40th Vibration Institute Annual Training Conference 2016

Asheville, North Carolina, USA
15 - 17 June 2016

ISBN: 978-1-5108-2991-6

Printed from e-media with permission by:

Curran Associates, Inc.
57 Morehouse Lane
Red Hook, NY 12571

Some format issues inherent in the e-media version may also appear in this print version.

Printed by Curran Associates, Inc. (2016)

For permission requests, please contact the Vibration Institute
at the address below.

Vibration Institute
2625 Butterfield Road
Suite 128N
Oak Brook, IL 60523-3415
USA

Phone: (630) 654-2254
Fax: (630) 654-2271

information@vi-institute.org

Additional copies of this publication are available from:

Curran Associates, Inc.
57 Morehouse Lane
Red Hook, NY 12571 USA
Phone: 845-758-0400
Fax: 845-758-2633
Email: curran@proceedings.com
Web: www.proceedings.com

Table of Contents

ACADEMIC PAPER SUBMISSIONS

HOW OPERATING DEFLECTION SHAPES CAN HELP YOU DIAGNOSE MACHINERY PROBLEMS

Mark H. Richardson
Vibrant Technology, Inc.
Aurora, CO 80016

Abstract: In this paper, it is shown how the Operational Deflection Shape (ODS) of a rotating machine can be used to detect and quantify rotor unbalance and shaft misalignment. An ODS derived from multiple accelerometer signals acquired at various points on the machine will be used to diagnose these two common machine faults.

The emphasis of this approach is a numerical comparison between the ODS of the machine when properly balanced and aligned with its ODS during an unbalance or shaft misalignment condition. These results point toward the implementation of a reliable numerical method for determining unbalance, shaft misalignment and other machine faults in an on-line machine health monitoring system.

Key Words: Auto power spectrum (APS), Cross power spectrum (XPS), Fourier spectrum (DFT), Frequency Response Function (FRF), Modal Assurance Criterion (MAC), ODSFRF, Operating Deflection Shape (ODS), Shape Difference Indicator (SDI)

Introduction: Vibration measurements have been used for decades as a means of assessing the health of rotating machinery. The use of vibration as a health metric is analogous to a medical doctor examining your electro-cardiogram in order to ascertain the health of your heart. Just as a doctor looks for small abnormalities in the time waveforms of the electronic signals from your chest, vibration signals also carry tell-tale signs that mechanical changes have taken place in a rotating machine.

Three of the most common health problems in rotating machinery are ***unbalance*** of rotating parts, ***misalignment*** of rotating parts, and loose mountings or an overly compliant foundation, called ***soft foot***.
It will be demonstrated that a form of data mining, using a ***numerical comparison*** between a ***current ODS*** and an ***archived ODS***, can be used to not only detect these types of faults but also to locate and quantify them.

Two assumptions are made about rotating machines;

1. All rotating machines are excited by ***internal cyclic forces*** which ***cannot be directly measured***

2. Order-related ***vibration levels acquired from multiple locations*** on a machine can be ***directly correlated*** with specific fault conditions

<u>**Hypothesis:**</u> ***When a rotating machine becomes unbalanced or misaligned, its order-based ODS will change.***

To validate this hypothesis, tests were performed on two different machinery fault simulators under various operating conditions. Accelerometers were attached to the motor, bearing blocks, and base plates of the simulators, and acceleration time waveforms were acquired using a multi-channel data acquisition system. Since both unbalance and misalignment produce dominant motion at the running speed & harmonics of a rotating machine, spectral data only from these frequencies was used to create order-based ODS's.

<u>**Significant ODS Change:**</u> It is well known and documented that unbalance and misalignment both cause a change in vibration levels. But; **"*What constitutes a significant change in vibration level*?"**

This question will be answered by applying two different metrics to the ODS's of the simulators, and using a popular ***"rule of thumb"*** to ascertain whether or not a ***significant change*** in the ODS has occurred. One of the metrics will prove to be more reliable as a measure of change than the other. Furthermore, it will be shown how one metric, the Shape Difference Indicator (SDI), can be used to uniquely identify machine faults by comparing a current ODS with the archived ODS of a known fault.

<u>**Output-Only Frequency Spectra:**</u> In operating machinery where the excitation forces cannot be measured, three types of ***output-only*** frequency spectra can be calculated from acquired time waveforms. The three types are a Fourier spectrum, Cross spectrum, and an ODSFRF.

A **Fourier spectrum** is simply the Digital Fourier Spectrum (DFT) of a digital time waveform. It is calculated with the FFT algorithm. A **Cross spectrum** is calculated between two simultaneously acquired responses. It is the DFT on one signal multiplied by the complex conjugate of the other. An **ODSFRF** is the Auto spectrum of a ***roving*** response combined with the phase of the Cross spectrum between the ***roving*** response and a ***(fixed) reference*** response.

<u>**What is an ODS?**</u> An ODS is defined as any ***forced motion of two or more points on a machine or structure***. Specifying the motion of two or more points defines a shape. Stated differently, a ***shape*** is the ***motion of one point relative to all others***. Motion is

a vector quantity, It has location and direction. Motion measured at a point in a specific direction is called a ***Degree of Freedom, or DOF***.

An ODS can be defined from any forced vibration, either from several time waveforms at a moment in time, or from several spectra at a specific frequency. An ODS can be created from different types of time waveforms, be they random, impulsive, or sinusoidal. An ODS can also be created from many different types of frequency spectra, including Fourier spectra (FFTs), Auto & Cross spectra, ODSFRFs, Frequency Response Functions (FRFs), Transmissibility's, etc. Three different types of ODS's are; ***time-based ODS's***, ***frequency-based ODS's***, and ***order-based ODS's***.

Time-Based ODS: A time-based ODS is simply the values of two or more simultaneously acquired or calculated time waveforms at the same time sample. A time-based ODS contains the ***true overall responses***, typically accelerations, that were simultaneously acquired at a moment of time from an operating machine.

Frequency-Based ODS: A frequency-based ODS is the values of two or more frequency domain functions or spectra at the same frequency sample. All frequency domain functions (except an Auto spectrum) are complex valued (having both magnitude & phase), and consequently all frequency-based ODS's are complex valued, containing the magnitude & phase of each response DOF relative to all other DOFs.

Order-Based ODS: In a rotating machine, the excitation forces are dominant at multiples of the machine running speed, called orders. An ***order-based ODS*** is assembled from the ***peak magnitude & phase*** values at one of the order frequencies in a set of output-only frequency spectra. When displayed in animation on a 3D model of the machine, an order-based ODS is a convenient way to visualize distributed relative vibration levels [5].

The components of an order-based ODS can also be used for monitoring the health of the machine by comparing their values with pre-defined warning levels. Three commonly used warning levels are termed ***alert, alarm, & abort***.

Measuring an ODS: In general, an ODS is created with ***magnitude & phase*** for each DOF on a machine or structure from which data is acquired. In order to measure the ***magnitude & phase*** for each DOF, either all responses must be ***acquired simultaneously***, or they must be acquired so that all of the ODS components have the correct magnitude & phase relative to each another

Simultaneous measurement requires a multi-channel acquisition system that can simultaneously acquire all responses of the machine. Sequential acquisition requires that a ***(fixed) reference response sensor*** be used during all of the acquisitions, so that all DOFs of the resulting ODS have the ***correct relative magnitudes & phases***.

Data Acquisition From a Rotating Machine: To create order-related ODS's, operating data was acquired from the machinery fault simulator shown in Figure 1. Tri-axial accelerometers were attached to the top of both bearing blocks and to the motor. A tri-axial and 2 uni-axial accelerometers were also attached to the base plate. These accelerometers provided a total of 14 vibration signals, which were simultaneously acquired with a 16 channel data acquisition system.

Figure 1. Rotating Machine with Accelerometers Attached

A set of ODSFRFs was calculated between each of the channels of data and a ***single reference*** channel. An ODSFRF is a ***"hybrid"*** cross-channel measurement, derived from both an Auto and Cross spectrum. It is formed by combining the ***phase*** of the Cross spectrum between a roving and reference signal with the Auto spectrum (***magnitude***) of the roving response signal.

The magnitude of an ODSFRF is the Auto spectrum of a response, and is a measure of the true magnitude of the machine's response at the location & direction of the sensor. Since the phase of an ODSFRF is the phase of the Cross spectrum between a response and the reference response, this phase is a measure of the ***relative phase*** between each response and all other responses.

Data was acquired from the machine at an operating speed of 2000 RPM, under seven different unbalance conditions. A typical ODSFRF is shown in Figure 2. It is clear that the dominant peaks in the ODSFRF are at the running speed and its higher orders (4000, 6000 RPM, etc.).

Order-based ODS's were created by saving the peak values in the ODSFRFs at each of the first three orders.

NOTE: An order-based ODS is the ***peak magnitude & phase*** values from a set of ODSFRFs at ***one of the machine orders***

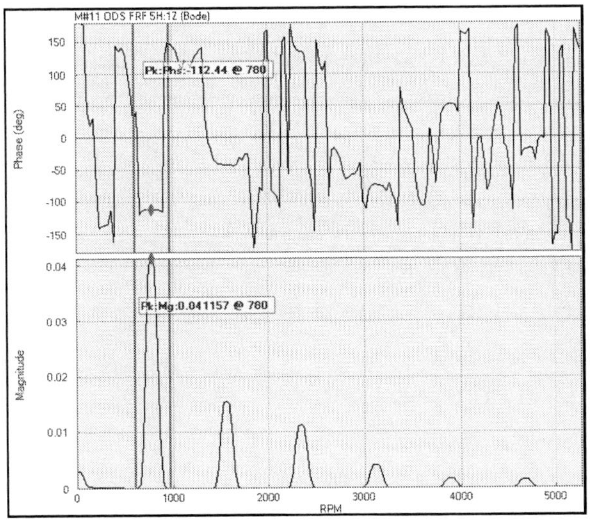

Figure 2 ODSFRF Showing Peaks at Machine Orders

<u>Unbalance Simulations:</u> Four tri-axial and two uni-axial accelerometers where attached to the rotating machine shown in Figure 1. The peak values at the running speed (or first order) of the machine were assembled together into an ODS with 14 shape components (or DOFs) [7].

However, ***only operating data from the two tri-axial accelerometers on the bearing blocks*** (indicated with red arrows in Figure 1) was used, thus providing order-based ODS's with only 6 DOFs in them. This minimal amount of data was used not only to prove the viability of ODS comparison as a method for locating and quantifying unbalances, but also to show that a ***minimal amount of instrumentation*** is required to implement this method on rotating equipment in a process plant.

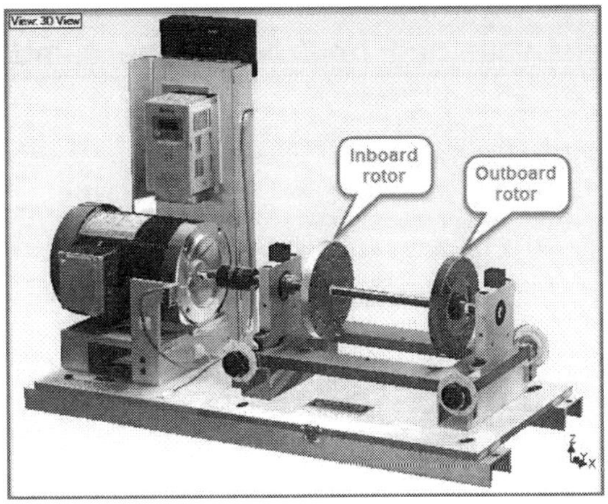

Figure 3. Unbalance Screws Added to Rotors

Seven Unbalance Cases: Vibration data was acquired from the machine when it was considered to be balanced, and under seven different unbalance conditions. Unbalance was created by adding weights (small screws) to either or both of the rotors, as indicated in Figure 3. Data was acquired for the balanced case and seven unbalance cases;

S1. Balanced
S2. Small weight (11.25 grams) - Inboard rotor
S3. Small weight - Outboard rotor
S4. Large weight (22.5 grams) - Inboard rotor
S5. Large weight- Outboard rotor
S6. Two large weights - 0 degrees apart
S7. Two large weights - 90 degrees apart
S8. Two large weights - 180 degrees apart

For case S2, a small unbalance weight (11.25 grams) was added to the inboard rotor, closest to the motor. For case S3, the same small unbalance weight was added to the outboard rotor, farthest from the motor. Cases S4 & S5 were the same as cases S2 & S3, but a larger unbalance weight (22.5 grams) was used.

In cases S6, S7 & S8, a large unbalance weight was added to both rotors, but the weights were added in different positions. In case S6, they were added at the same radial position on both rotors (with 0 degrees difference between them). In case S7, they were added at 90 degrees apart from one another, and in case S8 they were added at 180 degrees apart.

Modal Assurance Criterion (MAC): The Modal Assurance Criterion (MAC) [3], [4] is used for numerically correlating two shapes, but it has two limitations,

1. MAC only indicates the ***co-linearity*** of two shapes, not their difference
2. MAC requires shapes with ***two or more*** shape components. Its value for two scalars is ***always* 1**

MAC does not measure the difference in the actual values of the shapes themselves, only the difference in their ***"shapes"***. If two shapes lie together on the same straight line, **MAC = 1**. If two shapes do not lie on the same straight line, **MAC < 1**.

Shape Difference Indicator (SDI): Since MAC does not measure the difference between the component values of two shapes, a new measure was introduced [8], which does measure the difference. SDI has proven to be more useful than MAC for numerically comparing two ODS's. SDI values also range between **0 & 1**. If two shapes have ***identical shape components***, **SDI = 1**. If two shapes have ***different shape components***, **SDI < 1**.

MAC & SDI Applied to Order-Based ODS's: The following Figures 4 to 8 are bar charts of the MAC and SDI values between order-based ODS pairs for the balanced and the seven unbalance cases previously listed. Each bar chart depicts an 8 by 8 matrix of MAC or SDI values. Each ***diagonal bar*** (with value equal to **1.0)** is the MAC or SDI of each ODS paired with itself. Each bar chart also shows the ***worst case*** among the ***off-diagonal bars*** in the box at the top.

Rule of Thumb: If a MAC or SDI value for a pair of ODS's is ***less than* 0.9**, the ODS's are ***significantly different*** from one another.

NOTE: If ***all of the off-diagonal bars*** in the following MAC and SDI bar charts are ***close to zero***, that metric can be used to ***uniquely identify*** each machine fault

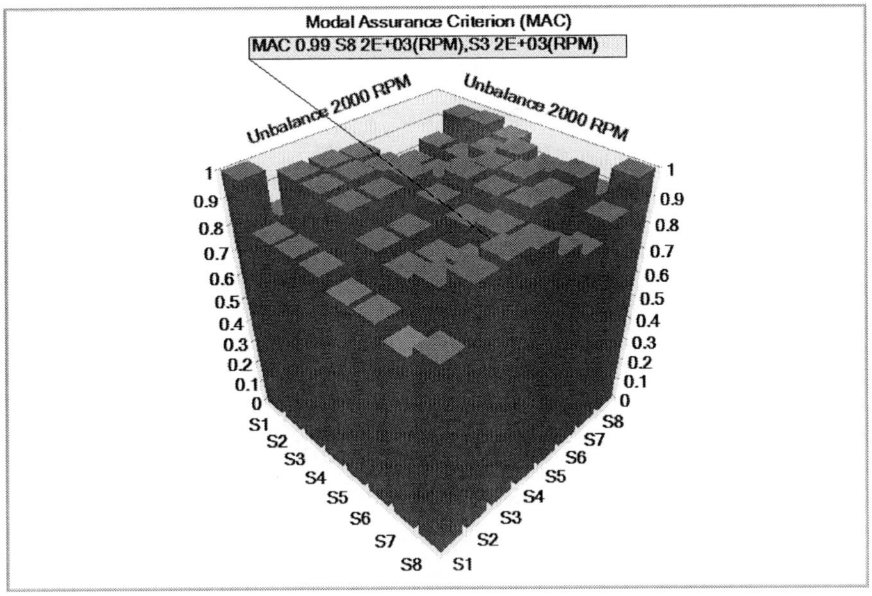

Figure 4. MAC for First-Order ODS Pairs

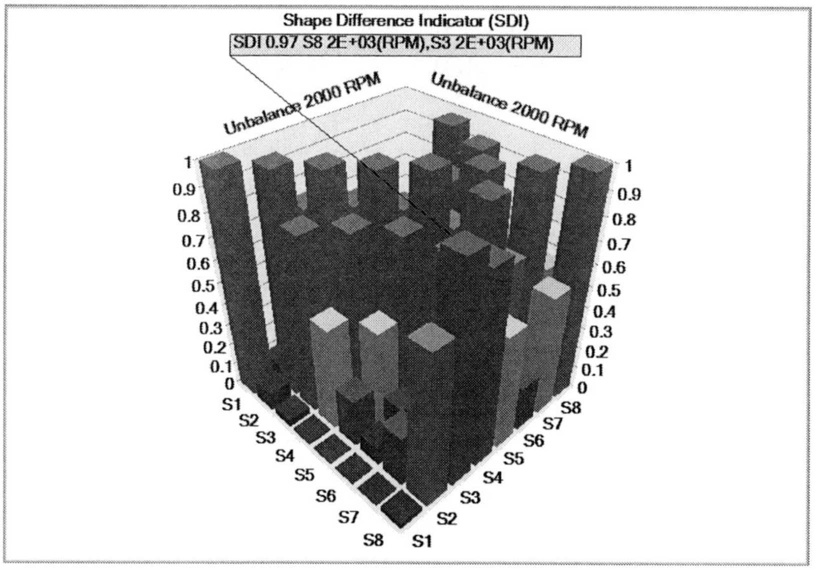

Figure 5. SDI for First-Order ODS Pairs

Figure 4 shows the MAC values for all eight first-order ODS pairs. Many of the MAC values for the off-diagonal pairs are ***close to* 1.0**. This means that although the vibration levels among the cases may be different, the ***ODS's*** for many of the pairs ***are essentially co-linear*** with one another. This bar chart verifies that with many MAC off-diagonal values ***close to* 1.0,** MAC cannot be used with first-order ODS's to uniquely distinguish between each of the eight cases.

Figure 5 shows the SDI values for all eight first-order ODS pairs. These values are closer to zero than the MAC values, but an SDI value ***equal to 0.97*** for the worst case pair (shapes **S8** & **S3**) indicating that they are closely correlated. Comparing the bar values in Figures 4 & 5, even though SDI appears to be a stronger metric than MAC for distinguishing differences between the first-order ODS's, SDI would still be an unreliable method for discriminating among all of the seven unbalance cases and the balanced case.

Increasing the SDI Sensitivity: The ***sensitivity*** of SDI becomes greater as one of the shapes contains values that are ***closer to zero***. Details of this, and a plot of the SDI curve are given in [9].

A simple modification of the ODS data can be used to increase the sensitivity of SDI. One shape is replaced with constant values that are ***closer to zero***, and the other shape is replaced with the ***difference between the two shapes*** added to the constant value of the first shape.

In order to distinguish the difference between a pair of ODS's, a sensitivity value should be chosen which ***drives all SDI value closer to zero***.

Figures 6, 7, & 8 contain bar charts of the SDI values with ***sensitivity* = 0.1**. These bars are the SDI values for the first-order (2000 RPM), second-order (4000 RPM), and third-order (6000 RPM) ODS pairs. The highest off-diagonal SDI for the first-order ODS pairs is ***0.19 (cases 7 & 5)***. The highest off-diagonal SDI for the second-order ODS pairs is ***0.85***, again for ***cases 7 & 5***. The highest off-diagonal SDI for the third-order ODS pairs is ***0.80 (cases 7 & 4)***.

These three Figures are graphic evidence that ***SDI with increased sensitivity*** provides a clear discriminator between a balanced and seven different unbalance conditions in the fault simulator.

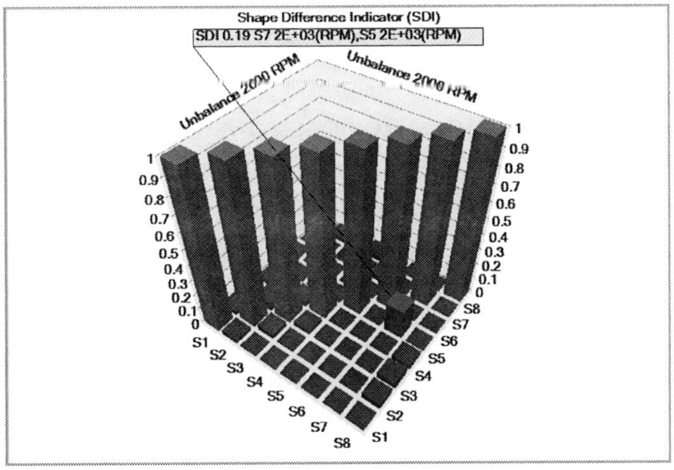

Figure 6. SDI for First-Order ODS Pairs

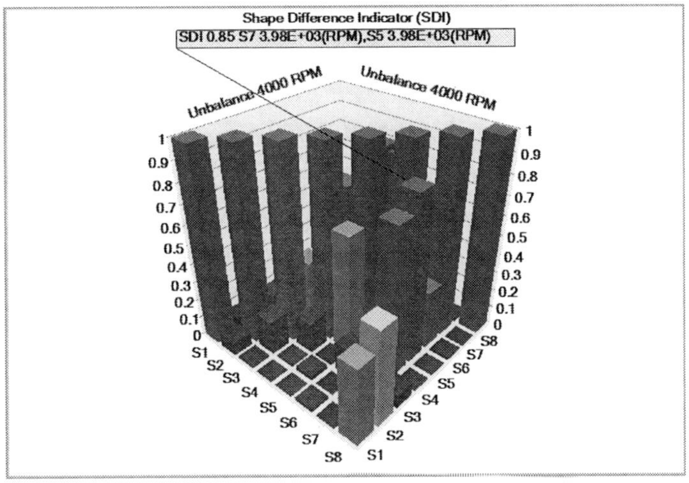

Figure 7. SDI for Second-Order ODS Pairs

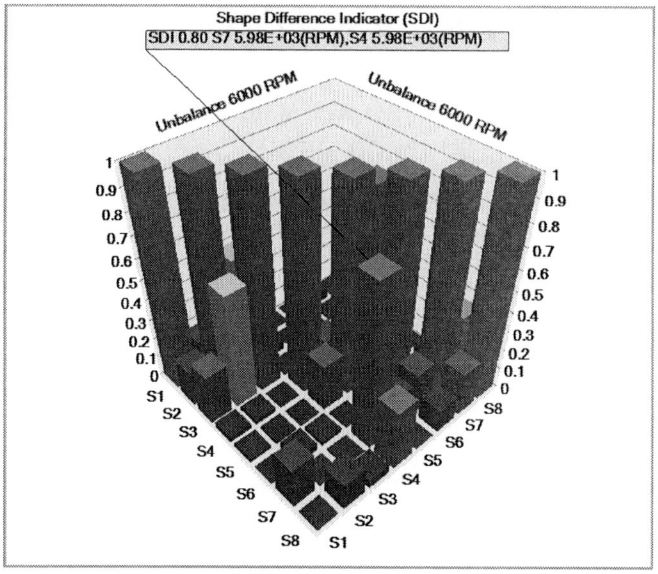

Figure 8. SDI for Third-Order ODS Pairs

Shaft Misalignment: A poorly aligned machine can account for up to 30 percent of a machine's downtime [1]. Downtime is expensive because of lost production, which translates directly into lost revenue. Misalignment also uses more energy, and requires a larger inventory of replacement parts, both of which add to costs.

A survey of the literature [2] on misalignment reveals that;

1. Misalignment produces significant vibration levels
2. Machine speed and coupling stiffness strongly influence vibration levels
3. Level profiling of a single-point vibration spectrum does not provide a reliable indication of misalignment
4. A machine can have parallel misalignment without exhibiting significant 2X vibration levels
5. Observation of spectra at several locations in several directions, and at varying speeds, may be required to effectively diagnose misalignment

Traditionally, vibration signatures (level profiling of single-point vibration spectra), and orbit plots have been the preferred tools for detecting and diagnosing machinery misalignment. Moreover, orbit plots require data from proximity probes, which must be imbedded in a machine to measure its fluid bearing displacements.

Although these tools may be effective when used by an expert vibration analyst, using a simple metric like SDI on ODS data offers a simpler, more straightforward approach to detecting misalignment. Misalignment is more easily characterized by a numerical and visual comparison of a current ODS with an archived ODS.

<u>Misalignment Data Acquisition:</u> To verify our fault hypothesis, tests were performed using the rotating machine fault simulator shown in Figure 9. Accelerometers were attached to the top of both bearing blocks, the motor, and the base plate of the machine. Operating data was acquired using a 16 channel acquisition system. A tachometer signal was acquired on channel 1 to measure the machine RPM, and 15 accelerometer signals were acquired on the remaining channels.

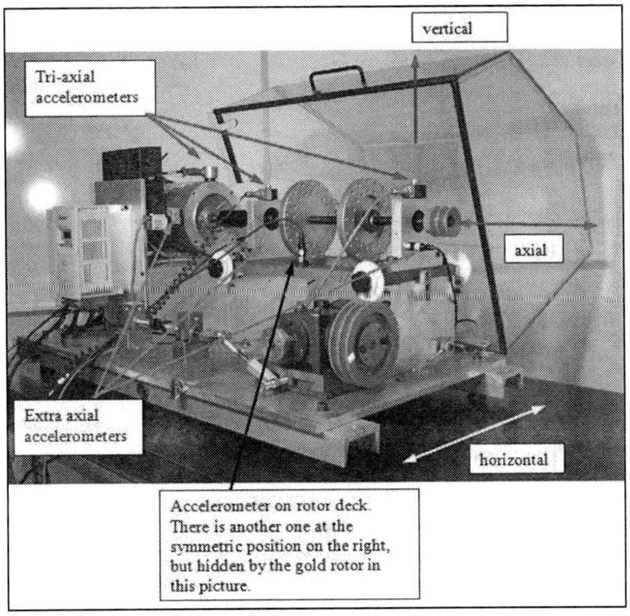

Figure 9. Fault Simulator with Accelerometers Attached.

Only data from the tri-axial accelerometer on the motor and the tri-axial accelerometers on the two bearing blocks was used to create ODS's. Again, this was done to demonstrate that only a minimal set of measurements (9 DOFs) is required to distinguish misalignment using the MAC & SDI metrics.

Figure 10 is a comparison of the first-order and second-order ODS's when the machine is aligned, using data only from the motor & bearing sensors. It is clear from the very low MAC & SDI values shown in the upper right-hand corner that these ODS's are very different from one another. The difference in deformation is better observed with an animated display of the two shapes. This shows that the first-order and second-order *ODS's* of a rotating machine are significantly different, even when it is properly aligned.

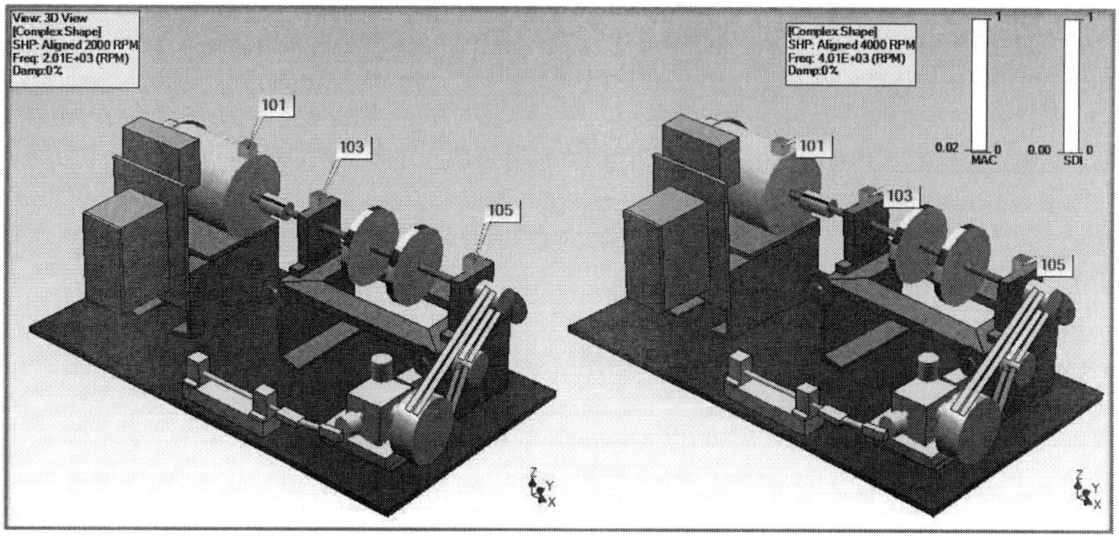

Figure 10. Aligned Machine First-Order versus Second-Order ODS

Parallel Misalignment: Figure 11 depicts both the parallel and angular misalignment that were simulated with the machine fault simulator. Parallel misalignment is depicted in Figure 11b. To simulate a parallel misalignment, the center of the rotor shaft was offset from the motor shaft in the simulator by ***25 mils at both bearing blocks***.

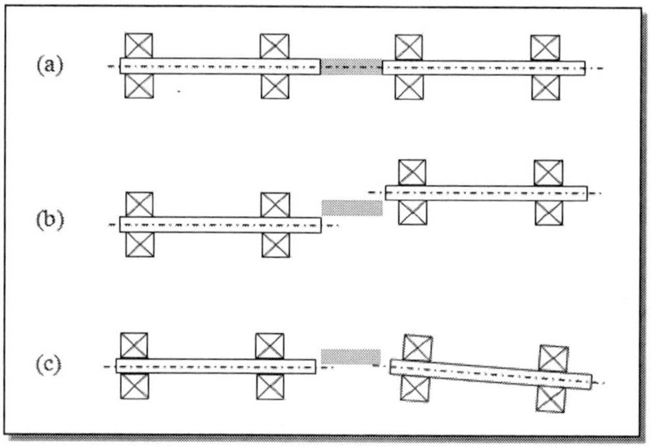

Figure 4. Parallel (b) & Angular (c) Misalignment.

Angular Misalignment: Angular misalignment is depicted in Figure 4c. To simulate angular misalignment, the inboard and outboard bearings were offset by ***6.2 mils and 26 mils*** respectively to obtain ***about 0.1 degree*** of rotor shaft misalignment [6].

ODS's were created for the following three alignment cases using data only from the motor and bearing blocks,

S1. Aligned
S2. Parallel Misalignment
S3. Angular Misalignment

Figure 11 shows MAC bars of the ***first-order ODS*** pairs for the three alignment cases. The ***worst case off-diagonal*** bar is **0.34**. Therefore, ***MAC could be used*** with the first-order ODS's to identify each of these three alignment cases. Figure 12 shows SDI bars of the first-order ODS pairs for the three alignment cases. It is clear from Figures 11 & 12 that SDI is a stronger metric than MAC when used on first-order ODS's.

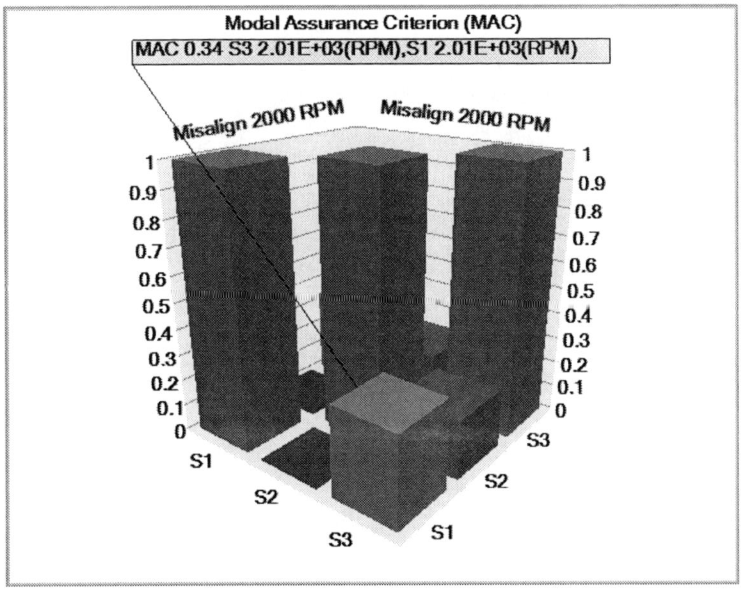

Figure 11. MAC for First-Order ODS Pairs

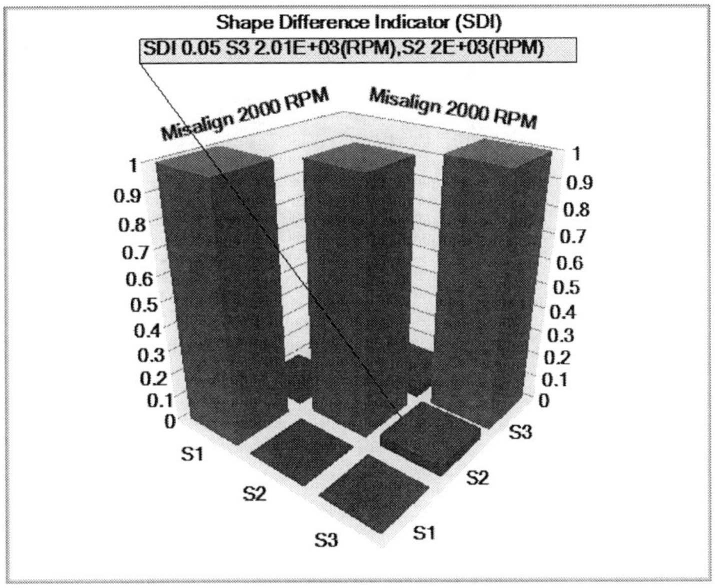

Figure 12. SDI for First-Order ODS Pairs

Figure 13 shows MAC bars of the second-order ODS pairs for the three alignment cases. The ***worst case off-diagonal*** bar is **0.95**. Therefore, ***MAC could not be used*** to discriminate between these three cases using second-order ODS's. Figure 14 shows SDI bars of the second-order ODS pairs for the three alignment cases. SDI would clearly identify these three cases without increasing its sensitivity.

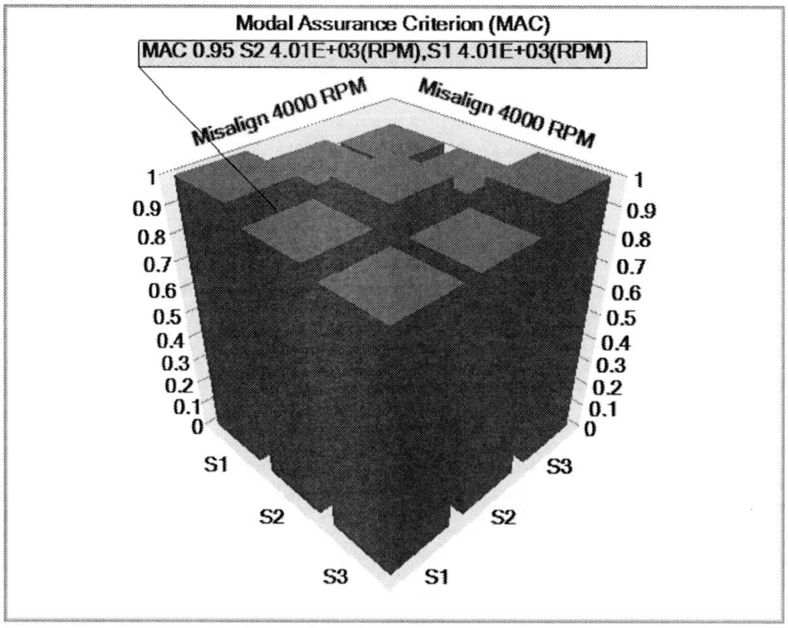

Figure 13. MAC for Second-Order ODS Pairs.

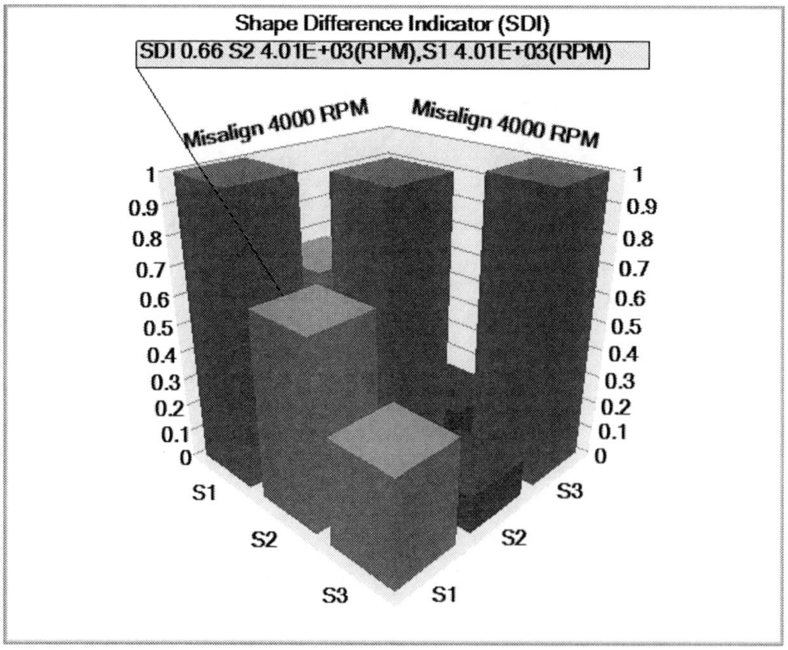

Figure 14. SDI for Second-Order ODS Pairs.

Fault Correlation Tools (FaCTs): When SDI is used together with an archival database, where archived ODS's have been associated beforehand with known unbalance or misalignment conditions, SDI bars can be used to identify those specific faults. This capability has been branded as **FaCTs™**, an acronym for **Fault Correlation Tools**, in the Vibrant Technology ME'scope software.

FaCTs can be used in a number of ways as part of the post-processing software in an online machine health monitoring system. In addition to correlating current ODS data with archived data that is associated with a known fault condition, FaCTs can also be used to graphically indicate any change in a machine's operating condition by displaying a real-time SDI bar of current versus operator chosen baseline ODS data. FaCTs can also be used as a ***pass/fail indicator*** in a qualification testing system where any type of engineering data stored in the form of a ***"shape"*** can be used for SDI calculations.

Conclusions: The purpose of this paper was to show that ODS data acquired only from the bearing blocks can be post-processed to detect and identify specific unbalance conditions in a rotating machine. Furthermore, ODS data acquired only from the motor & bearings can be used to detect and identify specific misalignment conditions in a rotating machine.

Two different metrics of shape correlation were used, MAC & SDI. Both metrics indicate the likeness of a pair of shapes by giving values between 0 & 1. A value ***equal to 1.0*** means that the shapes are the same, and a value ***less than 1.0*** means that they are different. A value ***less than 0.9*** has traditionally been used as a ***rule of thumb*** for declaring a ***significant difference*** between two shapes.

A unique property of the SDI metric is that as the values of one shape approach zero, the sensitivity of SDI to the difference between it an another shape increases. By increasing its sensitivity, it was shown that SDI more clearly identified all of the unbalance cases presented here.

Other machine faults, such as ***soft foot***, ***bearing oil whirl***, and ***loose connections*** might also be detected through the application of the SDI metric to ODS data. The simplicity of this approach to machinery fault diagnosis makes it a strong candidate for implementation in an online machine health monitoring system.

REFERENCES

1. S. Ganeriwala, S. Patel, and H. Hartung "The Truth Behind Misalignment Vibration Spectra of Rotating Machinery" Proceedings of the International Modal Analysis Conference, 1999
2. J.Piotrowski, "Shaft Alignment Handbook" Third Ed. Marcel Dekker, Inc. New York, 2006
3. R. J. Allemang, D.L. Brown "A Correlation Coefficient for Modal Vector Analysis", Proceedings of the International Modal Analysis Conference, 1982
4. R. J. Allemang "The Modal Assurance Criterion (MAC): Twenty Years of Use and Abuse", Proceedings of the International Modal Analysis Conference, 2002
5. B. Schwarz, M.H. Richardson, "Measurements Required for Displaying Operating Deflection Shapes" Proceedings of the International Modal Analysis Conference, January 2004
6. S. (Suri) Ganeriwala, Zhuang Li, M. H. Richardson "Using Operating Deflection Shapes to Detect Shaft Misalignment in Rotating Equipment" Proceedings of the International Modal Analysis Conference, February 2008
7. S. (Suri) Ganeriwala, B. Schwarz, M. H. Richardson "Using Operating Deflection Shapes to Detect Unbalance in Rotating Equipment" Proceedings of the International Modal Analysis Conference, February 2009
8. S. Richardson, J. Tyler, P. McHargue, M. H. Richardson, "A New Measure of Shape Difference", Proceedings of the International Modal Analysis Conference, February 2014
9. S. Richardson, M. Richardson, J. Tyler, P. McHargue "Using Operating Data to Locate & Quantify Unbalance in Rotating Machinery" Proceedings of International Modal Analysis Conference, January 2016

BIOGRAPHY

Mark Richardson, Vibrant Technology, Inc.

Dr. Richardson founded Vibrant Technology, Inc., Scotts Valley, California in 1991, and is currently its President & CEO. Vibrant is the developer of the ME'scope post-processing software, which is used in a wide variety of applications, including ODS & Modal testing and analysis, noise & vibration troubleshooting, continuous machine & structural health monitoring, environmental monitoring, and qualification testing.

As product manager at the Hewlett Packard Co. during the 1970's, Mark directed the development of the first commercially available modal testing software, which was part of the HP5451 FFT analyzer product line. Later he also directed the development of the first dedicated modal testing instrument, the HP5423 Structural Dynamics Analyzer.

In 1979, Dr. Richardson co-founded Structural Measurement Systems, Inc (SMS) where he was President & CEO until it was sold to GenRad in 1989. SMS developed and marketed the very popular Star Modal^{TM} software, used for analyzing structural noise & vibration problems.

Mark has authored over 60 technical papers and magazine articles on structural testing, digital signal processing, modal analysis, and machine fault detection. He has traveled worldwide conducting tests, giving seminars, and teaching short courses.

A NEW PARADIGM FOR PDM CONDITION MONITORING INSTRUMENTATION

Ken Piety
Vice President, Technology, Azima DLI, KPiety@AzimaDLI.com

There has been a paradigm shift in the instrumentation deployed for predictive maintenance (PdM) programs. Instrument vendors cannot keep up with the tablet computer and cellphone industries. Wireless communications are readily available and transferring and storing large amounts of data in the cloud is a common practice. These factors are affecting what will work best in your plant's PdM program. This presentation will discuss data collectors and analyzers, the functions they can provide, the cost of ownership, proper setup and training. Attend this session so you can make an informed decision when purchasing instruments to monitor the health of your plant's assets.

Examples of the effect of new paradigm:

1. More flexibility in ergonomic design
2. Lower lifecycle costs
3. Evergreen technology
4. Instrument can be used for other purposes
5. Flexibility in controller selection
6. Remote support and operation of instrument
7. Protected and shareable data via the cloud
8. Improved access to expertise
9. Better utilization of available expertise
10. More flexible delivery of technology
11. More sustainable PdM programs due to adaptable design
12. Improved access to PdM program information

Keywords: Condition based monitoring; Data collection instruments; Data processing; Lower life-cycle costs; PdM; Predictive maintenance; Vibration analysis; Vibration analyzers

Do you remember your first cellular phone? If you were an early adopter, it was likely a bulky device that flipped open and only had one function – to make calls.

Today, your cell phone looks nothing like your first one. It is thinner, lighter, and much, much more powerful. Besides making phone calls, you can text, surf the web, answer emails, manage your calendar, and take photos. For those who have teenage children, making phone calls is probably the least used feature! Today's mobile phones are literally powerful computers in the palm of our hands.

Vibration analyzers are following a similar path of evolution, albeit at a slower pace. The change is part of a paradigm shift in the PdM industry that is moving toward new technologies, cloud computing, data sharing, visibility, and program metrics.

You've Come a Long Way, Baby!

Industrial instruments were initially designed as monolithic, purpose-built meters that collected data. They were plain, heavy-duty, rubber-booted boxes from which some wires extruded (possibly attached to a sensor) to attach to a measurement location. These box-like meters made specific measurements, such as voltage or resistance. If you needed to make a different type of measurement, you needed a different meter.

While they remained purpose-built and monolithic, industrial instruments underwent two major technology changes over the past 50+ years. In the 1960s and 1970s, their design shifted from all analog components to digital technology. In the 1980's, instrument manufacturers began embracing microprocessors to control and expand the functionality of portable meters. The rise of personal computers, also in the 1980's, enabled data to be easily exchanged with a local PC workstation, thus resulting in a new generation of "smart" instrumentation. Vibration instrumentation followed this same path and vendors who did not see or appreciate the paradigm shift disappeared from the market.

Image 1 – Data collection devices have come a long way from these single-measurement instruments.

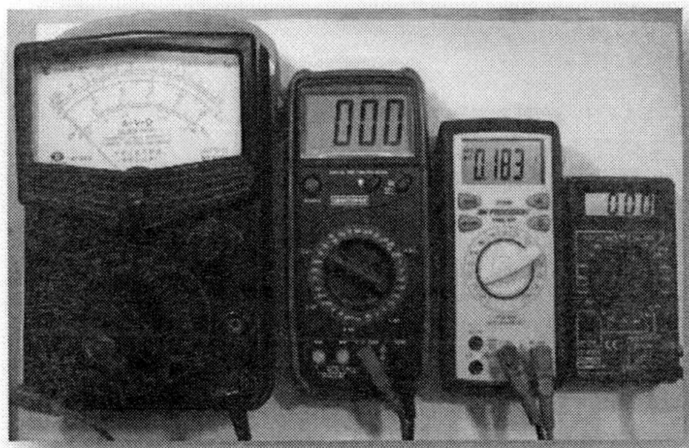

While the underlying technology changed, the operational process remained the same. A PdM analyst or technician collected vibration data with the analyzer, and then returned to his office to unload the data to a computer so it could be manually analyzed using proprietary desktop software.

Some vendors continue to operate that way today. However, others have recognized the technology advances made by tablet computer and mobile phone manufacturers and have found ways to capitalize on them, to the benefit of their instruments and their customers' PdM programs.

Thought leaders in the industry have adopted an "if you can't beat them, join them" attitude. These vibration analyzer vendors are incorporating commercial-off-the-shelf (COTS) tablet computers and smartphones into their designs, allowing their focus to be on the development of the data acquisition and diagnostic capabilities, not the development of another computer and user interface.

The results are evident in today's market. There are vibration analyzers with features that were unheard of in the industry 20 years ago. Touch screens, built in cameras and video recorders, GPS, wireless communications, remote access, and voice control abilities are readily available.

Analyzers built upon tablet computers have the added benefit of being part "regular" computer. Besides being able to collect and analyze vibration data, the tablet can be used for other purposes. Got a lubrication application that you usually bring to the shop floor on a PDA? Why not load it on your analyzer instead? The possibilities are only limited by the tablet's space and processing power – and the plant's IT security policies.

Need to access hazardous locations? No problem! Many vendors offer rugged, Class 1 Div 2 tablets and Class 1 Div 1 smartphones that meet hazardous use requirements.

Surprisingly, the ruggedized versions of commercially available tablets and smartphones are holding up well in harsh industrial environments. It seems the devices that many considered novel toys when they first emerged are actually tough. After all, they have to hold up under the daily abuse of teenagers.

Image 2 – Many mobile phone and tablet manufacturers have ruggedized devices that can hold up to harsh plant environments.

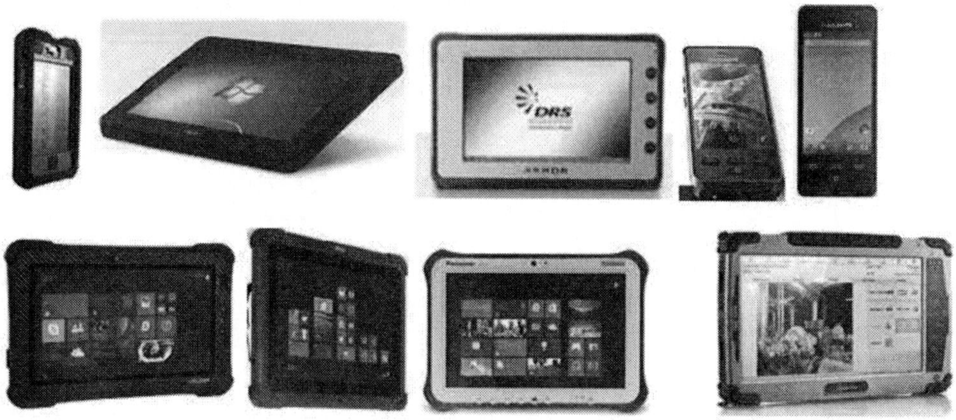

Physical Design Matters
Ergonomics, efficiency, and comfort are key factors in almost everything we use to accomplish our work.

A computer programmer, for example, will be less effective if his desk chair is uncomfortable and too low. Sore muscles, headaches, and stiff joints will likely distract him from his work and negatively impact his productivity.

The same is true for vibration analyzers. The weight of the device, how it is carried, and how it operates have a direct impact on the efficiency of the person collecting data.

Monolithic, purpose-built analyzers are typically heavier and have a poor display-area-to-weight ratio, though some vendors' latest offerings have proven to be a bit lighter. The bulkiness and weight of the meter should be an important factor considered when making a purchase. After all, the user will be carrying it around for hours at a time and may be climbing and reaching to access measurement points.

New technology and capabilities have given rise to a two-component design aimed at making the unit less taxing on muscles and joints. Using a COTS tablet that communicates wirelessly via Bluetooth with a separate data processing unit to trigger data collection has a number of advantages including lower cost of ownership and safer form function. An accelerometer is connected to the small data processing unit, which can be worn on a belt or placed near the machine. The user only needs the tablet—weighing less than 3lbs-- in his hand to initiate collection.

This design not only reduces the weight and bulkiness of the analyzer, but allows the operators to distance himself from the machine. The user can connect the accelerometer to the machine and leave the data processing unit near the machine, but step away to use the cable-free tablet to trigger collection.

When moving between machines, sensors and cables can be completely holstered in belt pockets, leaving both of the technician's hands free. No longer does the person collecting data need to be concerned with sensors and cables dangling out of the top of the instrument, forcing him to secure them with one of his hands.

Image 3 – The two-component design enables a technician to wear a small data acquisition device on a belt, leaving his hands free to trigger collection via a tablet computer.

Mobile phones and smartphone apps are slowly emerging as alternatives for technicians to communicate data to a remote data center, perform operator or lubrication rounds, or make helpful vibration-related calculations.

Many new instruments are being introduced:

- A rugged smart phone or tablet coupled to a separate data acquisition device.
- A self-contained smart sensor that communicates wirelessly to an iPad or smartphone.
- A smartphone that pushes data to the cloud from a traditional-looking instrument that makes the measurements.

Lowering Lifecycle Costs

The physical design of a data collector directly affects its lifecycle cost.

Most facilities replace their purpose-built, monolithic vibration analyzers every 5 to 10 years. This means buying a whole new device for $15K - $35K, after paying annual support fees between $1,500 and $3,000 over the life of the "old" analyzer.

For the two-component design, separating signal processing from the human interface lets vibration analyzer vendors take advantage of advances in computer technology. As faster, more powerful tablets and smartphones emerge, vendors can offer them to customers at a much lower cost (usually less than $3K). Typically, new technology makes the "old" model obsolete in about two or three years' time. Consider your last mobile phone. How long did you have it before upgrading to the shiny new model?

The data processing unit of a two-component design can be replaced much less frequently than the tablet or smartphone component, about every 10 years or longer. This is because the signal processing requirements for portable vibration data collection are very mature and well understood. The following is unlikely to change or be improved in any way that would necessitate a new design for route-based analyzers:

- Four parallel channels of data acquisition with frequency ranges up to 40 kHz and lines of resolution up to 25,600.
- Flexible sampling and triggering modes on tachometer or vibration signals.
- Ability to accept accelerometers, proximity probes, and other types of sensors.

For purpose-built, monolithic meters, most of the design work is focused on staying current with computer component advances. However, the customer base is not large enough to support rapid redesigns. This means that once you buy, you are likely frozen at that level of computer technology for 5 to 10 years. Think about the analyzer currently used in your plant. How long ago was it purchased?

The following chart shows the 12-year, lifecycle cost of the two-component design compared to two other traditional models.

Figure 1: Comparison of lifecycle costs

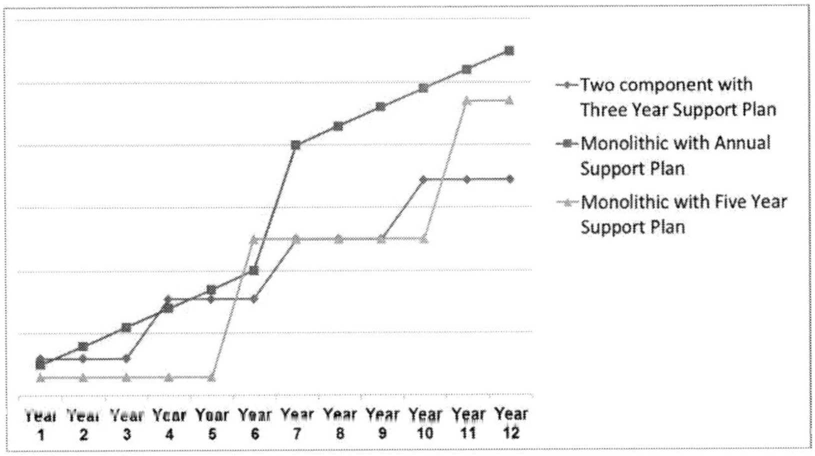

Keep in mind that as the costs of tablets and smartphones drop, so would the lifecycle costs for the two-component design. The estimated lifecycle cost of two-component systems is between 20% and 30% less than the lifecycle cost of monolithic systems, assuming an average replacement cycle for traditional data collectors, even with earlier and more frequent computer tablet refresh cycles.

Increased Availability & Enhanced Support

Being without your vibration analyzer can be difficult, especially if you have bad actors in your fleet and production deadlines to meet.

If your analyzer is on the fritz and technical support cannot help you over the phone or via email, you will likely ship it to the vendor for repair or replacement. Historically, this meant you would be without your analyzer for weeks, unless your vendor offered a loaner program and you were up to date with your support plan payments.

The two-component design does a lot to alleviate this pain:

- The tablet or smartphone of a two-component analyzer can be repaired or replaced for $2K - $4K, in a matter of days. (The tablet and smartphone industries are expected to deliver this type of turnaround to almost any location around the world.)
- You can purchase a spare tablet or smartphone component as a backup, to maximize the availability of your unit.

- The data acquisition unit of a two-component analyzer has a very simple, rugged design and is less expensive to replace or provide loaners for as compared to a purpose-built, monolithic analyzer.
- Repair or replacement of either component of two-component design is less costly than a monolithic device. Plus, it is highly unlikely that both components would fail at the same time.

Internet-enabled analyzers built upon a tablet or smartphone also provides an important advantage relative to technical support. Imagine a two-component analyzer that can be accessed remotely by its technical staff over the Internet because the tablet is a Windows-standard, internet enabled device. The tech support staff can troubleshoot the device or assist the customer with a difficult situation, such as setting up an advanced test. In many cases, this negates the need to ship the analyzer to the vendor for troubleshooting.

Bring Data to the Analyst
The paradigm shift in PdM instrumentation plays directly into an increased need to communicate predictive maintenance information and the concept of "cloud computing."

To a maintenance team responsible for ensuring maximum machine uptime, it begs the question: What's the benefit of cloud computing to *me* and *my* program?

The benefit of the cloud to a maintenance team is huge. Storing vibration data in the cloud enables access by a broader audience. This can be multiple analysts inside a single plant or across many plants, or a third-party PdM provider who is geographically remote. The cloud enables data to be delivered to people, rather than having to transport people to the location where the data is being generated.

Another major benefit of cloud storage is having the data backed up securely. Cloud storage means your data has little to no risk of loss, especially if your data collector and/or analysis software synchronizes with the cloud. Synchronization ensures both locations have the same information. This means that if the analyzer or the computer housing the diagnostic software fails, an exact copy of your entire database is backed up in the cloud. (The only risk is losing the vibration data collected just before the failure – if it has not been synched to the cloud.) Most cloud-storage providers also back up their servers, adding an extra layer of protection.

To reap the benefit of cloud storage, an instrument must be designed to transfer data over the Internet. Some vibration analyzers on the market today make it seamless, with just a click of a button and an Internet connection. Other OEMs including both industrial asset manufacturers and metering manufacturers have deployed smartphone applications as a transfer tool.

However, many of today's analyzers are not Internet-enabled at all. Some are designed for sharing via a LAN, which is limiting since everyone you want to access the data must be on the same LAN.

For vendors that have embraced the cloud, it is also important to consider the diagnostic software needed to analyze the data. The cloud concept supports a variety of new software configurations – all of which promote collaboration and data sharing:

- **Local Installation:** Diagnostic software is installed on the analyst's desktop or on a network server. A local database synchronizes to a database in the cloud. Multiple analysts can have individual copies of the diagnostic software, all of which synchronize to the cloud to enable sharing of information and collaboration.
- **Cloud Installation:** Diagnostic software is installed in the cloud with the data. Analysts access it via a Web subscription over the Internet. A synchronized copy of the cloud database resides on each analyst's vibration analyzer.
- **Analyzer Installation:** Diagnostic software is installed on the analyzer itself. A keyboard, mouse, and monitor (for those who prefer a larger screen for diagnostics) can be connected– essentially eliminating the need for a separate diagnostic computer. Data from the analyzer is synchronized with a copy of the database in the cloud so others can access it, too.

The cloud concept drives efficiency no matter how your program is set up.

For an in-house program with a single analyst, PdM in the cloud enables the analyst to stay on top of things, regardless of his location. He can go on vacation and be able to analyze machine data collected by his staff from anywhere with an Internet connection. If a company has more than one analyst located at different sites, they can collaborate on analysis. Without centralizing the data in the cloud, each analyst's data would be accessible only on an individual's computer after being unloaded from their meter or over a LAN.

For outsourced programs, the third-party PdM provider can send a technician to collect data (or train someone in-house) and then have an expert analyst remotely perform diagnostics on the data. This model saves the plant money, as the expert analyst does not have to travel to the site. Even if the PdM vendor sends a tech to collect the data, it is much less expensive than a visit from the expert analyst. It also saves time. No longer does the plant have to wait nervously wait while an analyst makes a long drive or hops a on a plane to come analyze problematic equipment. Simply collect and send the data to him via the cloud!

The design of today's analyzers plays a key role in the success of the PdM in the cloud model.

When the purpose-built, monolithic data collectors were designed originally, there were no formal standards relative to how the meter should operate from a hardware perspective. If you switched vendors, you had a brand new device to learn. One did not operate like the next.

The incorporation of COTS tablets and smartphones removes this usability barrier. Today's tablets and smartphones are designed to be easy-to-use right out of the box and tend to follow similar usability standards. Most children can pick up any tablet or smartphone and figure out how to use it within a few minutes.

By reducing the learning curve, a technician can be taught how to collect quality data in a very short time. Separating collection from diagnostics lets analysts focus on remote analysis with the confidence that quality data has been collected.

An Example:

Imagine a PdM provider that services large industrial companies with 100 or more sites distributed throughout the United States. The onsite staff has been trained on how to collect quality data.

Each month, all 100 sites collect and send data via the Internet to the cloud. It is reviewed by the vendor's analysts, who produce comprehensive reports for each plant that describe the faults found and recommend corrective actions.

Corporate-level managers at the company log on to a web Portal to see an overview of machinery health across all of the plants and can drill down into the Portal for more details. All of the reports are posted on the Portal for easy access.

This visibility shines a light on sites where collections are skipped, as it is tracked on the Portal. Correlations can be made between sites with higher rates of machine failures and low data collection compliance. Corporate managers are able to clearly see the value of their PdM program.

Visibility into the PdM Program

PdM in the cloud is NOT new. Companies have been using it for more than a decade. What has changed is its use from basic data sharing to something much more.

As mentioned previously, the cloud enables data to be shared with a broader audience and drives visibility into PdM programs.

A plant's PdM budget needs to be justified each budget season to continue. Showing a return on investment is not always easy—especially if you have a mature, strong program that catches problems early. Many maintenance managers can relate to hearing: "We've had very little downtime this past year, so why do we need this line item for a PdM program?" It is often the first thing cut in a bad economy.

PdM Web Portals that work in conjunction with vibration analyzers support the tracking of metrics, from how many faults were detected, to their severity, to the time it took to repair. There are even ways to track costs savings from avoided downtime.

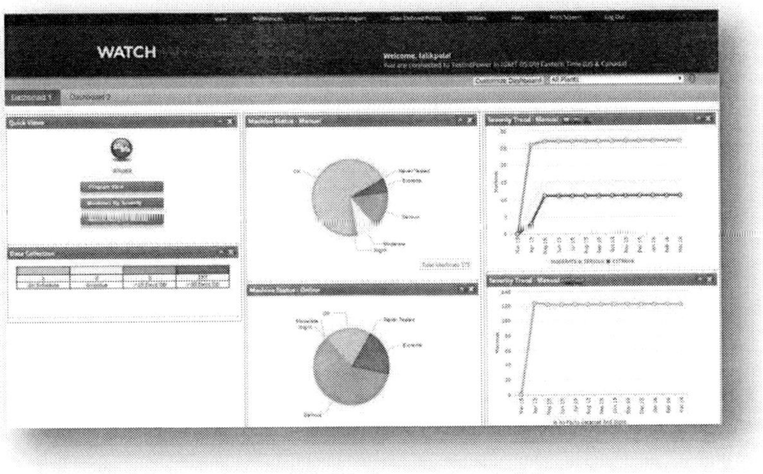

Image 4 – Web portals aggregate information from various data collection sources and provide visibility across one or many plants via data-driven metrics.

Having this type of information readily available arms the maintenance manager with the facts and figures he needs to show just how effective his program is and tout how much downtime cost was avoided.

For a maintenance manager in charge of multiple facilities, these metrics also help him better manage his vibration program by tracking whether data collection is occurring on schedule and by comparing the time to repair across sites. A Web Portal can also show the current statuses of machines across many plants so the manager gets a complete picture of corporate-wide machinery health, rather than a piecemeal view.

Solves the Expertise Shortage Problem

Many studies have shown that the available diagnostic expertise within plants is dwindling. Many skilled analysts are nearing retirement and there are not enough new analysts to fill their shoes.

The need to solve the expertise shortage problem is one of the driving factors behind the paradigm shift in the PdM industry, including changes in instrumentation. Remote analysis enabled through cloud computing and Web-enabled vibration analyzers lets one expert analyst service many geographically dispersed customers. If the analyst had to make monthly or quarterly visits to collect and analyze data on-site, he would not be nearly as effective or be able to help as many disbursed sites.

For those who do not outsource, having an instrument that integrates with a Web Portal means historical machine health information is not lost when the in-house analyst leaves or retires. While other systems can offer a new PdM manager a database full of data and a slew of past reports, a Web Portal aggregates and displays an intuitive picture of the site's history, which saves lots of time and headaches.

Many PdM programs simply die off when their champions leave. When the program is brought back, usually after a string of unexpected machinery failures, it starts from scratch – sometimes even with a new brand of vibration data collector and software. The new paradigm of PdM helps keep that from happening. Not only can a Web-based PdM system be easily outsourced when talent is lost, it can also easily be re-started by a new hire since all of the data and history is available on the Portal and the data collection unit is intuitive and easy to use.

Example Case Studies

The following three case studies demonstrate the impact the new paradigm is having on the efficiency and effectiveness of reliability programs.

Case Study #1:

Analyst Detects Bearing Failure from 2,300 Miles Away

When a bearing in a roll stand gearbox failed without warning, a major US specialty steel producer escaped major gear damage and lengthy downtime thanks to cloud computing.

This East Coast steel mill collects its own vibration data on a regular schedule and then uploads it over the Internet to a skilled analyst, located in Seattle, WA. In this case, the mill uploaded vibration data from two, eight-stand rolling modules on a Sunday. The mill relies on the 16 critical roll stands to produce various size bars of specialty steel.

The analyst received an email notification of the upload and logged in over the Internet from his Seattle home that same day for an initial review. To his surprise, vibration from one of the vertical stand's planetary gearboxes was so random, it could only be described as "noise." For gearboxes, this is almost always a sign of bearing failure.

The analyst immediately notified mill personnel and strongly recommended the machine be shut down to minimize consequential damage. An undetected bearing failure can completely destroy a gearbox.

After shutting the module down, the mill inspected the suspect gearbox and found that one of the bearings had completely failed and destroyed the cage. None of the gears were damaged.

This failure was unusual because the previous set of data from the gearbox – collected just a few months earlier – was normal. Generally, bearing wear is a gradual process that is clearly evident in vibration readings as it degrades. In addition, oil samples analyzed around the time of the failure did not show any trace metal increases. This is also unusual because as a bearing degrades, tiny metal particles usually rub off into the oil for several months prior to a failure. The absence of vibration anomalies and no increase of trace metals in the surrounding oil suggest the bearing failure was sudden.

The mill had a spare gearbox which they were able to rebuild and swap out with the damaged one in just 50 hours. Since the damage was confined to one of the two rolling modules, the other was able to swap to a different size bar to keep production rolling. In total, the mill suffered just eight hours of downtime.

The ability to share data via the cloud made all of the difference. Without a vibration data collector that could collect data and transmit it to the cloud, the analyst would have had to travel to the plant—likely on Monday to review data. Even if the analyst was locally based, it may have been too late.

Case Study #2:
Remote Analysis Catches Imminent Failure Just In Time

Vibration data collected from a newly installed motor on a major industrial gas plant's critical main air compressor showed unusually high readings. This was of great concern because the motor was the site's only spare and had been installed because the previous motor was out for repair.

An analyst in South Carolina was remotely reviewing the plant's data via the cloud. Data had been collected in North Carolina with a vibration analyzer that transfers data via the cloud for analysis. The analyst noticed that at higher speeds, the amplitude of the vibration was much larger than it should be. The frequency of the energy was twice the running speed of the machine, a common indication of misalignment.

To confirm his diagnosis, the analyst asked the plant's staff to immediately collect two more sets of data with the handheld vibration data collector and upload them to him for review. He requested one set be taken with the motor running uncoupled and one with it running coupled to the compressor, but with no load. In a misalignment situation, the removal of load usually restores vibration to normal ranges.

The analyst reviewed the data as soon as it was posted and saw what he expected – no sign of the misalignment. As a final test, the analyst asked for more data, this time with the compressor fully loaded. If there was an alignment issue, the load would trigger the same vibration increases he saw when he first made his diagnosis.

As he expected, the new data confirmed the misalignment and the analyst recommended the motor be aligned promptly. Plant staff followed his advice and found the motor to be substantially misaligned. The motor would most likely have failed within 24 hours without the alignment.

New data showed the repair eliminated the high vibration with a full load. As a precaution, the analyst recommended plant staff send him weekly data sets for the next month. If the misalignment had damaged the bearings, the data would start revealing such damage over the next few weeks.

No bearing issues were detected and the motor remains problem free.

In this case, time was of the essence. The analyst received multiple sets of data and diagnosed the root cause within a 24-hour period. The plant spent an estimated $50,000 in parts and labor to perform the alignment and suffered another $50,000 in lost production during the downtime needed for the repair.

However, had the problem not been identified quickly, the motor would have failed at a much higher cost. It would have taken several days to locate, deliver, and install a new one. It is estimated that the new equipment, associated downtime, trucking costs, and customer fees for missed deliveries could have topped $1 million.

Technology was the key to the analyst being able to work quickly and remotely. Plant staff uploaded data via the cloud and the analyst reviewed it and issued timely, actionable advice from miles away. If there had been no alternative to bringing an analyst on site, the machine may have failed with the analyst in transit.

This intervention on a single piece of equipment saved the company more than the total annual cost of its reliability program across 100 of its plants.

Case Study #3:
Remote Access to Instrument Enables Analyst to Nurses Critical Expander

Today's vibration analyzers can not only bring data to the analyst, but they can also bring the analyst to his instrument.

This is what happened in Pennsylvania when an expander at a large industrial gases facility began exhibiting serious problems during the December holidays.

The expander uses proximity probes to detect high vibration. The probes are attached to a transmitter, which converts the signal to a 4-20 output and connects directly to a DCS system. The DCS readings had been gradually increasing throughout the fall. As luck would have it, the readings suddenly soared in December, nearly tripping the expander. On-site employees replaced that expander with a spare. However, the new expander impeller started to come in contact with the shroud during startup, causing damage to the impeller. The company replaced the expander with yet another spare. Meanwhile, they sent the original, faulty expander to the repair shop.

The inspection revealed no condition that could have caused the increase in vibration.

The on-site person who normally collects vibration data at the plant was on vacation and had taken his analyzer with him, so the third-party, remote analyst for the site was called in to assist. This analyst brought his analyzer to the site and, after collecting data from the expander and transmitting it to the cloud, left his unit there--connected and ready to take data.

He went home and over the next several days accessed his vibration analyzer over the Internet and collected various tests under a variety of conditions. At one point, the plant had an unexpected power outage and he was able to capture valuable startup data. His analysis was done via the cloud, right from his home.

It is important to also note another hurdle that posed little trouble for the instrumentation. Strict IT policies at the plant prohibit access to the Internet via its LAN/WAN. To resolve this, the analyst showed the plant's staff how to use a mobile hotspot so the data collector could connect to the Internet without exposing the network.

After comparing many sets of data to what the DCS was showing, the analyst determined that that the high vibration was not coming from the expander at all. It was a faulty DCS sending the high vibration.

The analyst continued to collect data remotely from the expander for six weeks, while new proximity probes and transmitters were ordered and installed. This was important to the plant, as the expander was critical to production.

Without the ability to remotely access his instrument, the analyst would have had to stay on site for a lengthy amount of time to capture all of the data needed at a large expense to the plant. (The analyst is a third-party analysis provider.)

By capitalizing on the vibration analyzer's underlying technology, the analyst was able to get the right data at the right time and solve the problem.

The ability to nurse equipment in this way was unheard of with the analyzers of yesteryear. With the new paradigm of instrumentation, however, it is becoming the norm.

In Summary

Ten years from now, young analysts will look at today's purpose-built, monolithic vibration analyzers and consider them as old fashioned as we consider those clunky flip phones. The new generation of PdM personnel will arrive on the job already tablet and smartphone savvy, expecting and embracing the latest technologies.

Today's paradigm shift to Internet-enabled instruments and cloud-based storage will eventually become the norm. Management will increasingly demand visibility into program metrics on a corporate scale as it realizes the efficiency gains possible through "big data."

Whether we embrace these changes -- or go kicking and screaming – there's little doubt that technology improvements will continue to drive PdM and its instrumentation into the cloud.

While a lot is changing, one thing remains a constant: The need to sustain a solid PdM program that ensures maximum uptime and gives maintenance staffs ample warning of needed repairs.

# # #

About the Author

Ken Piety is a pioneer in vibration analysis and other condition assessment technologies. Ken holds more than thirty patents related to predictive and proactive maintenance technologies.

He is currently the vice president of technology for Azima DLI continuing his vision of innovation in product development. He was a co-founder of Computational Systems Inc. and a key contributor at CSI for nearly twenty years holding a number of senior-level positions in marketing and engineering.

Ken's skills have been put to work for General Electric, Technology for Energy Corporation, the Oak Ridge National Laboratory, and other leading organizations.

He holds a Ph.D. in Nuclear Engineering from the University of Tennessee.

GAS TURBINE INTRODUCTION AND VIBRATION DIAGNOSTIC BASICS

Timothy S Irwin, P.E.

Mitsubishi Hitachi Power Systems Americas, Inc.
Gas Turbine Service Engineering Manager
2287 Premier Row
Orlando FL, 32809
Timothy.Irwin@MHPowerSystems.com

Abstract: Today's large frame power generation gas turbines are complex machines with a large number of components and systems that affect their operating characteristics. There are many similarities to other large frame turbines (such as steam), but understanding the mechanical details of these machines will improve the vibration diagnostic analyst's efficiency and efforts to accurately assess the machine's condition and potential issues. A thorough understanding of the details of the rotor assembly structure, stationary component structure, casing and structural support, bearing design, rotor and casing seal interfaces, and auxiliary system operation will improve the analyst's ability to determine specific vibration characteristics and fault conditions. This paper will describe the general physical characteristics of a large frame gas turbine and vibration data formats that are typically used for diagnostic purposes on these machines.

Keywords: Diagnostics; Faults; Gas; Plots; Turbines; Vibration

Introduction:

There are several thousand of large frame power generation gas turbines presently in use in the United States. The technical concepts of both steam and gas turbines have been around for hundreds of years and there are many books and articles available that document the history and development of these machines. Technological advances in recent decades have allowed the gas turbine to become the preferred land based power generation method for base load and load following duties of today. The advances in metallurgy and technical concepts have allowed the large frame gas turbine to exceed 40% efficiency in simple cycle mode and to exceed 60% efficiency in combined cycle mode. The quick response in startup and load following of the most recent designs are a very good match for balancing load demands with renewable power sources.

There are benefits and advantages to understanding the mechanical assembly of large frame gas turbines for personnel that operate and maintain these machines. There are many mechanical similarities to other machines that result in similar operational behavior. There are also many mechanical differences to these machines that may result in similar operational behaviors to other machines, but are driven by different conditions. Understanding the mechanical assembly and desired/typical operational behaviors will make an analyst's condition assessment and subsequent recommendations much more accurate improving equipment and overall plant operation.

Gas Turbine Components and Assembly: Figure 1 below shows a general cutaway of a large frame gas turbine. Figure 1 will be referenced for the discussion of the major components and assembly of a gas turbine. The cutaway referenced shows the inlet to the left and the exhaust to the right.

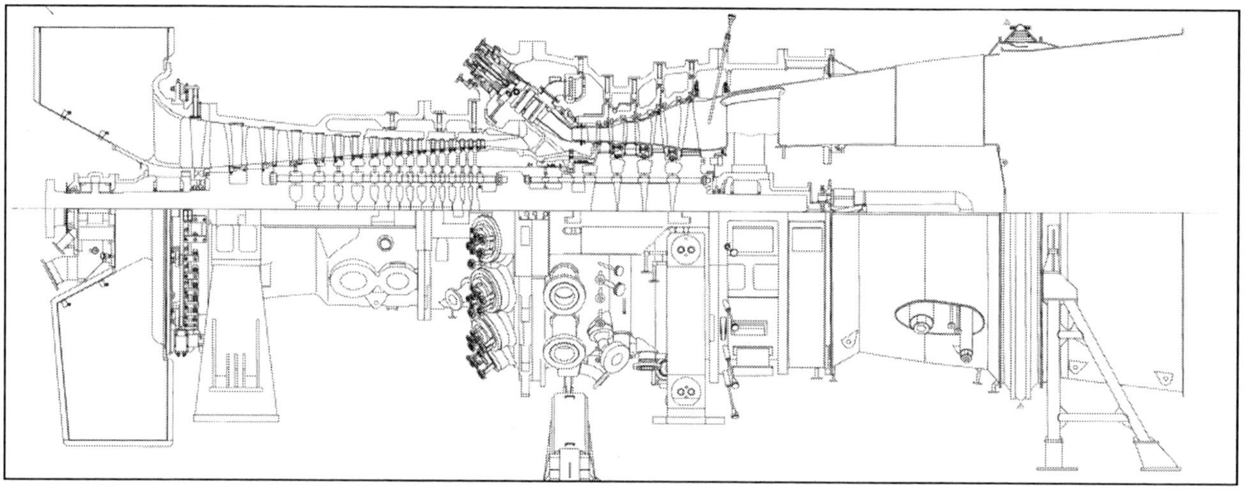

Figure 1 - M501G Large Frame Gas Turbine

Stationary Components: The inlet duct provides the compressor inlet air flow drawing air flow through the silencers and inlet filters to the inlet bell of the compressor casing. The inlet bellmouth assembly includes the thrust bearing, the compressor end journal bearing, inlet guide vanes, and various shaft seal assemblies. The inlet bellmouth is mounted to the compressor casing assembly. The compressor casing houses and includes the compressor diaphragms, compressor blade rings, and compressor structural support. The compressor casing is then bolted to the compressor/combustor (CC) casing that includes two more compressor blade rings, fuel nozzles, baskets, transition pieces, torque tube cover and torque tube seals. The CC casing is bolted to the turbine casing that includes the bypass ring, stationary seal assembly, turbine structural supports, and four turbine blade rings including inner stage seals. The turbine casing is bolted to the exhaust casing which includes dead air space seals, exhaust diffuser assembly and bearing tunnel, exhaust bearing struts, turbine journal bearing and associated shaft seals, and mechanical overspeed trip assembly. The turbine casing is bolted to the exhaust manifold that includes access to the exhaust bearing tunnel, associated auxiliary piping, and an expansion joint connection to the downstream duct work.

Rotor Structure: Starting from the inlet the compressor stub shaft includes the coupling, thrust bearing collar, compressor journal, and first three rows of discs all as part of a single shaft section; the remainder of the compressor section is made up with 14 individual discs that are all bolted together between the stub shaft and the torque tube. The center section of the machine is the torque tube and it is bolted to the seal disc. The seal disc is then bolted as an assembly with the four turbine discs. The row 4 turbine disc also includes the exhaust bearing journal section.

After the general overview of the whole machine, details of the various sections will be reviewed to understand the mechanical interfaces between the rotor and the casings.

The mechanical interfaces between the rotor and stationary components include the following items of each section:

- Inlet bellmouth assembly – various shaft seals, thrust bearing, and journal bearing.
- Compressor section – blade tips to the casing and blade rings; and the diaphragm shaft seals to the compressor discs.
- Torque tube and seal disc section - shaft seals.
- Turbine section - blade ring inner stage seals, exhaust bearing air and oil seals, and exhaust journal bearing.

The above interfaces can also be classified as intended or managed contact (bearings) and unintended contact (various shaft seals).

Knowing the bearing configuration is very important to understanding typical and non-typical vibration signatures. Figure 2 below shows three common bearing configurations used on today's large frame gas turbines and their pressure profiles. Vibration signatures between a sleeve style bearing and a tilt pad bearing will typically show some different characteristics under the same operational and load conditions. Reactions during transient conditions and shaft criticals may also be different between bearing designs due to different damping characteristics of the different bearing types.

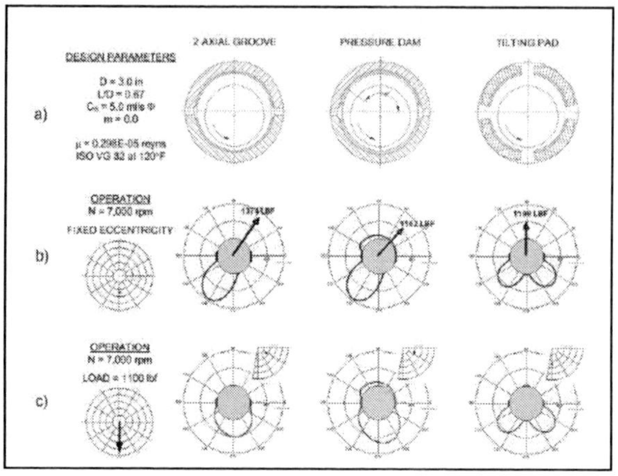

Figure 2 - Typical Bearing Configurations [ref 1]

Knowing whether the shaft seals are tooth type seal strips, honeycomb sections, or floating type rings, or bush seals provides an advantage to the person performing the diagnostics. When diagnosing vibration signatures understanding the various contact types and their locations within the machine will improve the analysis.

For the operational and auxiliary system considerations please refer to Figure 1.

Operational considerations: The primary flows through the gas turbine are the inlet air and fuel that is combined and ignited creating the combustion gas. The fuel and air are mixed in the combustor baskets and ignited in the transition piece that directs the combustion gas flow to the turbine section. The combustion gas provides the torque and motive force for the compressor to supply the necessary air supply along with the additional power to drive the generator. The compression of the air provides axial load on the rotor in the upstream direction, the exhaust gases provide axial load in the turbine section in the downstream direction. The resultant axial force will vary in magnitude and may change directions dependent on the operational speed and load.

Auxiliary operational considerations: The auxiliary systems that can have a direct influence on the vibrational characteristics of the machine are the fuel system, bearing lubrication system, the vapor extraction system, the seal air system, the compressor bleed air system, the vane cooling air system, and the turbine cooling air system.

There are several physical configuration and operational differences that need to be considered when performing vibration diagnostics.

Large frame gas turbine rotors are considered flexible rotors and typically operate between their 2nd and 3rd shaft lateral criticals.

The pedestal and bearing support structures are very different than typical turbomachinery (steam turbines, generators, compressors, etc.) structural supports. The bearings are not mounted to pedestals that sit directly on a large concrete foundations; the gas turbine casing is mounted to vertical steel supports (trunnions) that are then mounted to the foundation. The exhaust bearing is supported by tangential struts to the outer casing and is axially downstream from the trunnions; so the exhaust bearing is not vertically located over the concrete pedestal. Similarly the compressor bearing is also supported within the inlet assembly and is supported by radial struts that are axially upstream from the casing support to the foundation.

Once the machine has reached operational speed (3600 rpm Full Speed No Load) the machine is not truly idling, but is at less than 50% of full power through the shaft train. The compressor consumes this power from the turbine to provide the necessary air flow for combustion. Therefore at 'idle' the vibration signature is not as 'lightly' loaded as other rotating machinery, but achieving thermal stability is appropriate before completing the analysis.

Rotor System Vibration Parameters: The rotor system parameters and nomenclature that are used to analyze and describe rotor dynamics and vibration are reviewed below. Rotor system parameters are mass, stiffness, and damping. The equation below describes how displacement, force, and dynamic stiffness are related. It is important to note that as described in the equation the parameters are not linearly related as it is shown as vector and summation quantities (they are not linear).

$$\overrightarrow{\text{Displacement}} = \frac{\sum \overrightarrow{\text{Forces}}}{\sum \overrightarrow{\text{Dynamic Stiffness}}}$$

A change in displacement (vibration) is due to a change in the force applied or stiffness of the system. The equation below shows the relationship between frequency, stiffness, and mass.

$$F_n = \frac{1}{2}\Pi \times \sqrt[2]{\frac{k}{M}}$$

F_n = frequency, Hertz
k = stiffness, N/m
M = mass, kg
Π = 3.14159

Using the above equations as the foundation for vibration, force, mass, stiffness and response, the following equation was developed to explain the motion of a rotor relative to the forces on it.

Rotor Equation of Motion (for fluid film bearings):

$$F_s + F_t + F_d + F_p = M\ddot{r}$$

Defined as - Spring stiffness + Tangential stiffness + Damping force + Pertubation force = mass x acceleration:

$$-Kr + jD\lambda\Omega r - D\dot{r} + mr_u\omega^2 e^{j(\omega t+\delta)} = M\ddot{r}$$

The above method is more fully described in "Fundamentals of Rotating Machinery Diagnostics – Charles T. Hatch & Donald E. Bently" [ref 2]

Forces that Act on the Machine: Continuing with the basic understanding of rotor dynamics, the forces that actually act on the machine must be defined. There are many variations of categorizing or grouping the forces that act on a machine. One example is below:

Internal influences – interaction with parts of itself:

- Support forces in bearings
- Forces resulting from shaft deflection
- Fluid dynamics and forces

External influences – forces that are applied to the rotor system:

- Axial Restraint – Rub
- Casing distortion – Rub
- Static radial loads - Rotor Misalignment

Unbalance – Is listed separately, but is typically considered an external force

Fault Conditions:

The forces that are categorized above are driven by specific conditions. Below is a list of the most typical fault conditions for large frame gas turbines:

- Instrumentation Error
- Unbalance
- Misalignment
  - Rotor to rotor
  - Casing to rotor

    - Radial
    - Axial
- Bearing Damage or excessive clearance
- Rotor Bow
- Looseness
- Fluid Induced Instability
- Shaft Crack
- Resonance

The above items and the various details within each are what should be considered and evaluated whenever reviewing vibration and operational data for machinery condition analysis. Knowing the mechanical details and auxiliary system operational conditions will allow the best opportunity for the analyst to perform detailed and accurate diagnostics of a specific condition.

Data Acquisition Parameters: Dynamic vibration data collection is typically performed by connecting a temporary dedicated data acquisition system to either permanently mounted vibration transducers or temporary transducers on the machine in the desired locations. The desired locations are to have two transducers located at each bearing at 90 degrees to each other. It is also necessary to have a Keyphasor® installed and data collected for a phase reference.

When performing vibration data acquisition there are additional operating data parameters that should also be collected and reviewed:

- Machine operating conditions including megawatts, megavars, air inlet temperatures, blade path temperatures and rotor speed.
- Lube oil system operating conditions including, supply pressure, supply temperature, and drain temperatures
- Bearing metal temperatures

Defining additional terms:
Steady State Conditions: The machinery is operating under steady conditions i.e. steady load, speed, etc.

Transient Conditions: The machinery is operating under changing conditions i.e. changing speeds (startup, coast down), changing loads

Static Data: can be represented by a single number or status. Examples include vibration amplitude filtered vibration amplitude, filtered vibration phase lag, transducer gap voltage, etc. Plot examples: Tabular List, Trend, Spectrum, etc.

Dynamic Data: the actual vibration waveform. Sampling of the waveform and processing to provide both time domain (Orbit or Timebase plots) and frequency domain (Spectrum plots). Dynamic data cannot be characterized by a single number. Plot Examples: Bode, Polar, Shaft Centerline, Timebase, Orbits, Cascade, Waterfall.

Slow roll compensation is another critical parameter that is necessary when using eddy current proximity probes. A data sample is collected at an rpm of approximately 10% of design operating speed. This data sample is the 'noise' in the signal from these types of probes that includes shaft surface defects, mechanical run-out, and residual magnetism (electrical run-out). This data sample can then be used to subtract this 'base' noise from the signal at normal operating speeds.

Plot Examples: Trend plots are used for tracking the amplitude and amplitude changes of a variable relative to time. These changes can occur suddenly or gradually and are typically evaluated in relation to alarm or trip levels that require some kind of corrective action.

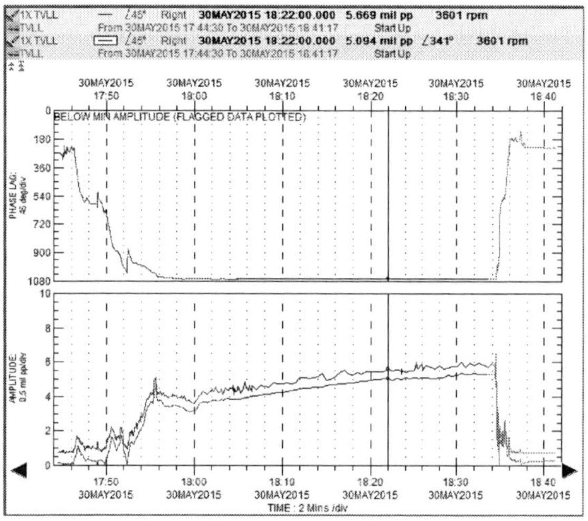

Figure 3 – Trend Plot 1

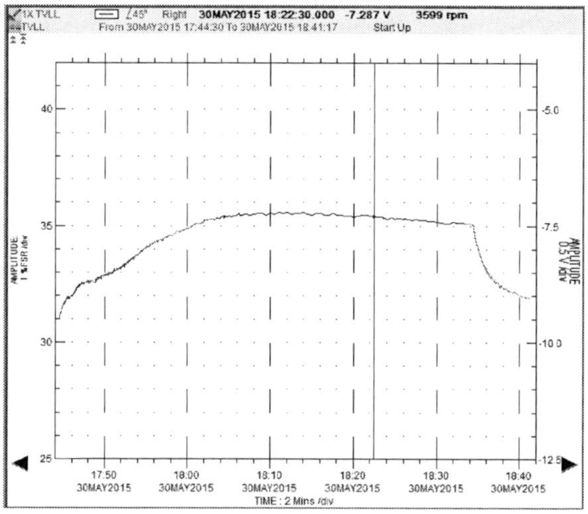

Figure 4 – Trend Plot 2

Figure 3 above shows a trend plot of an overall amplitude, 1X amplitude, and the 1X phase relative to time during a startup, loaded operation and a coast down. The amplitude levels in trend plots are limited in diagnostic details, but are a good indication of overall machine condition and indicate when conditions may have changed. Figure 4 above shows a gap voltage trend during startup, loaded operation, and coast down.

Bode plots are used for enhancement of the amplitude and phase response to shaft rotative speed. Using this information a rotor/bearing system parameter called "synchronous amplification response" can be calculated. They are also used to determine the slow roll vector that should be used for compensation. The Bode 1X and amplitude plots are filtered synchronously with shaft speed. The amplitude and phase data is plotted relative to shaft rotational speed.

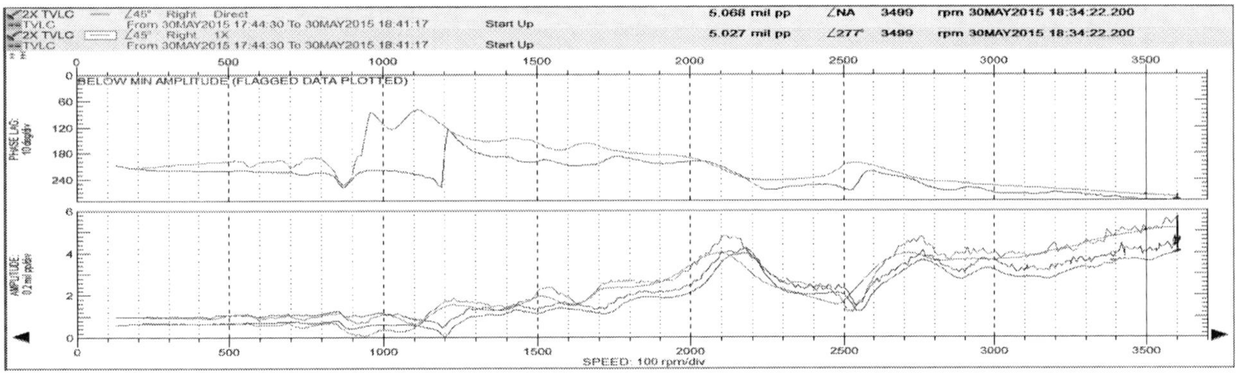

Figure 5 - Bode Plot 1 – Uncompensated

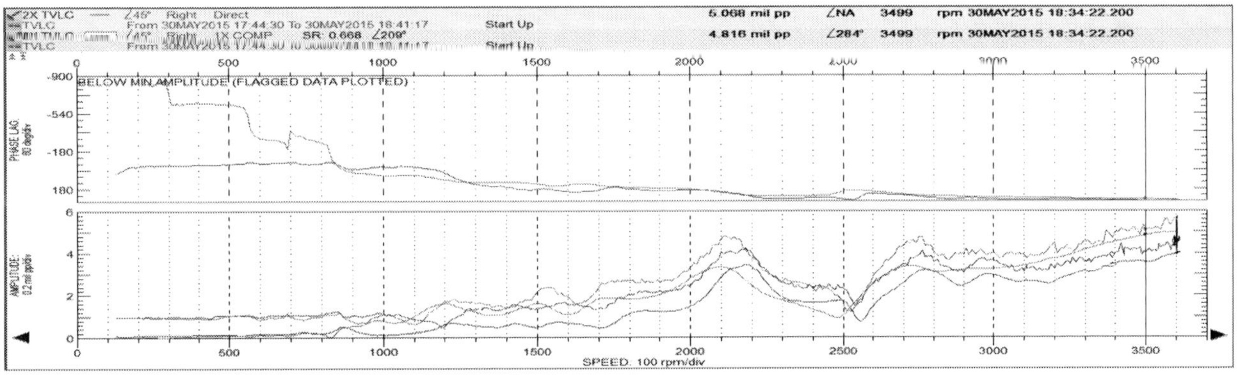

Figure 6 - Bode Plot 2 - Compensated

The two Bode plots above in Figures 5 and 6 show a startup, loaded operation, and coast down. The data includes overall, 1X and 1X phase. The two plots show the variation of the 1X compensated and uncompensated data.

Polar (Nyquist) plots use the same data applied to Bode plots, but are represented in a different format. The plot format is a circular plot (polar coordinates) that shows the vibration vector (phase angle and amplitude) typically referenced to speed or time.

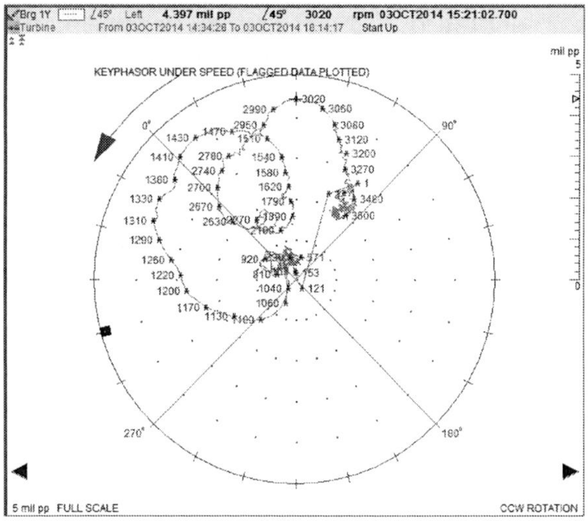

Figure 7 - Polar Plot 1

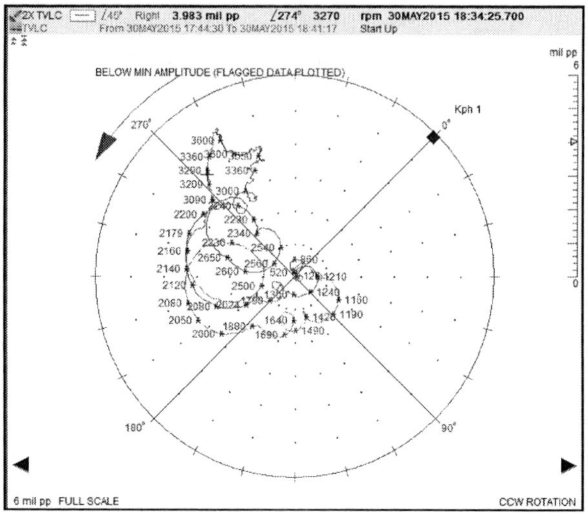

Figure 8 – Polar Plot 2

Figures 7 and 8 above are plots from two different bearings showing some of the differences that can be noted. Polar plots are very useful for performing balancing work. The pattern and phase relationship of the 1st and 2nd shaft criticals along with determining if a shaft critical is being approached are very appropriate to know when to perform balancing changes. Comparing startup to coast down data can provide information was to whether a machine is thermally sensitive or other influences are changing the vibration signature from the startup condition. Patterns and loops within the plot can show structural resonances.

Timebase (Waveform) plots show the actual signal that provides the information that will be processed into the other plot formats. The unfiltered and filtered plots allow examination of the 'form' of the vibration. The form is evaluated for whether the signal is clean or if there is a lot of frequency content, is the form repetitive, is there a preload, etc.

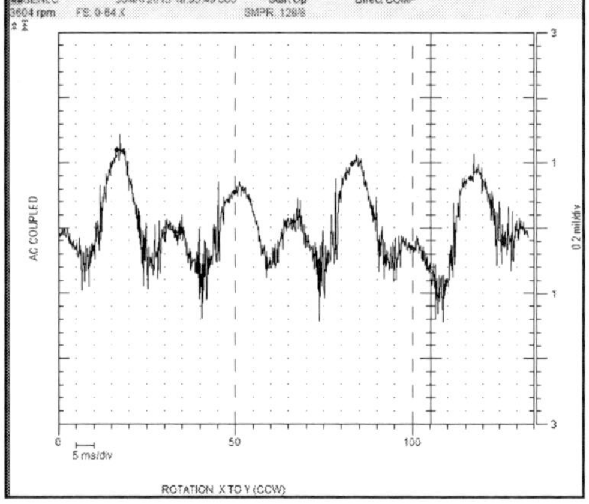

Figure 9 – Timebase Plot 1

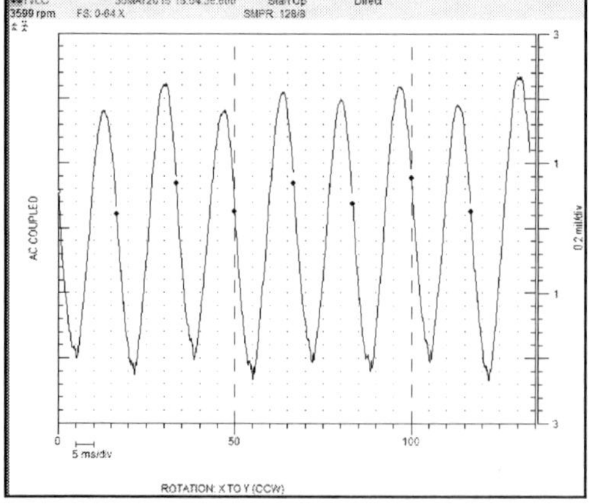

Figure 10 – Timebase Plot 2

Figures 9 and 10 above show examples of Timebase plots unfiltered and filtered. The high frequency content and uneven peaks of the unfiltered plots indicates that slow roll data and filtered analysis is appropriate. The filtered plot shows much more consistent amplitudes and phase consistency but there is still some variation rotation to rotation.

Orbit plots are processed from the information in the Timebase plots but in a different format. This plot format shows the two dimensional motion of the shaft centerline perpendicular to the shaft centerline or in the axial view of the shaft. Orbit plots can also be filtered and unfiltered for evaluating the different components of the vibration motion.

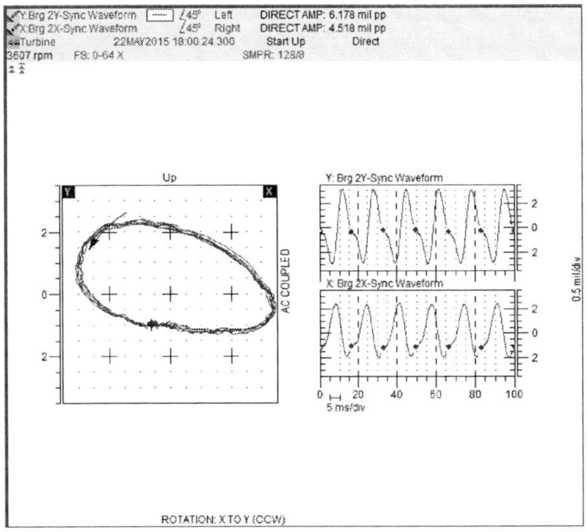

Figure 11 – Orbit Timebase Plot

The Orbit plot to the right, Figure 11, shows the unfiltered Timebase channels that formed the unfiltered Orbit plot. This Orbit shows a mostly horizontal attitude angle with a flattened movement through the bottom of the shaft movement indicating a possible preload.

Shaft average centerline plots are used to view the changes of the average shaft centerline position within the bearing's radial clearance. A change in the shaft position within the bearing clearance can be an indication of a change in alignment, bearing wear, the amount of preload and the dynamic stability of the bearing.

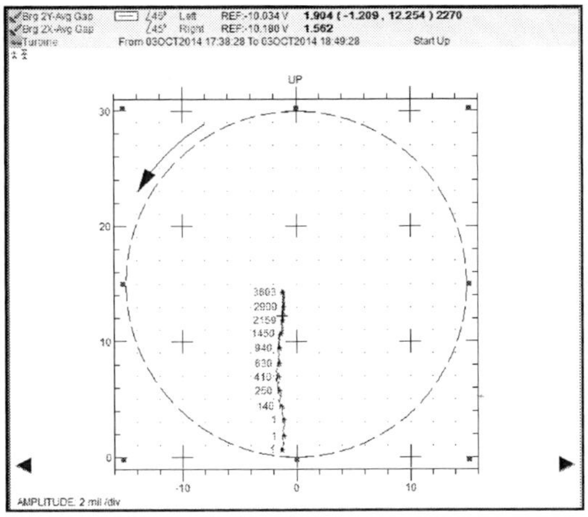

Figure 12 – Shaft Centerline Plot 1

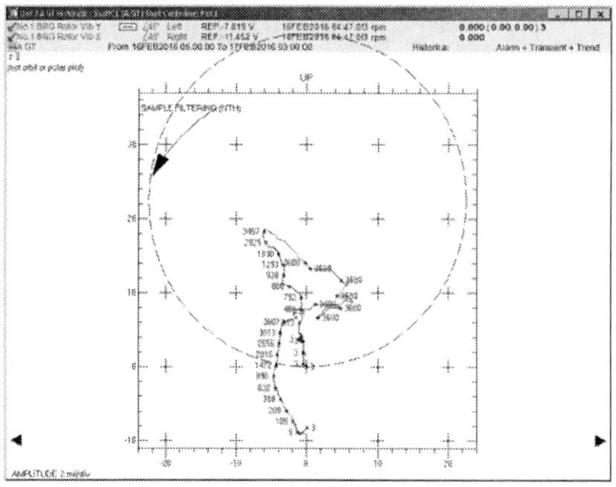

Figure 13 – Shaft Centerline Plot 2

Figure 12 above shows a typical shaft average centerline movement for a tilt pad bearing with the shaft rising from bottom dead center of the bearing to near the center of the bearing clearance. Figure 13 above shows what appears to be an abnormal condition, but in reality this type of movement (from startup to shutdown) is not unusual due to the thermal changes the casing, components, and probes incur during the startup for this location.

Full spectrum plots including single sample spectrums, waterfall spectrums, and cascade spectrums show both the forward precession and reverse precession components of the vibration signal. These plots are a spectral breakdown of the vibration signature that is shown in the timebase/orbit plots. The forward precession components of the signal represent the vibration that rotates in the direction of rotation; the reverse precession components of the signal represent the vibration that rotates opposite the direction of rotation.

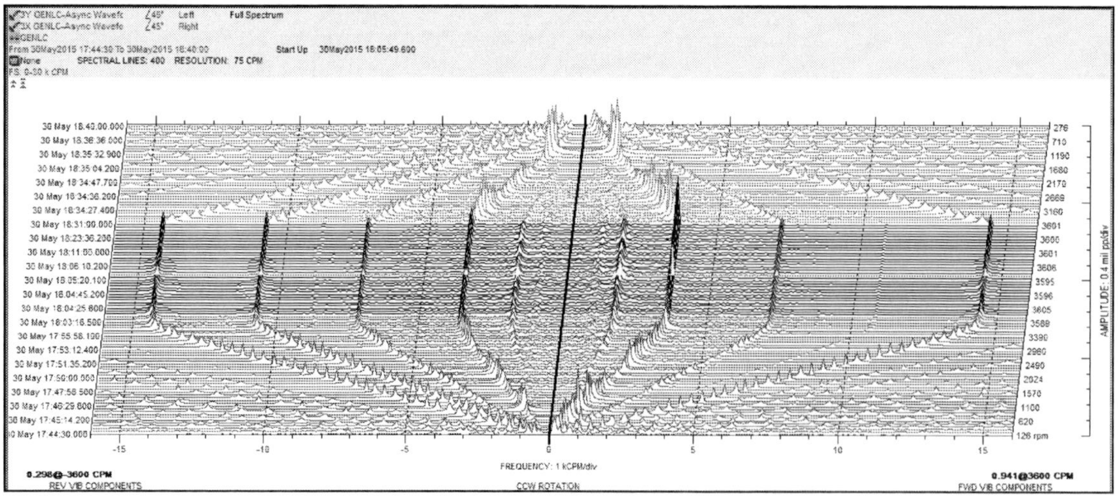

Plot 14 – Full Spectrum Waterfall Plot

Figure 14 above shows an example of a full spectrum waterfall plot. You can note the behavior of the sub-synchronous, synchronous, and synchronous multiples. Comparing the forward precession, (right) side of the plot, to the reverse precession, (left) side of the plot, shows that the dominant energy during the majority of the operational run was the forward precessed vibration. During the 3600 rpm run time the dominant energy was synchronous (1X) with some influence of sub-synchronous and synchronous multiples. During the transient conditions the synchronous forward and reverse precession energy is dominant.

Summary:

In essence, the more known about a machine, the various components, and their failure or fault modes, the more thorough a condition analysis can be. Understanding the potential faults and what can drive them allows a more comprehensive examination of the machine's operating parameters including typical operating conditions and vibration data. Analyzing the vibration data and evaluating the vibration signature relative to typical fault conditions takes practice and experience. As knowledge of the machine increases and vibration analysis experience increases then the more thorough and accurate the analysis will be.

References:

1. M. He, C. H. Cloud, J. M. Byrne, Fundamentals of Fluid Film Bearing Operation and Modeling, Proceedings of the 34th Turbomachinery Symposium 2005, Pages 155-176
2. C.T Hatch, D.E. Bently, Fundamentals of Rotating Machinery Diagnostics, Bently Pressurized Bearing Press

Bibliography:

1. M.P. Boyce, Gas Turbine Engineering Handbook, Elsevier Science
2. Investigation of Vibration, Westinghouse Electric Corporation Technical Bulletin – PH9603.51
3. V. Wowk, Machinery Vibration – Measurement and Analysis, McGraw-Hill
4. ISO – 10816 – Mechanical Vibration
5. ISO – 7919 – Mechanical Vibration
6. API – 670 – Machinery Protection systems

Biography:

Timothy Irwin graduated from University of Illinois in 1985 and has spent his 30 year career providing technical assistance and guidance on rotating machinery. He has worked for large turbine (steam and gas) OEMs; as a Mechanical Engineer for Utilities with responsibility for both critical and balance of plant rotating machinery (turbines, generators, pumps, gearboxes, air compressors, etc.); and as a 3rd party consultant providing machinery diagnostic (vibration and process) and outage/overhaul related technical direction support for both Utility and Industrial Class machines.

He presently works for Mitsubishi Hitachi Power Systems Americas, Inc. as a Gas Turbine Service Engineering Manager and is responsible for providing Corporate Engineering support to Mitsubishi customers and equipment owners.

He has been a member of the Vibration Institute since 1993, was actively involved in the Piedmont Chapter from 2000 to 2011, and is presently vice chair of the Central Florida Chapter.

DYNAMIC ANALYSIS And FIELD BALANCING Of 100 MW UTILITY GAS TURBINE - GENERATORS

Edgar J. Gunter, Ph.D.
RODYN *Vibration Analysis, Inc.*
Professor Emeritus Mechanical and Aerospace Engineering
Former Director, Rotor Dynamics Laboratory, Univ. Of Virginia
Fellow ASME
DrGunter@aol.com

John Waite, Engineer III
Power Generation Engineering, Dominion Resources
Glen Allen,VA
John.k.Waite@dom.com

Abstract

This paper deals with some of the challenges in the dynamic analysis and field balancing of the W501D gas turbine-generator system currently used in a number of utilities. The current W501D model represents a power enhancement over the earlier Westinghouse 70 MW units which have been in service for many years. The utility gas turbines present a number of challenges as compared to conventional steam turbines. Because of the design of the gas turbine, the hot end of the turbine requires a very flexible support under the tilting pad bearing. This in turn causes a significant reduction of stiffness and particularly of damping. In addition, there is substantial rotor-casing interaction which is not present in conventional steam turbines. As a result the dynamic characteristics are totally unlike that encountered with conventional steam turbines. In 1985, papers were presented at Vibration Institute meetings on the dynamic analysis and field balancing of 70 MW gas turbine-generator systems. Extensive experimental data was collected on various units at that time. It was determined that computer assisted balancing was required to properly calculate the balance corrections necessary for field balancing. Unlike steam turbines, in which the casing is quite substantial, the W501 series of gas turbines have significant casing motion present in the operating speed range. The dynamics of the casing introduces rotor-casing modes which must be taken into consideration in order to perform proper field balancing. At the time of the earlier papers, it was not possible to compute the dynamics of the system including casing effects. Currently, with the modern finite element computer codes, the casing can be included in the dynamic analysis of the turbine-generator to produce a more realistic representation of the system. In the earlier analysis of the 70 MW gas turbine system, the transfer matrix method of rotor dynamics was employed. This has proven to be insufficient for the analysis of the gas turbine-generator systems including casing and foundation effects to compute stability and predicted unbalance response. It is only with the finite element approach that a complete analysis of the system can be accurately made. A significant amount of field data is presented and descriptions on recommended methods of balancing the W501D are presented.

Introduction and Statement of The Problem

Figure 1 Represents the Cross-Section of a modern W501D gas turbine which has been upgraded to 100 MW from the earlier Westinghouse 70 MW design. With the current emphasis on the reduction of CO_2 emissions from utilities, the W501D gas turbine generator systems have become very popular with various utilities. Although the W501 series of gas turbines have been very successful, and in use for many years, they are not without their problems.

One of the difficulties encountered with the gas turbines is the requirement for a very flexible support under the tilting pad bearing at the gas turbine hot end. This not only causes a reduction of the effective bearing stiffness, but can lead to over a 90% reduction in the tilting pad bearing effective damping. This can then cause the turbine to be very sensitive to unbalance conditions and thermal bowing in the turbine and distortion in the generator. In addition, there is considerable casing motion that interacts with the rotor that is not present in your typical steam turbine design. This then can lead to additional critical speeds in the operating speed range. This has led to difficulties in many instances of balancing.

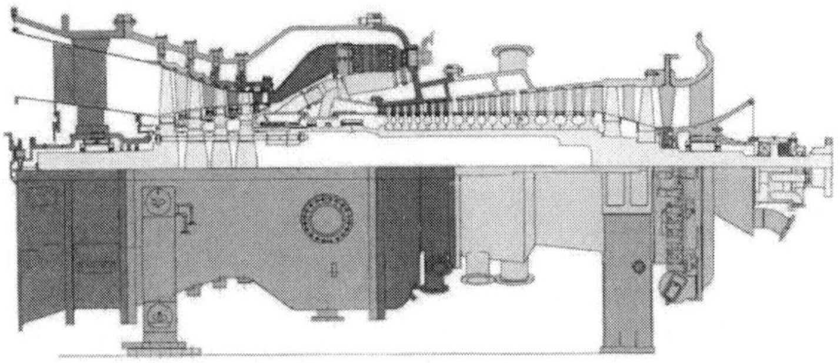

Fig. 1 Schematic Drawing of 100 MW Utility Gas Turbine

Figure 2 represents a W501D turbine which has been removed for maintenance. Visible in this picture are the turbine center plane balance holes. In the photograph are shown the four turbine stages and 17 smaller axial turbine blades. Additional planes available for balancing are located at the compressor inlet, the last large outboard turbine station, and also at the coupling.

Fig. 2 100 MW Gas Turbine Being Removed For Servicing

Background on Analysis and Balancing Earlier 70 MW Gas Turbine-Generators

Figure 3 shows a cross-section of the 70 MW gas turbine showing the balance planes. The Rotor Dynamics Laboratory at the University of Virginia was contacted in 1983 by Carolina Power & Light to investigate the difficulties encountered with the field balancing of several of their W501 70 MW gas turbine-generators.

In particular one field engineer had spent three months in an attempt to balance one of the gas turbines. He was instructed by field service to perform a simultaneous two plane balance. However, in the field, he would place a balance weight on the turbine which in turn would cause high vibrations on the compressor side. A correction weight placed at the compressor balance plane would then in turn cause high turbine end vibrations.

Gas turbine balancing runs can be extremely time-consuming. For example, after a weight has been placed at the hot end of the gas turbine, it usually takes a day for the system to cool down before another balance weight may be applied at that location.

It is particularly embarrassing if a balance weight should be dropped into the gas turbine balancing location. This then requires a dismantling of the turbine section in order to remove the missing weight. Failure to do so can cause the missing balance weight to be ingested by the blades leading to extensive turbine damage.

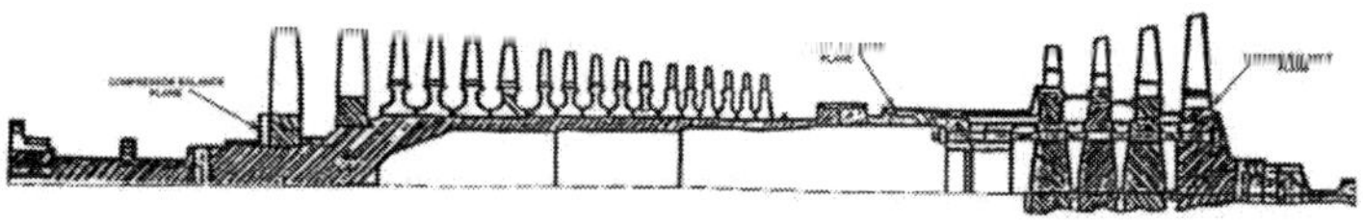

Fig. 3 Cross Section of 70 MW Gas Turbine Showing Balance Planes

The Bently Nevada Corporation, in support of the University of Virginia Rotor Dynamics Program, supplied instrumentation for the turbine generator System. Bill Pryor (currently PdM Solutions) was assigned to collect the data. Figure 4 shows the five locations along the turbine generator system where Bently 7000 series dual probes were installed on the system. The Bentley dual probes system uses a proximity probe for shaft relative motion and the velocity probe for the casing motion.

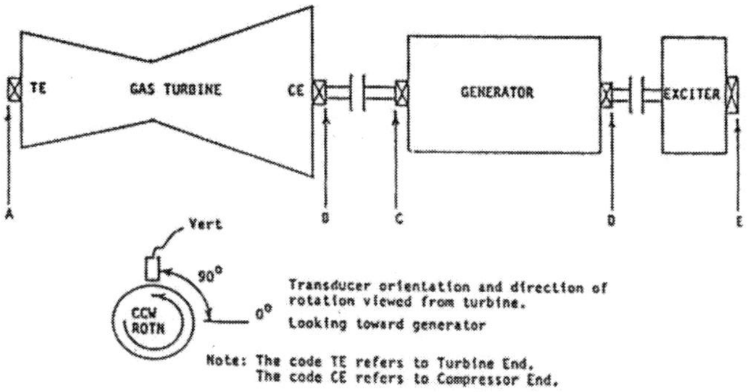

Fig. 4 Gas Turbine - Generator System Showing Measurement Planes

The casing velocity motion was integrated to determine displacement. The casing motion could then be added to the relative shaft motion to determine the absolute shaft motion in space. An additional probe referred to as a key phaser was provided to determine relative phase of motion. Accurate synchronous amplitude and phase are required for the balancing calculations.

70 MW Turbine-Generator Motion With Large Compressor Plane Unbalance:

Figure 5 represents the absolute vertical turbine motion at the hot end. Distinct turbine and generator modes are seen in the experimental data. The predicted turbine second critical speed is approximately 2700 RPM. It is noticed from the experimental data that the highest amplitudes are around 3100 RPM and 3448 RPM. These modes were not originally computed because of the inability of the transfer matrix computer program to include casing effects.

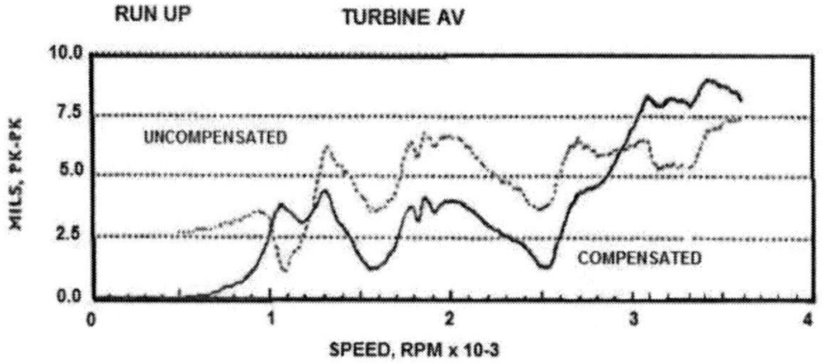

Fig. **ınce**

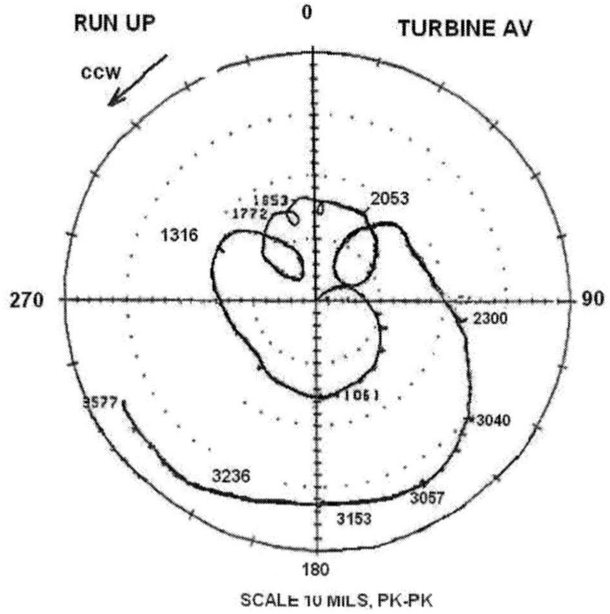

Fig. 6 Turbine Vertical Absolute Motion Polar Plot With Large Compressor Unbalance

Figure 6 represents the polar plot of the turbine vertical absolute motion with large compressor unbalance. From the examination of the polar plot, we can distinctly see various resonance frequencies in the system such as at 1051 RPM, 3016 RPM and 3057 RPM.

Of particular interest is the phase angle change that occurs from approximately 3000 RPM to 3600 RPM. This phase angle change is approximately 120°. This large phase angle change also helps to explain the difficulty in balancing the gas turbine at speed. If one considers balancing data only at running speed, then the mode at 3100 RPM will be excited.

Thus single plane balancing at the turbine location will cause high vibrations in the 3100 RPM speed range. This will result in excessive vibrations in the 3100 - 3200 RPM speed range such that it will not allow the system to come up to speed.

This mode is due to the interaction of the casing with the turbine. The mode generated is referred to as a bifurcated second critical speed. With bifurcated critical speeds, the casing can interact causing the second critical speed to split into higher and lower mode. Review of these modes will be shown in detail at a later section..

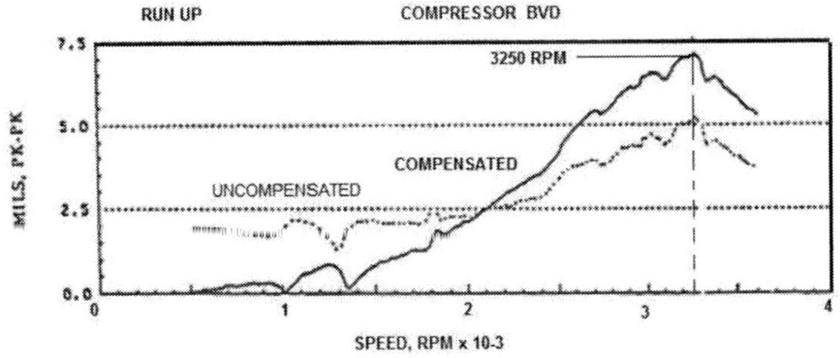

Fig. 7 Compressor Vertical Motion With Large Compressor Unbalance Showing Strong Resonance Frequency At 3,250 RPM

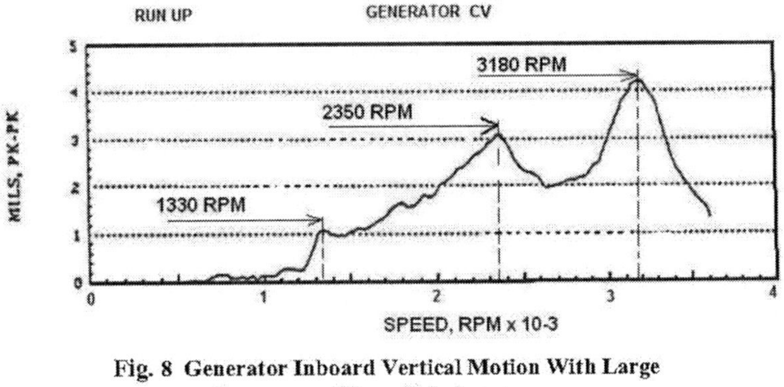

Fig. 8 Generator Inboard Vertical Motion With Large Compressor Plane Unbalance

Figure 8 represents the generator inboard vertical motion under the influence of a large unbalance weight at the gas turbine compressor inlet. The turbine rotor-casing mode is clearly seen in the vibration data at 3180 RPM. The response at 2350 RPM is the generator second mode. Also observed is the turbine mode at 1330 RPM. The computed critical speeds, including casing effects, and the corresponding mode shapes will be shown in The next section.

Problems with Generator Effecting Gas Turbine Balancing

In the course of attempting to balance the CP&L 70 MW gas turbine with a two plane balance correction, we were informed by Bill Pryor that an unusual occurrence had happened with the generator casing response. The vibration data indicated erratic behavior as well as a three times super synchronous vibration being generated. It was apparent at the time that an internal rub had occurred in the generator.

This was further verified later by examination of the amplitude and polar plots for the generator motion at station D. The turbine-casing mode at 3,200 RPM had sufficient response on the generator that it caused a horizontal generator rub. This caused a highly nonlinear jump as can be seen in Fig. 10 and Fig. 11. Balancing is not possible under these conditions and the generator must be realligned.

The problem is due in part to the poor 120° arc bearings used which provide little horizontal support.

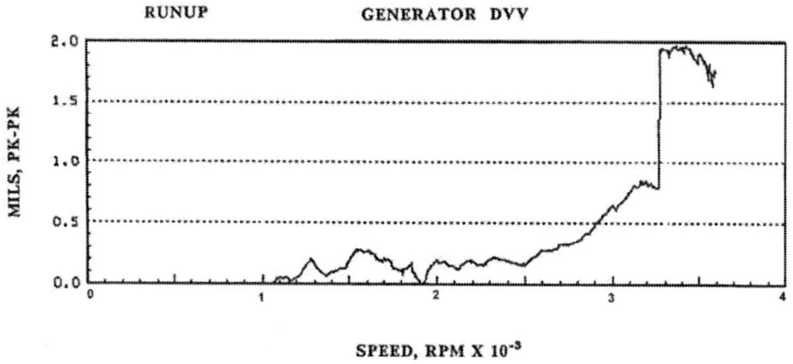

Fig. 9 Generator Casing Motion Showing Large Nonlinear Jump at 3,300 RPM

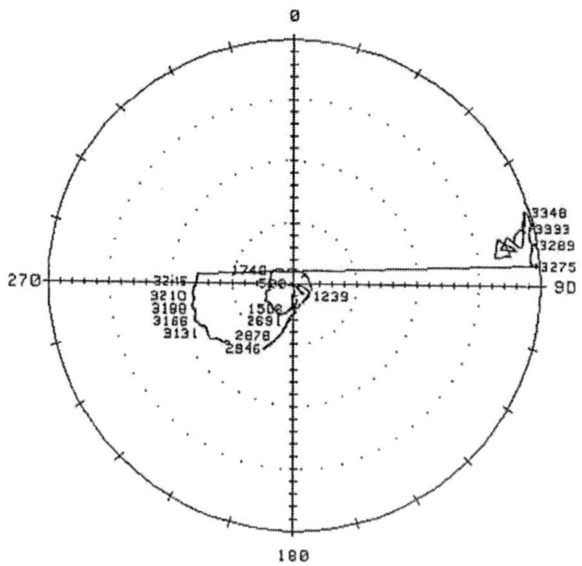

Fig. 10 Polar Plot of Generator Outboard Casing Motion Showing Large Nonlinear Jump at 3,300 RPM

Critical Speeds Of a W501D Gas-Turbine Generator

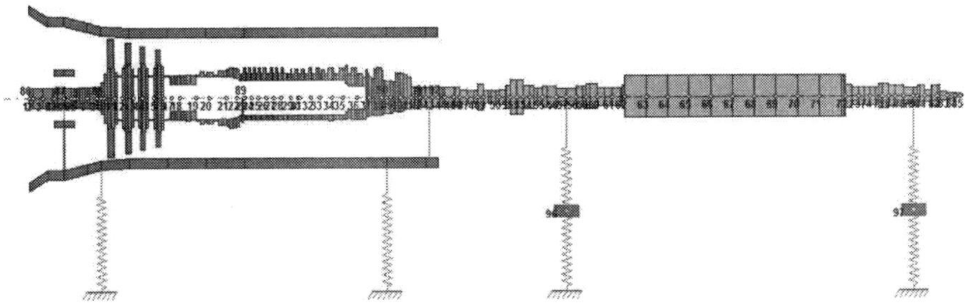

Fig. 11 Computer Model of the W501 Gas Turbine Generator

The model as shown in Figure 11 above has 92 elements and 378° of freedom. The solution of such a model would have been impossible with transfer matrix methods. In earlier reported papers on the W501, the critical speeds of the turbine-generator system were analyzed without casing effects. Other problems that are encountered with transfer matrix methods are the occurrence of missing and unconverged modes in the system. This is not the situation with the finite element method. There are no modes that are skipped in the system.

The gas turbine is supported by 2 4-pad tilting pad bearings. The generator is supported by 120° partial arc bearings. These generator partial arc bearings have been a continual source of problems as they provide low horizontal stiffness which can lead to seal rubs, provide low system damping, and are unstable with self excited whirling at running speed. Generator manufacturers are reluctant to change away from a bearing design that they have been using for over 100 years.

What is unusual about the gas turbine is the flexible webbing supporting the turbine end tilting pad bearing. In a previous paper presented on the W501 series gas turbines, it was shown that the flexible supporting web can reduce by as much as 95% the effective damping at the turbine end bearing. This combination of low turbine end damping with a flexible support and casing interaction leads to a number of unusual dynamical modes in the system.

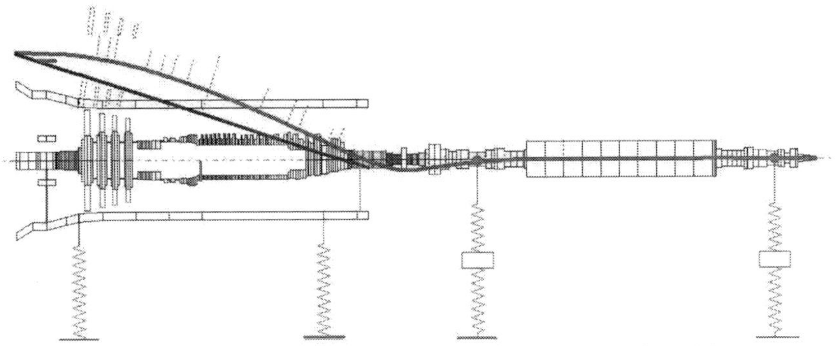

Fig. 12 Turbine-Generator 1st Undamped Mode, M1 = 688 RPM

Figure 12 represents the turbine-generator first mode at 688 RPM. These Critical Speeds are based on the average bearing horizontal and vertical stiffness values. For example, the horizontal resonance RPM could be as low as 430 RPM while the vertical could be 850 RPM.

The casing and rotor are moving in phase. This mode is caused by the low stiffness of the support at the hot end of the gas turbine.

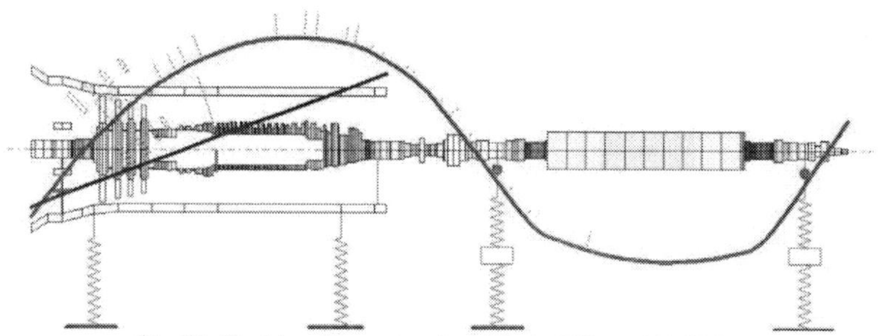

Fig. 13 Turbine-Generator 2nd Mode, M2 = 1131 RPM

Figure 13 represents a combined turbine-generator mode at 1131 RPM. In this case the casing is moving in phase with the turbine first mode. The effect of the casing has caused a bifurcated mode in which the casing at the lower speed is moving predominantly at the turbine end. In this mode the turbine end casing motion is reduced and the predominant motion as at the compressor location.

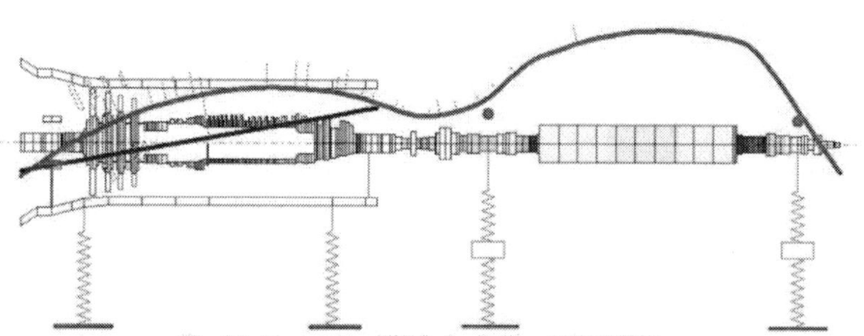

Fig. 14 Generator 1st Mode, M3 = 1171 RPM

Figure 14 for mode three is basically a generator mode at 1171 RPM. Note the close proximity of the frequencies between modes three and two. These two modes would be virtually impossible to compute by means of the transfer matrix method because of their close proximity. Closely spaced modes do not present a problem for the finite element method of computation.

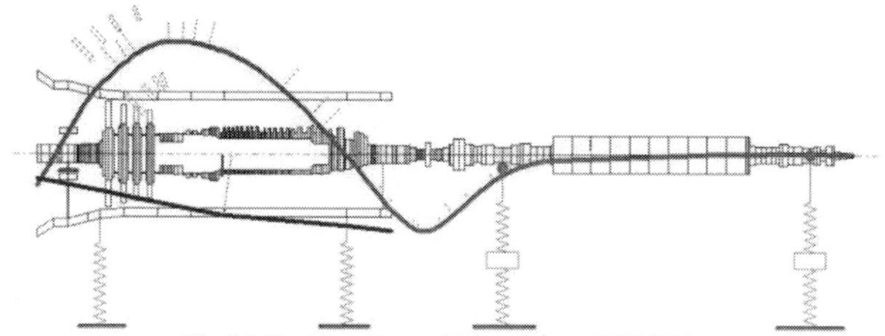

Fig. 15 Turbine-Casing Mode, M4 = 1618 RPM

Figure 15 for mode four at 1618 RPM is a bifurcated mode of the turbine mode shown as shown in Fig. 13, mode two at 1131 RPM. In the case of M4, the casing is moving out of phase with the first turbine mode. Without the influence of the casing, the first turbine mode would be somewhere between 1131 RPM and 1618 RPM.

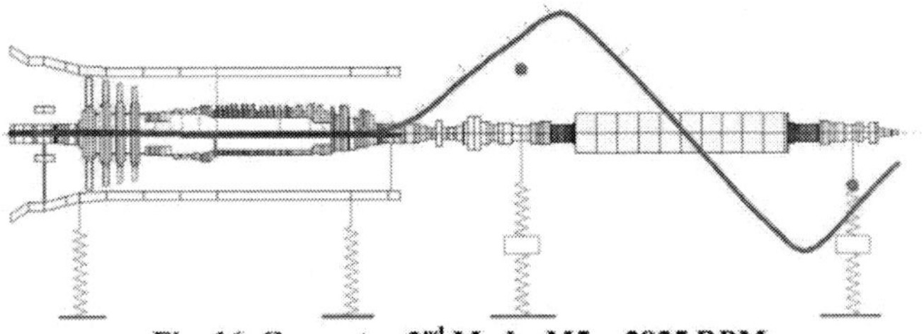

Fig. 16 Generator 2nd Mode, M5 = 2925 RPM

Fig. 16, system mode M5 at 2,925, represents the generator 2nd critical speed. It is little effected by either the turbine or the turbine casing.

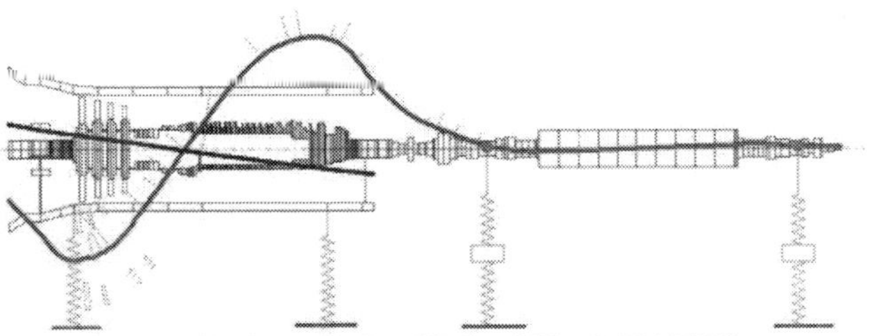

Fig. 17 Turbine-Casing Mode, M6 = 3,164 RPM

Fig. 17 represents the principal mode of concern with respect to the dynamics and the balancing of the gas turbine system. Due to the influence of the casing, the interaction of the turbine and casing causes a strong dynamical mode to fall in the operating speed range. Attempts to the balance the turbine by taking into consideration only the vibration at the operating speed of 3,600 RPM can result in an unbalance correction that will excite the system rotor dynamics response in the 3,200 RPM speed range. This may be to the extent that the turbine will trip off line before reaching operating speed.

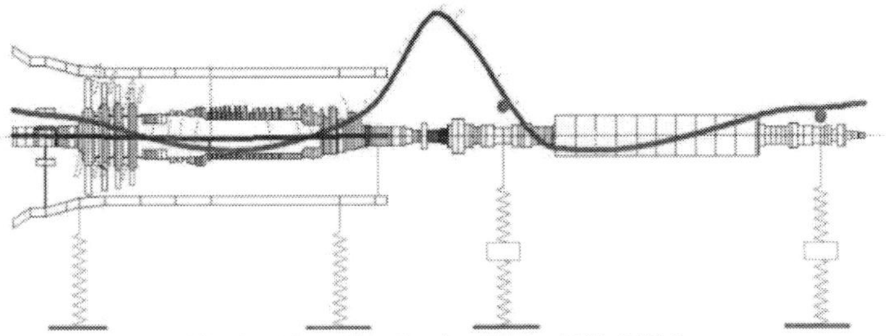

Fig. 18 Coupling Mode, M7 = 4651 RPM

Fig. 18 represents the response of the coupling supported between the turbine and generator bearings.

Gas Turbine - Generator Damped Eigenvalues

The analysis of the system undamped critical speeds is useful for the understanding of the mode shapes and the effective planes for balancing. In the following section, the system damped eigenvalues will be presented. In the evaluation of the damped system natural frequencies, all eight bearing stiffness and damping coefficients are taken into consideration. For example, for the case of the undamped critical speeds, the average of the horizontal and the vertical bearing stiffness coefficients was assumed for each bearing.

In this case, the bearing stiffness and damping characteristics at 3,600 RPM are used for the damped eigenvalue analysis.

The gas turbine is supported by 2-4 pad bearings with load between pads. The generator bearings are supported by 120° partial arc bearings. Fig. 19, as shown below represents the first system eigenvalue M1 at 485 CPM. This low frequency mode is a horizontal oscillation with the gas turbine and casing moving together. It is caused by the soft turbine horizontal casing support.

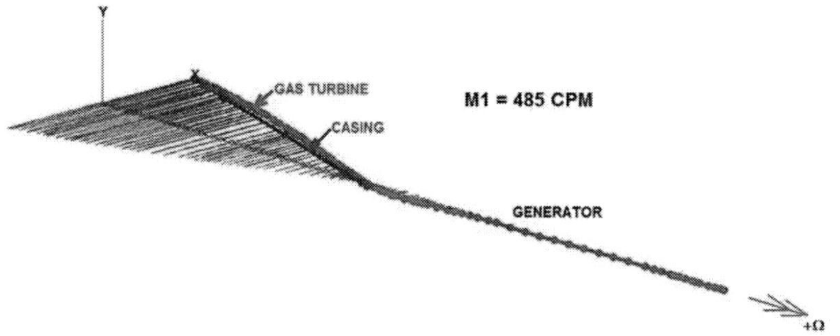

Fig. 19 Horizontal Gas Turbine-Casing Mode, M1= 485 CPM , Log Dec = 0.0038

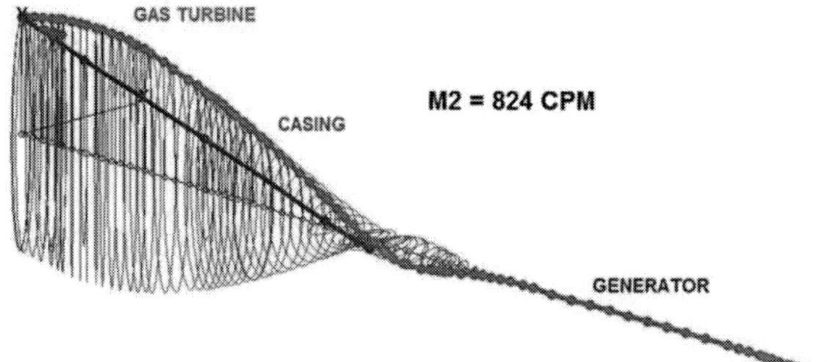

Fig. 20 Vertical Gas Turbine-Casing Mode, M2 = 824 CPM , Log Dec = 0.02

Figure 20 represents the second system mode M2 at 824 CPM. This mode is basically a vibration of the gas turbine and casing moving together in the vertical direction. The undamped critical speed for this mode was computed to be approximately 624 CPM since the stiffness in the vertical and horizontal directions were averaged.

Figure 21 represents the generator first critical speed. The damped eigenvalue analysis shows that the generator in 120° arc bearings is unstable. The system mode M4 at 3600 RPM is predicted to be at 1091 CPM. It would be highly desirable to have the generator bearings redesigned for tilting pad bearings. This would provide additional horizontal support and added system modal damping.

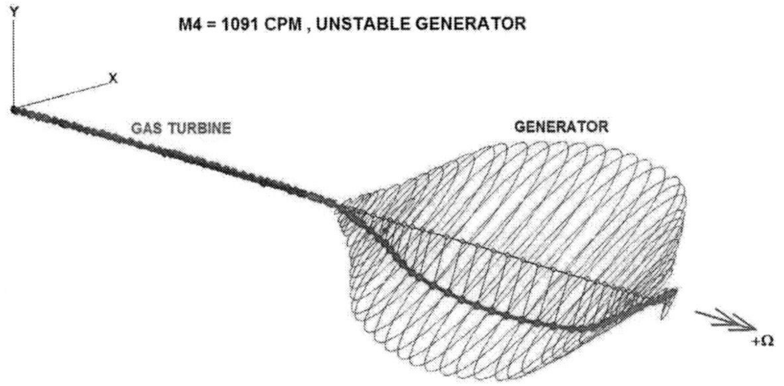

Fig. 21 Unstable Generator Mode M4 = 1091 CPM , Log Dec = -0.073

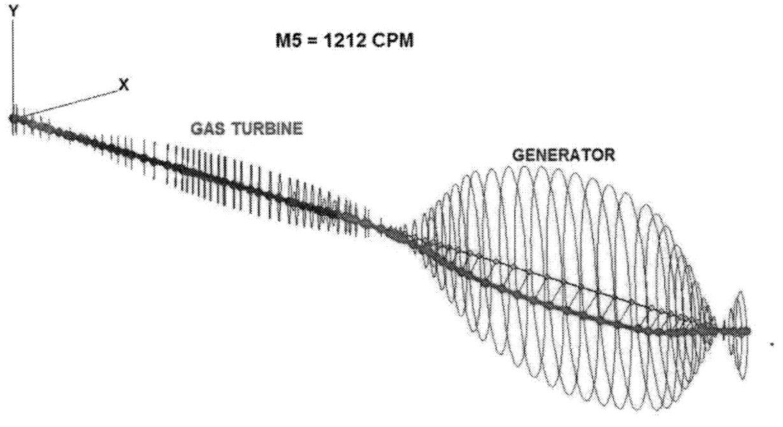

Fig. 22 Generator Vertical 1st Critical Speed M5 = 1212 CPM, Log Dec = 0.132

Figure 22 represents the generator vertical first critical speed. Because of the 120° partial arc bearings, the resonance frequency in the horizontal and vertical directions are different. The mode in the horizontal direction is unstable, however the resonance frequency in the vertical direction is stable because of the higher vertical bearing stiffness and damping values. Well-designed tilting pad bearings would be of benefit for the dynamical characteristics of the generator system. Generator manufacturers for years have resisted the call to employ superior bearings for their generators. This is in contrast to the well-designed tilting pad bearings employed on the gas turbine rotor.

Figure 23 represents the turbine vertical first critical speed mode M6 at 1274 CPM. Note that the casing motion is conical with the casing end displacements moving in phase with the turbine bearing motion. There is also some generator activity shown in this mode. The predicted amplification factor for this mode is over 40. Therefore this mode should be discernible in the vibration data.

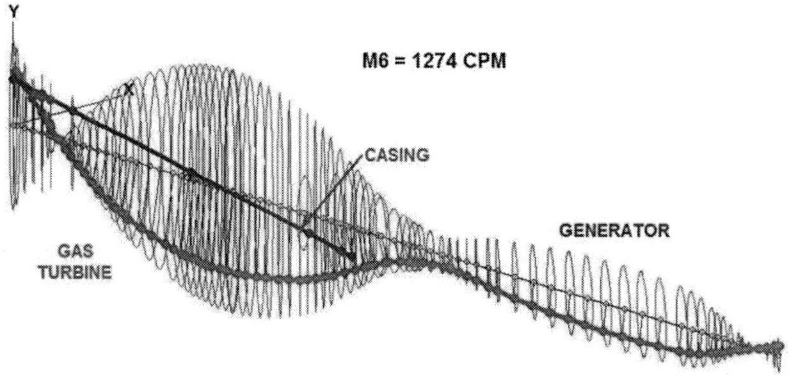

Fig. 23 Vertical Turbine Casing Mode M6 = 1274 CPM, Log Dec = 0.078

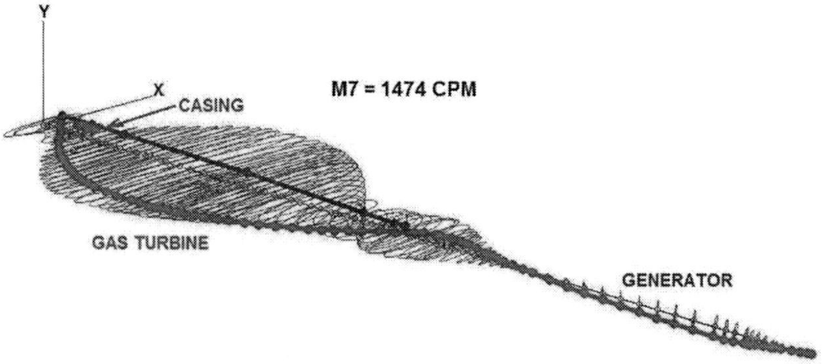

Fig. 24 Gas Turbine-Casing Horizontal Critical Speed, M7=1474 CPM, Log Dec = 0.145

Figure 24 represents the gas turbine horizontal critical speed mode M7 at 1474 CPM. In this mode the casing is moving out of phase to the turbine There is also some generator motion associated with this particular mode. The amplification factor for the horizontal mode is half the value as predicted for the vertical mode.

Figure 25 represents the turbine second critical speed mode M8 at 1798 CPM. In this mode, the casing displacement is following the gas turbine bearing motion. The log dec for this mode is 0.063 which is equivalent to an amplification factor of the almost 50.

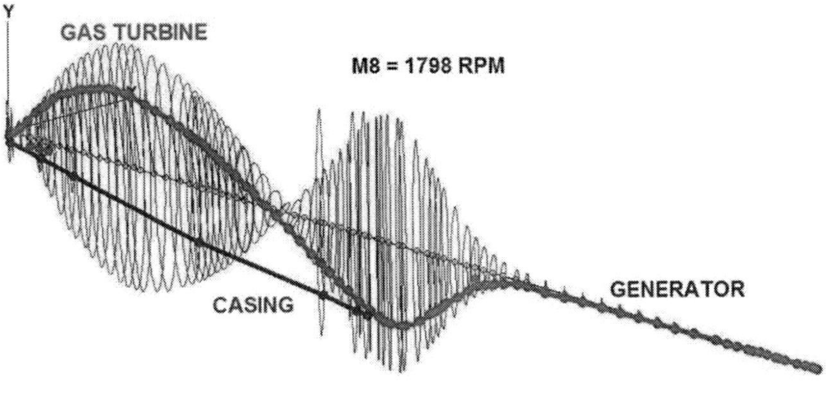

Fig. 25 Gas Turbine Vertical 2nd Critical Speed, M8 = 1798 CPM, Log Dec – 0.063

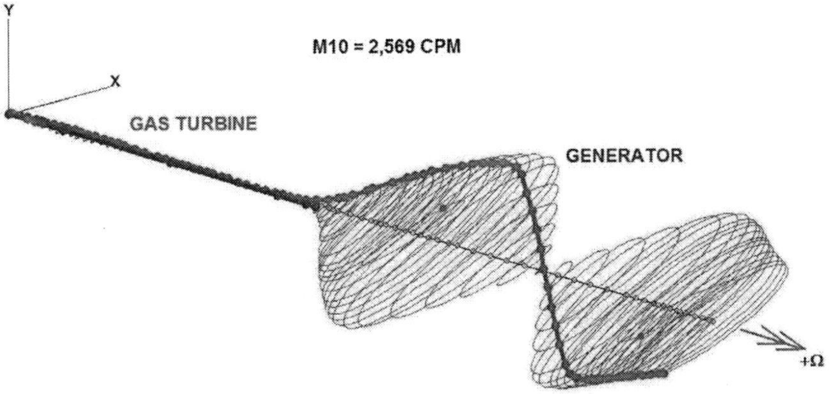

Fig. 26 Generator 2nd Critical Speed, M10 = 2569 CPM, Log Dec = 0.227

Figure 26 represents the generator second critical speed, M10 at 2569 CPM. The log dec of 0.227 is equivalent to a rotor amplification factor of approximately 14. In this mode, there is little gas turbine or casing participation.

Figure 27 represents the generator vertical second critical speed, M 12 equals 3163 CPM. Again this generator second critical speed has very little interaction with the gas turbine rotor and casing.

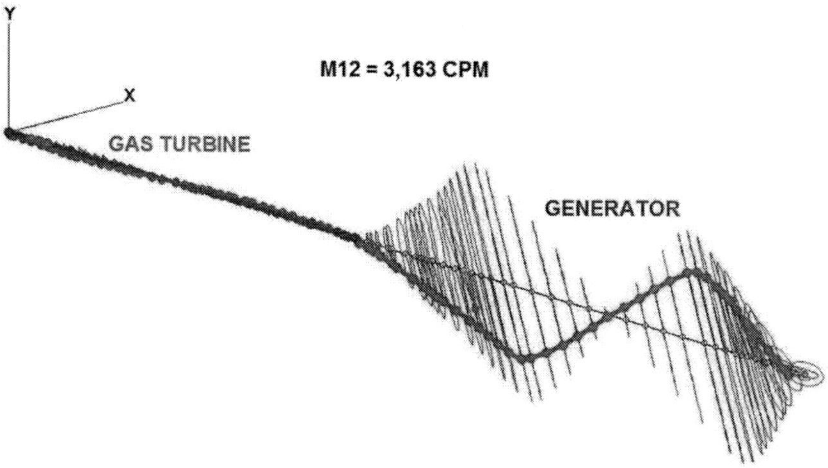

Fig. 27 Generator Vertical 2nd Critical Speed, M12 = 3163 CPM, Log Dec = 0.246

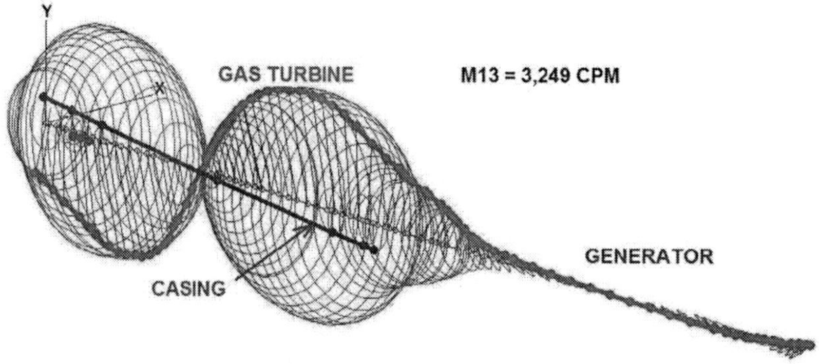

Fig. 28 Gas Turbine Casing Mode M13 = 3249 CPM, log Dec = 0.613

In Figure 28 represent one of the most significant modes of the system. This mode is actually a bifurcated turbine-casing mode and is related to the turbine second critical speed M8 at 1798 CPM. The action of the casing causes a pair of modes to occur due to the interaction of the casing acting in phase in mode M8 and out of phase to the turbine in mode M13. Without the influence of the casing, this mode M13 cannot be computed. Hence one would not know that this mode lies just below the operating speed range.

This mode plays a significant role in the difficulty of balancing the W501 series of gas turbines. For example, if balancing is applied only to vibration data at the 3600 RPM speed range, then this mode will be excited to the point that the turbine may not be able to reach operating speed due to the excitation of this mode at the lower speed.

Experimental W501D Vibration Data

Figure 30 as shown below represents the W501D with the turbine and compressor end probe locations identified. These are the locations at which the relative and absolute motion of the gas turbine may be identified at these two planes. Also identified in this figure are three planes that are available for field balancing. These are at the turbine exhaust, the center plane and the compressor balance planes. For the situation with the balancing of the earlier 70 MW gas turbines of CP&L, the center plane was not available due to the interference of the piping.

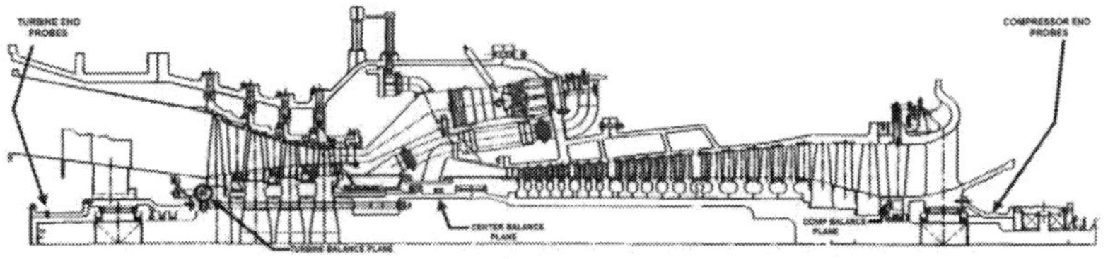

Fig. 29 Gas Turbine Showing Field Balancing Planes and Probe Locations

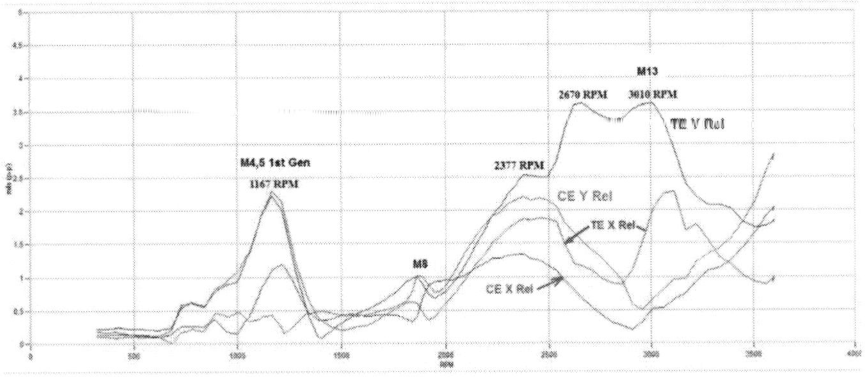

Fig. 30 Relative Probe Measurements at Compressor and Turbine Locations

Figure 30 above represents the relative vibrations measured at the compressor inlet location and at also at the turbine exhaust position. The compressor locations are referred to as CE-X and CE-Y. The two turbine locations are labeled TE-X and TE-Y. Figure 30, shows a substantial vibration at 1167 RPM. This vibration is labeled as M4 and M5 as it represents the generator first critical speed.

The most substantial relative vibrations are observed at TE Y at 2670 RPM and at 3010 RPM. The large response at 3010 RPM is believed to be the mode M13 which is the interaction of the gas turbine second mode with the rotor casing. Note that around 2800-2900 RPM, the compressor motion is at a minimum. As the speed increases above 3000 RPM, it is seen that the compressor relative motion increases while the gas turbine exhaust end reduces.

Thus in the balancing attempts of the 70 MW gas turbines, if single plane balancing was applied at the hot end of the turbine, then excessive vibrations at the compressor end would not allow the unit to come up to speed.

Attempts to apply a single plane balancing correction to the compressor end would then result in excessive vibrations at the turbine exhaust end.

Figure 31 represents the relative and absolute motion at the turbine probe location TE–X.
When the casing motion is included to determine the turbine end absolute motion at plane TE-X, it is seen that there is a very substantial vibration of six mils at 3272 CPM. This again is an example of mode M13 in which there is an interaction between the exhaust casing and the gas turbine second critical speed mode. As previously mentioned, this mode had not originally been predicted for the earlier 70 MW units or for the current improved 100 MW units.

There is a significant phase change between the vibrations around 3200 RPM and at the operating speed of 3600 RPM. It is therefore readily apparent that if a single plane balancing were attempted at speed on the turbine hot end, then mode M13 around 3200 could be significantly excited. The two plots as shown in Figures 30 and 31 then lead to some significant conclusions as to the procedure for field balancing the gas turbines.

It is apparent that single plane balancing at either the gas turbine exhaust or the compressor inlet end will be insufficient because of the occurrence of the strong rotor-casing mode that is encountered just below running speed. Therefore it is necessary to perform a two plane balancing in which vibration measurements are taken at running speed at both the compressor and turbine locations and also at the lower speed range around 3200 RPM. Balancing can then be accomplished using a lease squared error computer program.

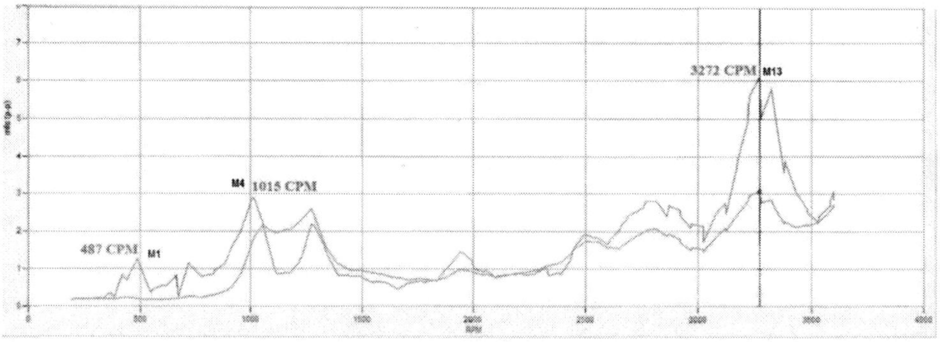

Fig. 31 TE-X Shaft Relative and Absolute Motion on Run Up

There is a small vibration observed around 487 CPM. It is shown that this is the rotor-casing mode M1. In this mode , the turbine and casing are moving in essentially a horizontal plane pivoting about the compressor bearing.

Figure 32 represents the frequency spectrum at the turbine exhaust location. The spectrum on the left represents the relative probe motion and the spectrum on the right is the absolute probe motion which includes the vibration of the casing.

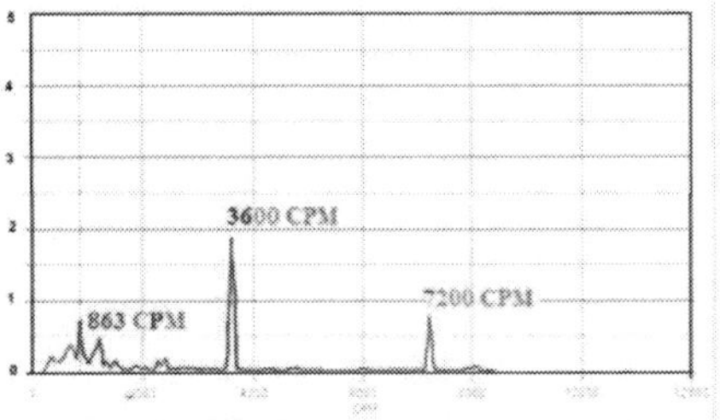

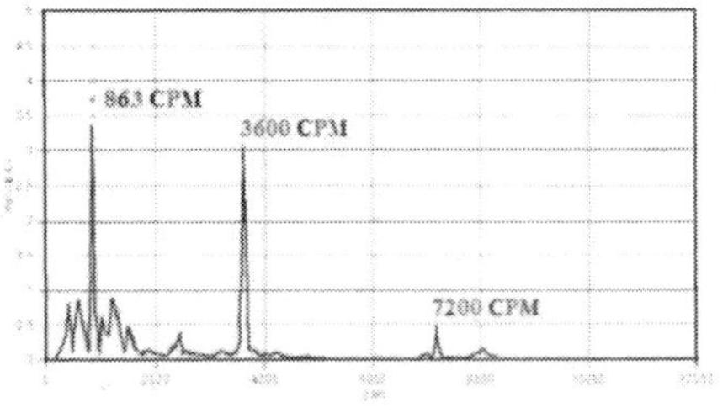

Fig. 32 TE-X Frequency Spectrum At 3600 RPM Showing Subsynchronous Whirling
A. Relative Motion Spectrum **B. Absolute Motion Spectrum**

Of particular significance in the vibration spectrum is the occurrence of the sub-synchronous component observed at 863 CPM. There is also a two times excitation as seen at 7200 CPM. This is quite typical for 3600 CPM turbine-generators. However, the occurrence of the 863 CPM resonance is quite unusual and was not observed earlier in the spectrum data of the 70 MW turbines.

At first, it was considered that this whirl frequency was the excitation of the generator first critical speed which was predicted to be unstable at a whirl frequency 1091 CPM. The observed frequency of 863 CPM is much lower than this and could not correspond to the generator first critical speed. What is different with the current W501D design as compared to the earlier 70 MW turbines is the increase in the horsepower.

In a paper written by Joel Alford of GE in 1965, all gas turbines can generate self excited rotor dynamic cross coupling forces leading to rotor instability. To simulate cross coupling, additional bearings were added in which Kxy equal to -Kxy of 50,000 Lb/In was applied to several turbine stations. The damped eigenvalue analysis now predicts the turbine to be unstable in the vertical mode M2 with a whirl frequency of 826 CPM.

In this mode, the turbine and casing are moving together in essentially a vertical plane as shown in Fig. 33. It would be of interest, therefore, to review the sub synchronous whirl component on the turbine at various power levels.

Figure 33 shows the damped eigenvalue of the system with aerodynamic cross coupling applied.

The stability analysis shows that the turbine is unstable with the applied level of aerodynamic cross coupling.

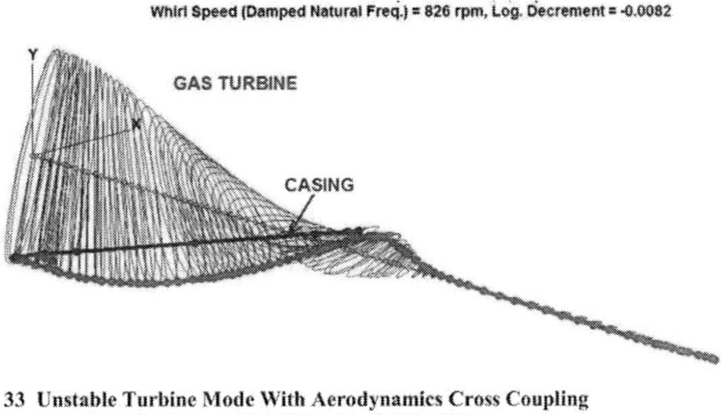

Fig. 33 Unstable Turbine Mode With Aerodynamics Cross Coupling Showing Whirling At 826 CPM, N = 3,600 RPM

Review of Field Balancing a CP&L 70 MW Gas Turbine-Generator

In 1982, the Rotor Dynamics Laboratory at the University Of Virginia was contacted by the research division of CP&L to investigate the balancing problems encountered with one of their W501 gas turbine-generator units. When reviewing the balancing data, we were informed that the young field engineer from Westinghouse had spent three months attempting to balance one of their problem gas turbines.

In his attempts to balance the CP&L unit, when high vibrations were encountered at the turbine exhaust, correction weights were applied at the turbine balance plane. This would result in inducing high vibrations at the compressor end. When balance correction weights were applied at the compressor location to correct excessive compressor vibrations, excessive vibrations would be induced at the turbine end often causing the turbine to trip off line when coming up to speed.

Before attempting to balance the W501 gas turbine, a critical speed analysis was performed using the current rotor dynamics codes based on the transfer matrix procedures of Lund. A preliminary critical speed analysis is extremely important if one is to successfully balance a large turbine-generator which operates through multiple critical speeds. It is important to have an understanding of the system mode shapes and also as to which planes will be the most effective for balancing. It is also important to understand how many balance planes will be required to properly balance a particular rotor-bearing system operating through multiple critical speeds.

An important paper was presented a number of years ago by Kellenberger in which he discussed the N or N +2 method of balancing. He stated that if a rotor operates through N critical speeds then a minimum of N balance planes must be supplied. If one also wishes to minimize the bearing forces transmitted, then an additional two planes is required. As can be seen from the experimental data and past experience, a complex system such as the W501 gas turbine can not be success successfully balanced at speed by the use of a single balance plane.

Attempts to balance the gas turbine-generator combination at speed with a single balance plane can lead to exciting various modes at lower speeds. This can lead to the unit tripping off line when coming up to speed due to excessive vibrations. There were some cases experienced in which the monitors were turned off in order to allow the unit to reach the operating speed. This, in general, is not a recommended procedure to use!

Detailed vibration measurements were recorded on the problem W501 unit. As a contribution to the Rotor Dynamics Research Program at the University of Virginia, the Bently Nevada Corporation supplied instrumentation and the assistance of Bill Pryor for data acquisition. The actual data used for the balancing calculations were obtained with the assistance of Ed Shepherd of Mechtel Analysis using their portable Turbo Balancer (www.Mechtell.com/turbobalancer). A thermal printout for each run would be obtained at 50 RPM increments recording the rotor speed, amplitude, and phase for the W501 unit at 4 vertical locations along the turbine and generator.

In 1984, a 46 page paper entitled ***'Dynamic Analysis and Field Balancing of 70 MW Gas Turbine Generators'*** by E .J. Gunter and R. R. And Humphris was presented at the Vibration Institute. For more specifics of the dynamic analysis and the various balancing cases presented, the full paper may be obtained on download from the website www.RODYN.com/papers Several of the original balancing procedures for the 70 MW unit will be presented and reviewed.

The polar plot of Figure 6 shows the result of the large correction weight of 39.7 ounces placed at the compressor location. At that time, the rotor dynamics computer program based on the transfer matrix theory was not able to compute the apparent gas turbine rotor and casing mode observed around 3000 RPM. This mode appears to be closely related to the mode M13 as shown in Figure 28.

Table 5.1 Compressor End Balance Calculation of 70 MW Gas Turbine Using Turbine End T_e Data only at 3,600 RPM

RUN #7,-20 OZ @359 Deg LEAD AT COMPRESSOR END LOCATION ON GAS TURBINE
SINGLE PLANE BAL; 2 OF 4 BAL WEIGHTS(38.9 OZ) REMOVED AT Ce PLANE

NO. OF PROBES = 1 ; NO. OF SPEEDS = 1 ; NO. OF BALANCE PLANES = 1

THE SPEEDS FOR BALANCING DATA:

SPEED NO.	SPEED (RPM)
1	3600

INITIAL READINGS OF AMP & PHASE:

SPEED NO.	SPEED (RPM)	PROBE NO.	PROBE LOC	AMPLITUDE (MILS)	PHASE (DEG LEAD)
1	3600	1	Te	6.50	113.0

TRIAL UNBALANCES:

PLANE	MAGNITUDE	PHASE ANGLE (DEG LEAD)	BALANCE LOC	BALANCE (UNITS)	TRIAL WEIGHT AFTER RUN
1	20.00	359.0	COMP.	OZ	WEIGHT LEFT IN

READINGS AFTER A TRIAL WEIGHT HAS BEEN ADDED:

BALANCE PLANE	SPEED NO.	SPEED (RPM)	PROBE NO.	AMPLITUDE (MILS)	PHASE (DEG LEAD)
1	1	3600	1	3.10	99.0

THE BALANCE CORRECTION WEIGHTS:

PLANE	Ux (OZ)	Uy	MAGNITUDE ABSOLUTE VALUE	ANGLE (DEG) LEAD-CCW (LAG-CW)	MAGNITUDE TRIM (+TRIAL)	ANGLE (DEG) LEAD (LAG)
1	35.45	-8.26	36.40	346.9(13.1)	17.36	332.9(27.1)

SUMMARY TABLE OF THE SYSTEM INFLUENCE COEFFICIENTS

UNITS OF INFLUENCE COEF:MILS/OZ

BAL #	UBAL MAG	SPD #	SPEED (RPM)	PROBE #	AMP1 (MILS)	PHASE1 (LEAD)	AMP2 (MILS)	PHASE2 (LEAD)	INF. COEF (MAG)	(ANGLE) LEAD	(REL LAG)
1	20.00	1	3600	1	6.50	113	3.10	99	.1786	306	53

INITIAL READINGS AND PREDICTED COMP.RESPONSE

SPEED (RPM)	PROBE NO.	PROBE LOC	INITIAL AMP MILS	INITIAL PHASE LEAD	RESIDUAL AMP MILS	RESIDUAL PHASE LEAD	(LAG)
3600	1	Te	6.50	113	0.000	297	63

Table 5.1 was taken from the balancing paper presented in 1984. This table represents the compressor end balance calculations for the 70 MW gas turbine using only the turbine end data Te at 3600 RPM. In this balancing run, 20 ounce inches was removed from the compressor end balance location. After the weight was removed, the new amplitude at the gas turbine and has been reduced from 6.5 mils to 3.1 mils. Note that the influence coefficient for the response of the turbine at 3600 RPM for a balance weight at the compressor end is approximately 0.18 mils/oz.

The balancing calculation computes that the absolute magnitude of unbalance is 36.4 ounces @ 332°. Therefore, using only the turbine end data, it is seen that we would end up with approximately the same unbalance as was used in the original balancing case. From this calculation it becomes apparently clear that the young Westinghouse engineer was computing balancing based on turbine end vibration only at the operating speed of 3600 RPM.

Table 5.2 Compressor End Balance Calculation Using Compressor C_e and Turbine T_e Data at Various Speeds - U_c = 20 oz @ 359° - Initial Vibration Data

TRIAL WEIGHT LEFT IN PLACE AFTER RUN

NO. OF PROBES = 2 ; NO. OF SPEEDS = 6 ; NO. OF BALANCE PLANES = 1

THE SPEEDS FOR BALANCING DATA:

SPEED NO.	SPEED (RPM)
1	1066 TURBINE 1ST CRIT
2	1702 GENERATOR 1ST CRIT
3	2300 TURBINE 2ND CRIT
4	2756 TURBINE-GENERATOR CRIT
5	3600 RUNNING SPEED: COLD OMW
6	3600 RUNNING SPEED: HOT 70MW

Predicted Gas Turbine Response at Various Speeds Correction Based on T_e Data Only at 3,600 RPM

ORIGINAL WEIGHT REMAIN IN PLANE(S),TRIM WEIGHT(S) ADDED

ORIGINAL TRIAL WEIGHT IN PLACE,TRIM WEIGHT ADDED

TRIAL UNBALANCES:

PLANE (OZ)	MAGNITUDE	PHASE ANGLE (DEG LEAD)	BALANCE LOC	BALANCE (UNITS)
1	17.36	333.0	COMP END	OZ

INITIAL READINGS AND PREDICTED COMP.RESPONSE

SPEED (RPM)	PROBE NO.	PROBE LOC	INITIAL AMP MILS	INITIAL PHASE LEAD	RESIDUAL AMP MILS	RESIDUAL PHASE LEAD	(LAG)
1066	1	Ce	2.60	313	3.304	296	64
1066	2	Te	5.90	7	7.708 ←	339	21
1702	1	Ce	3.90	232	4.885	230	130
1702	2	Te	4.40	4	2.389	12	348
2300	1	Ce	4.20	160	4.961	167	193
2300	2	Te	2.80	340	2.391	357	3
2756	1	Ce	3.50	120	3.627	142	218
2756	2	Te	5.40	325	7.012 ←	333	27
3600	1	Ce	4.10	73	3.079	81	279
3600	2	Te	3.70	97	1.116	74	286
3600	1	Ce	1.50	141	2.300	168	192
3600	2	Te	3.10	99	.702	79	281

Table 5.2 shows the five speed values that were recorded. The cases five and six represent the gas turbine at zero and at the full power level of 70 MW at 3600 RPM. Table 5.2 shows the initial and predicted response using balancing data based only on the turbine vibration at running speed. The turbine vibration is reduced from 3.12 mils to 0.702 mils at the full power level.

Of interest it is seen that using only turbine vibration data at full power for the balancing calculations results in a predicted turbine amplitude at 1066 RPM of 7.7 mils. This is a significant increase from the original value of 5.90 mils. This vibration level would be enough to trip the machine off line. The turbine data at 2756 RPM, which is a coupled turbine-casing 2nd critical speed, would increase from 5.40 to over 7 mils.

There were a number of cases in the balancing runs by the Westinghouse engineer in which the monitors were disconnected in order to allow the machine to come up to speed.

Table 5.3 represents the compressor end balance calculations and influence coefficients using all of the compressor data Ce and the turbine data Te at the 6 speed conditions. In this case the absolute magnitude of the unbalance is 22.6 ounces as shown in Fig. 35. If the original trial weight is left in place,then only a trim balance of 5 ounces in hole 34 is required as compared to the excessive predicted trim weight of 22 oz to be placed in balance hole 37 in the compressor balance plane.

Of particular interest is the examination of the influence coefficients for example the placement of the balance weight at the compressor causes a larger response at the turbine and at speeds of 106 six and also at running speed of 3600 RPM.

The more balancing data that is applied to the lease squared error balancing program, the greater is the reduction in the computed value of balance correction. The reason for this is that all of the additional readings act as constraints upon the amount of balance weight that should be placed at the balance location.

Thus we see at the initial and predicted response that the turbine response at zero power is 6.5 mils and is reduced to three mils at full power the initial response was 5.7 and with the reduced value of balancing this value is reduced to 2.5 mils is important to note that it is not always possible to reduce the vibration at all speeds going through more than two modes with only two planes of unbalance. For example we see that the turbine amplitude at 106 six RPM has increased from 4.62 5.5 mils at the second turbine casing mode of 2756 RPM the amplitude turbine amplitude has increased from the original value of 4.7 mils to 5.94 mil. However this value is not enough to trip the machine off line

Table 5.3 Compressor End Balance Calculations and Influence Coefficients Using Compressor C_e and Turbine T_e Data at Various Speeds

RUN #7 SHOT AT Ce: 20 OZ @ 359 DEG ON COMPRESSOR BALANCE PLANE
TRIAL WEIGHT LEFT IN PLACE AFTER RUN

THE BALANCE CORRECTION WEIGHTS:

PLANE	Ux (OZ)	Uy	MAGNITUDE ABSOLUTE VALUE	ANGLE (DEG) LEAD-CCW	(LAG-CW)	MAGNITUDE TRIM (+TRIAL)	ANGLE (DEG) LEAD	(LAG)
1	22.09	-4.97	22.64	347.3	12.7	5.08	294.4	65.6

SUMMARY TABLE OF THE SYSTEM INFLUENCE COEFFICIENTS

UNITS OF INFLUENCE COEF:MILS/OZ

BAL #	UBAL MAG	SPD #	SPEED (RPM)	PROBE #	AMP1 (MILS)	PHASE1 (LEAD)	AMP2 (MILS)	PHASE2 (LEAD)	INF. COEF (MAG)	(ANGLE) LEAD	(REL LAG)
1	20.00	1	1066	1	1.70	339	2.60	313	.0653	279	80
1	20.00	1	1066	2	4.60	54	5.90	7	.2177	317	42
1	20.00	2	1702	1	2.80	226	3.90	232	.0577	248	111
1	20.00	2	1702	2	6.70	10	4.40	4	.1185	202	157
1	20.00	3	2300	1	3.90	145	4.20	160	.0549	228	131
1	20.00	3	2300	2	3.70	333	2.80	340	.0491	134	225
1	20.00	4	2756	1	4.50	103	3.50	120	.0771	242	117
1	20.00	4	2756	2	4.70	302	5.40	325	.1064	26	333
1	20.00	5	3600	1	5.40	74	4.10	73	.0651	258	101
1	20.00	5	3600	2	6.50	113	3.70	97	.1558	313	46
1	20.00	6	3600	1	1.98	98	1.50	141	.0676	230	129
1	20.00	6	3600	2	5.70	114	3.10	99	.1411	312	47

INITIAL READINGS AND PREDICTED COMP.RESPONSE

SPEED (RPM)	PROBE NO.	PROBE LOC	INITIAL AMP MILS	INITIAL PHASE LEAD	RESIDUAL AMP MILS	RESIDUAL PHASE LEAD	(LAG)
1066	1	Ce	1.70	339	2.567	306	54
1066	2	Te	4.60	54	5.520	357	3
1702	1	Ce	2.80	226	4.095	229	131
1702	2	Te	6.70	10	4.018	10	350
2300	1	Ce	3.90	145	4.479	160	200
2300	2	Te	3.70	333	2.819	345	15
2756	1	Ce	4.50	103	3.729	125	235
2756	2	Te	4.70	302	5.938	325	35
3600	1	Ce	5.40	74	3.948	77	283
3600	2	Te	6.50	113	3.037	104	256
3600	1	Ce	1.98	98	1.820	145	215
3600	2	Te	5.70	114	2.530	108	252

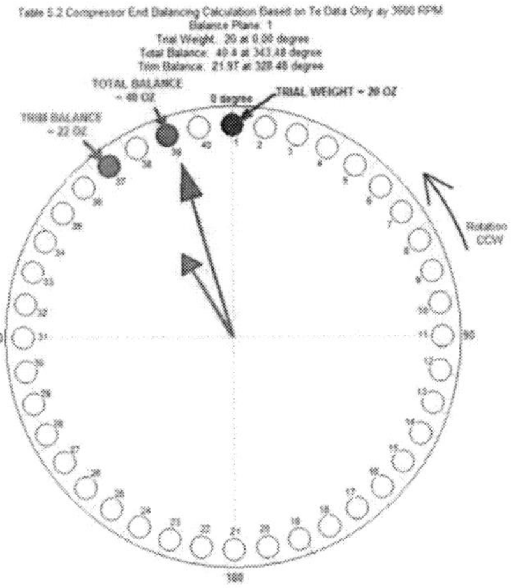

Fig. 34 Ce Balance Based on Turbine Data Only at 3600 RPM

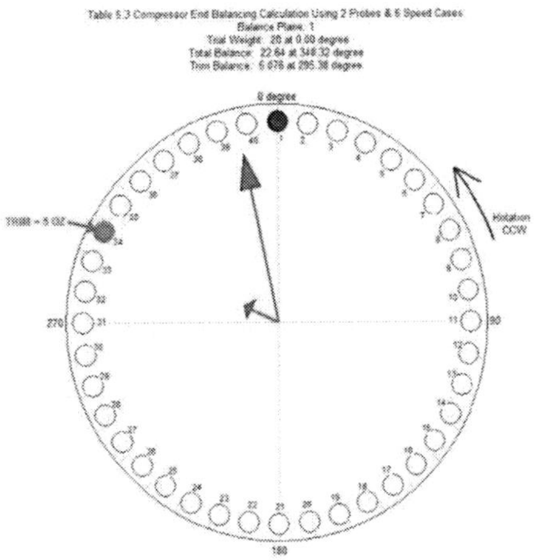

Fig. 35 Ce Balance Based on Compressor & Turbine Data at 6 Speeds

Figure 34 represents the balancing calculations based on only the Te data at 3600 RPM. In this case it is seen that a trim balance of 22 ounces would be applied at hole 37. In Table 5.3b below it is seen that at full power the amplitude of 5.7 mils would be reduced to zero and the amplitude at zero power of 6.5 mils would be reduced to 0.36 mils. However if this large correction weight were applied then it is seen that at the turbine first critical speed the amplitude would increase from 4.6 mils to 8.09 mils. This would trip the gas turbine off-line. With this large correction weight, the gas turbine could pass through the first critical speed.

Figure 35 shows the balance correction that would be added to the compressor balance plane if all of the data of the compressor and turbine were taken at the six speed cases. In this case the balance correction weight is reduced from 22 ounces to be placed at hole 37 to only 5 ounces placed at balance hole #34. As a result of this correction, the first critical speed amplitude increases only slightly to 5.5 mils. The amplitude at the second turbine critical speed is slightly increased to 5.9 mils which is acceptable for this initial single initial balance correction.

The important consideration is that at running speed at the zero power level, the amplitude is reduced to 3 mils and to 2.52 mils at full power. These values are acceptable compared to the original values of 6.5 and 5.7 mils observed at zero and full power.

Table 5.3 b

	Speeds ***(RPM)***	**Initial)** ***(Mils)***	**Balance Te Only** ***(Mils)***	**Balance With Ce & Te Data** ***(MILS)***
1	1066	4.60	**8.09**	5.52
2	1702	**6.70**	1.99	4.02
3	2300	3.70	2.42	2.82
4	2756	4.70	**7.54**	5.94
5	3600 ***0 PW***	**6.50**	0.36	3.04
6	3600 ***Full PW***	5.70	0.0	2.52

Table 5.5 Turbine End Balance Calculation and Predicted Response Based on T_e and C_e Data at Various Speeds

BALANCE DATA WITH 4 SPEEDS WITH Ce & Te MEASUREMENTS,10 OZ @ 270 DEG
SINGLE PLANE CORRECTION,3600 ,2850 ,1054,1107 RPM- 4 SPEED SETS

THE BALANCE CORRECTION WEIGHTS:

PLANE	Ux (OZ)	Uy	MAGNITUDE ABSOLUTE VALUE	ANGLE (DEG) LEAD-CCW (LAG-CW)	MAGNITUDE TRIM (+TRIAL)	ANGLE (DEG) LEAD (LAG)
1	7.72	-23.37	24.61	288.3(71.7)	15.44	300.0(60.0)

SUMMARY TABLE OF THE SYSTEM INFLUENCE COEFFICIENTS

UNITS OF INFLUENCE COEF:MILS/OZ

BAL #	UBAL MAG	SPD #	SPEED (RPM)	PROBE #	AMP1 (MILS)	PHASE1 (LEAD)	AMP2 (MILS)	PHASE2 (LEAD)	INF. COEF (MAG)	(ANGLE) LEAD	(REL LAG)
1	10.00	1	3600	1	4.10	73	2.60	101	.2178	309	321
1	10.00	1	3600	2	3.70	97	3.40	95	.0325	28	242
1	10.00	2	2850	1	3.40	114	1.40	91	.2181	39	231
1	10.00	2	2850	2	5.50	320	4.30	327	.1339	207	63
1	10.00	3	1054	1	2.60	313	1.80	319	.0831	210	60
1	10.00	3	1054	2	5.90	7	4.30	15	.1748	257	13
1	10.00	4	1107	1	2.50	284	1.80	292	.0760	175	95
1	10.00	4	1107	2	4.50	351	3.40	4	.1412	228	42

INITIAL READINGS AND PREDICTED COMP.RESPONSE

SPEED (RPM)	PROBE NO.	PROBE LOC	INITIAL AMP MILS	INITIAL PHASE LEAD	RESIDUAL AMP MILS	RESIDUAL PHASE LEAD	(LAG)
3600	1	CE#4BRG	4.10	73	1.803	199	161
3600	2	Te#5BRG	3.70	97	3.128	88	272
2850	1	CE#4BRG	3.40	114	3.113	3	357
2850	2	Te#5BRG	5.50	320	2.233	327	33
1054	1	CE#4BRG	2.60	313	.592	295	65
1054	2	Te#5BRG	5.90	7	1.607	12	348
1107	1	CE#4BRG	2.50	284	.631	287	73
1107	2	Te#5BRG	4.50	351	1.431	28	332

Table 5.7 Compressor and Turbine Two Plane Balancing Calculation Without Removing Trial Weights - Balance Correction Weights and Predicted Turbine Response

RUN #7 SHOT AT Ce: 20 OZ @ 359 DEG; RUN #8 SHOT AT Te: 10 OZ @ 270 DEG
TRIAL WEIGHTS LEFT IN PLACE AFTER RUN

THE BALANCE CORRECTION WEIGHTS:

PLANE	Ux (OZ)	Uy	MAGNITUDE ABSOLUTE VALUE	ANGLE (DEG) LEAD-CCW	(LAG-CW)	MAGNITUDE TRIM (+TRIAL)	ANGLE (DEG) LEAD	(LAG)
1	26.24	-5.71	26.85	347.7(	12.3)	8.23	319.3(	40.7)
2	-1.00	-20.62	20.65	267.2(	92.8)	10.67	264.6(	95.4)

INITIAL READINGS AND PREDICTED COMP.RESPONSE

SPEED (RPM)	PROBE NO.	PROBE LOC		INITIAL AMP MILS	INITIAL PHASE LEAD	RESIDUAL AMP MILS	RESIDUAL PHASE LEAD	(LAG)
1066	1	Ce		1.70	339	1.093	310	50
1066	2	Te		4.60	54	2.594	358	2
1702	1	Ce		2.80	226	2.875	257	103
1702	2	Te		6.70	10	1.803	41	319
2300	1	Ce		3.90	145	4.850	66	294
2300	2	Te		3.70	333	.734	317	43
2756	1	Ce		4.50	103	1.610	154	206
2756	2	Te		4.70	302	4.331	341	19
3600	1	Ce	O POWER	5.40	74	.542	343	17
3600	2	Te		6.50	113	2.112	114	246
3600	1	Ce	FULL PW	1.98	98	1.637	207	153
3600	2	Te		5.70	114	2.086	118	242

Table 5.5 represents the turbine and balancing in which 10 oz was placed at 270°. The data at both the turbine and compressor locations with the six speed cases was used to compute a final 2 plane balance. Thus we have achieved an amplitude at the compressor location from 4.1 mils to 1.8 mils. The amplitude at the turbine is only reduced slightly from 3.7 to 3.1 mils. However there is a significant reduction in the amplitude at the first critical speed from 5.9 mils to 1.6 mils.

In table 5.7, the compressor and turbine data has been combined to perform a two plane balancing calculation without removing of either of the 2 trial weights. As can be seen from the above chart, this has proven to be highly successful with the peak turbine vibrations of 6.5 mils and 5.7 mils respectively been reduced to slightly over 2 mils. Of considerable importance is that the 6.7 mill turbine vibrations observed at 1702 RPM has now been reduced to 1.8 mils.

It is very apparent that by combining the influence coefficients from the two planes with six different speed cases, that a successful balance may be achieved on the gas turbine.

Review of a Balancing Case History For a 501D 100 MW Gas Turbine

In this section, the field balancing data generated for a 100 MW 501D gas turbine will reviewed. Balancing calculations based on this field data will be presented for various assumed probe vibration measurements and single and 2 plane balancing solutions will be presented. The balancing predictions will be generated using the Weighted Least Squared Error Balancing program of Dyrobes. The least squared error balancing procedure is well known and was first reported on by Goodman of GE in 1965. The first case to be considered is that of a single plane balancing shot applied to the gas turbine exhaust end.

The instrumentation for the current gas turbines consists eight probes placed along the gas turbine and generator at 45° from the vertical axis. There are 4 probes placed at the turbine inlet and exhaust sections and 4 more probes placed on the generator. The turbine probes measure the rotor relative shaft to casing motion. Casing pickups in series with the turbine probes may be used to determine the turbine absolute motion. Absolute motion is obtained by integrating the casing motion with the relative prox motion. For the earlier balancing cases with the 70 MW 501 gas turbines, superior balancing was achieved by using absolute probe measurements.

Review of Turbine Exhaust Balance Calculations

Table 1 represents the input table for a least squared error calculation for a trial weight placed at the turbine exhaust. In this example, a trial weight of 18.9 ounces was placed in hole number 13. This balance shot represents two normal balancing steel weights of approximately 9.45 ounces each. The balancing radius for the turbine end is 18 inches.

In Table 1 as shown below are the initial vibration readings before application of the balancing weight and the resulting vibrations after the balancing weight was applied. Note that the highest amplitudes before balancing occur at the turbine and compressor Ty and Cy probes of over four mils. It is seen that the trial weight has reduced the four mil responses to under three mils along the turbine. The table as shown below is for a least squared error balancing calculation. It is possible with the least squared error program to add weighting functions to the various probe readings. In the first case a weighting function of one is applied to only the relative turbine Ty vibration. This is equivalent to a single plane - single speed balancing calculation. By placing 0 for the weighting factors for the other probes, there amplitudes do not come into consideration for the balancing calculation.

Number of Balancing Planes: 1 Number of Speeds/Cases: 1 Shaft Rotation: CCW (•) CW Phase: (•) Lag Lead 0 degree at: (•) Y-Up X-Right

Number of Measured Probes: 8 Runout Compensation: No

Comment: Turbine Balance Using Turbine Ty Measurement Only

Comment: 18.9 oz @108.4 deg at Turbine Exhaust

Comment: 3600 RPM Full Load

Balancing Holes: No. of Holes: 40 1st Hole Angle: 0 Numbering Direction: (•) CCW CW

	Condition	Speed	Description	Amplitude	Phase (deg)
1	Initial Readings	1	Ty	4.18	262
2	Initial Readings	1	Tx	3.9	15
3	Initial Readings	1	Cy	4.07	42
4	Initial Readings	1	Cx	3.15	146
5	Initial Readings	1	Gty	0.98	298
6	Initial Readings	1	Gtx	0.24	43
7	Initial Readings	1	Gey	1.84	128
8	Initial Readings	1	Gex	1.46	217
9	Trial Run < 1 >	---	Left-In Afterward	18.9	108.4
10	Response	1	Ty	2.46	262
11	Response	1	Tx	2.95	36
12	Response	1	Cy	2.7	78
13	Response	1	Cx	3.01	151
14	Response	1	Gty	0.63	183
15	Response	1	Gtx	0.11	275
16	Response	1	Gey	2.49	138
17	Response	1	Gex	1.32	132
18					
19					
20					

Weighting (Scale) Factors for probes and speeds

Probe	Factor
1	1
2	0
3	0
4	0
5	0
6	0
7	0
8	0
9	
10	
11	
12	
13	
14	
15	
16	
17	
18	
19	
20	

Speed	Factor
1	1
2	
3	
4	
5	
6	
7	
8	
9	
10	

New Open Save Save As Run Close

Table 1 Input table For Single Plane Turbine Exhaust Balancing With 8 Probe Inputs

Figure 36 shows the trial weight of 18.9 ounces placed at 108.4° which corresponds to balancing hole # 13. It is seen that a trim balance of 27 ounces is computed corresponding to the location of the placement of the trial weight at hole 13. It is therefore apparent that the original balancing calculation was based simply on the measurement of the vibration of 4.18 mils at 262°.

It will be shown later that if this computed trim balance weight of 27 ounces is applied, then the response at Ty of 4.18 mils will be reduced to zero amplitude. However it will be shown that this is an excessive amount of unbalance and would result in increasing the vibration amplitudes at the other points along the turbine and generator.

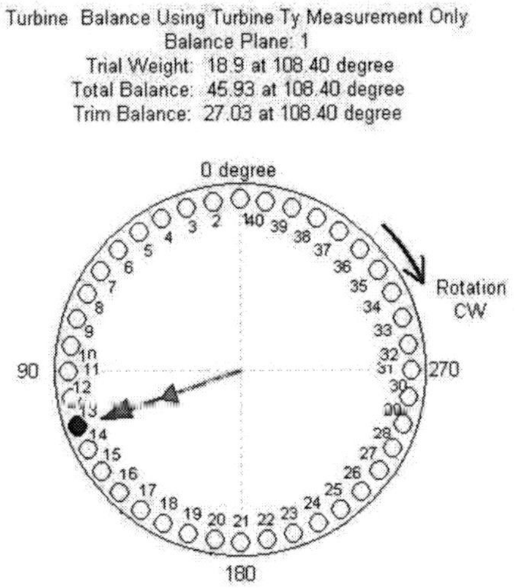

Fig. 36 Turbine Exhaust Balancing Based On Compressor Inlet Cy Measurement Only

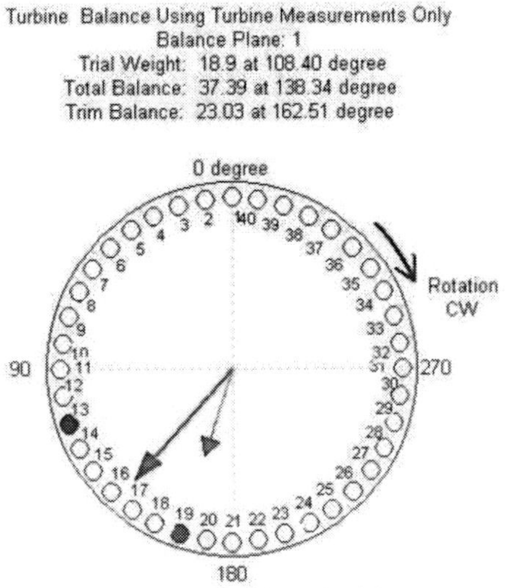

Fig. 37 Turbine Exhaust Balancing Based On Turbine and Compressor Inlet Measurements

Figure 37 represents the balancing calculation computed if the four turbine measurements are taken into consideration. This is done by assuming a weighting function of 1 for the 4 turbine probes which are labeled 1 to 4. By leaving the weighting factors for the generator probes at zero, they are not included in the calculations. By including the generator probes along for the ride, it will be seen later how much their amplitudes would change with the additional probe measurements taken into consideration.

By including the additional probe readings, each additional reading acts as a constraint on the total amount of balancing that will be computed. Thus the total unbalance is reduced from 45.9 ounces in the first case with only a single probe measurement to 37.4 ounces when the 4 turbine probes are included in the calculations. The trim balance, based on 4 turbine measurements at speed, has been now reduced to 23 ounces to be placed at balance location 19.

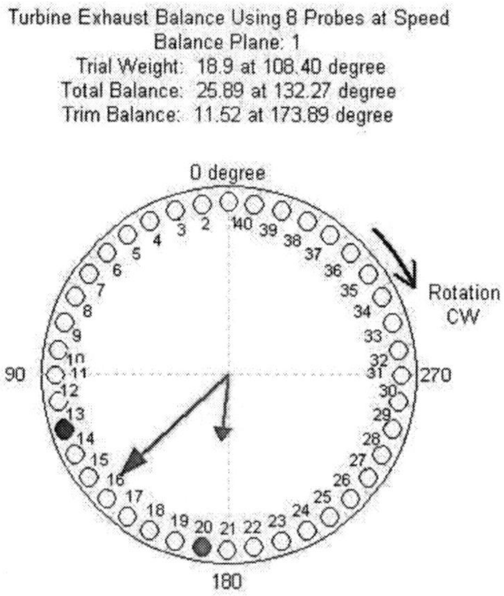

Fig. 38 Turbine Exhaust Balancing Based On 8 Turbine & Generator Measurements

Figure 38 represents the balancing calculation when all eight probes of the turbine and generator are taken into consideration. Again with the four additional generator measurements including in the balancing calculation, it is seen that the total balancing weight is reduced further to approximately 26 ounces and the correction trim weight is reduced to 11.5 ounces to be placed at balance hole number 20.

If we were to compare the difference in balancing based on a single measurement as compared to all eight probe readings, then it is seen that there is a substantial difference in the requested trim weight as shown between Figures 36 and 38. Using only a single plane measurement, a trim balance of 27 ounces is requested for hole 13 as compared to the greatly reduced value of 11.5 oz for hole 20.

Thus, when all eight turbine and generator measurements are taken into consideration, the trim weight is substantially reduced to less then one-half of the original value with the trim weight location shifted from balance hole 13 to balance hole 20. This represents a significant difference in the balancing strategy. This type of balancing calculation requires a computer simulation and can not be accomplished in the field by simple single plane vector plots.

Table 2 represents a summary table of the predicted final turbine and generator amplitudes of motion with the three various methods of balancing. In this table, the four turbine and four generator initial vibration values ,taken at speed under load, are shown along with the predicted results if the various computed correction weight were applied.

In the first balancing case in which only the Cy measurement was used, we see that if the full trim balance of 27 ounces is applied at hole 13, then the response at Cy would be reduced to zero.

The application of a single plane balancing to reduce a single vibration reading to zero is a highly undesirable procedure. Since the unit is operating above multiple critical speeds, the attempt to highly balance a single location will always result in higher amplitudes along the rest of the rotor. These increased amplitudes can be of such a significant magnitude that it may prevent the rotor from coming up to speed.

Of particular importance is that a high level of balancing applied at the turbine end will result in increased amplitudes of vibration at the other end of the machine on the generator. Note that the generator turbine end and exciter have increased to over 3.5 mils. In previous maintenance situations, the generator bearings have had to be re-Babbitted due to bearing over temperature. It is very easy to see that this damage may have been caused by excessive loadings placed on the generator bearings due to high exhaust turbine balance corrections.

In the second case, with four turbine measurements, we see that Tx is reduced to 2.65 mils and Cy is two mils, which is acceptable. The generator exciter motion Gex is unacceptable as its amplitude has increased to 3.6 mils.

When all eight turbine and generator measurements are taken into consideration, as explained before, this acts as a constraint and the maximum amount of requested trim unbalance is further reduced to 11 ounces. The turbine amplitudes are all acceptable as they are under three mils the generator amplitudes are now in the acceptable range even though the amplitudes have increased, they are still within 1.5 to 2.5 mils which is an acceptable range.

Table 2 Predicted Turbine-Generator Response Calculations

	(Trim Bal)	**27oz @ 108°**	**23 oz @ 163°**	**11 oz @ 174°**
Probe	**Initial, Mils**	**Cy only**	**Turbine-4**	**Turbine & Gen**
Tx	3.15	2.89	2.65	2.80
Ty	4.01	3.89	1.13	1.20
Cx	3.90	2.80	1.12	2.00
Cy	4.18	0.00	2.10	2.34
Gtx	0.24	0.55	0.50	0.30
Gty	0.98	2.32	2.80	1.40
Gex	1.46	**3.67**	**3.60**	2.45
Gey	1.85	**3.49**	2.8	2.53

Review of Turbine Compressor Inlet Balancing

In this field balancing case, after a balancing correction weight had been applied to the turbine exhaust end, a trial weight was added to the compressor inlet section of the turbine. Two balancing weights of 9.5 ounces were added to balancing holes 21 and 22. This represents a total balancing trial weight of 19 ounces at 185° as shown in Figure 39.

Table 3 as shown below represents a section of the input table used to compute the balancing at compressor balance plane. This balancing plane has 40 holes at a radius of 15 inches. The balancing calculations were based on 12 vibration measurements taken at three speeds. These speeds corresponded to system resonances observed at 3070 RPM and 3490 RPM. A third speed case was taken at 3600 RPM at full power level.

The 12 measurements consisted of the four relative probe measurements taken on the turbine exhaust and inlet sections and the four absolute vibration measurements which incorporate casing response of the turbine. Four additional measurements from the generator were also included in this particular balancing calculation.

Number of Balancing Planes: 1
Number of Measured Probes: 12
Number of Speeds/Cases: 3
Runout Compensation: No
Shaft Rotation: CCW / CW (selected: CW)
Phase: Lag / Lead (selected: Lag)
0 degree at: Y-Up / X-Right (selected: Y-Up)
Comment: Compressor Balance With 19 oz @ 185 Deg, 12 Measurements, 3 Speeds
Comment: Turbine relative & Abs Motion, generator motion
Comment: Speeds at 3070, 3490, & 3600 Full load
Balancing Holes
No. of Holes: 40
1st Hole Angle: 0
Numbering Direction: CCW / CW (selected: CCW)

	Condition	Speed	Description	Amplitude	Phase (deg)
1	Initial Readings	3070	Trx	8.1	54
2	Initial Readings	3070	Try	2	276
3	Initial Readings	3070	Tx	1.6	265
4	Initial Readings	3070	Ty	2.1	159
5	Initial Readings	3070	Crx	1.1	143
6	Initial Readings	3070	Cry	3.7	268
7	Initial Readings	3070	Cx	1.7	287
8	Initial Readings	3070	Cy	0.8	156
9	Initial Readings	3070	Gtx	1.5	250
10	Initial Readings	3070	Gty	2.9	161
11	Initial Readings	3070	Gex	0.6	176
12	Initial Readings	3070	Gey	1.9	101
13	Initial Readings	3490	Probe: 1	5.7	326
14	Initial Readings	3490	Probe: 2	1.7	238
15	Initial Readings	3490	Probe: 3	2.9	23
16	Initial Readings	3490	Probe: 4	1.4	252
17	Initial Readings	3490	Probe: 5	2.5	70
18	Initial Readings	3490	Probe: 6	1.9	216
19	Initial Readings	3490	Probe: 7	2.9	23
20	Initial Readings	3490	Probe: 8	1.9	216

Weighting (Scale) Factors for probes and speeds

Probe	Factor
1	1
2	1
3	1
4	1
5	1
6	1
7	1
8	1
9	1
10	1
11	1
12	1
13	
14	
15	
16	
17	
18	
19	
20	

Speed	Factor
1	1
2	1
3	1
4	
5	
6	
7	
8	
9	
10	

New Open Save Save As Run Close

Table 3 Section of Input Table For Turbine Inlet Balancing Calculations
12 Turbine and Generator Vibration Measurements at 3 Speeds

In the least squared error calculation method of balancing, weighting functions may be added to any particular probe or speed data set. This allows one to examine the situation when vibration data is ignored in the balancing calculations. For example if one wishes to calculate a simple single plane balance based on only the value of the relative turbine exhaust x probe motion of Trx of 8.1 mils, then the weighting factor for probes 2 to 12 would be set to zero.

After a balancing calculation is performed, the results should be reviewed as there are cases in which one may wish to modify particular vibration values. These values may either be amplified or reduced by the use of the weighting factors. Later balancing studies showed that the generator vibration values had little influence on the compressor balance predictions. Hence the weighting factor F is set to zero for the 4 generator vibration readings.

Figure 39 represents the computed balancing to be applied to the compressor inlet plane based on 12 vibration measurements at three different speeds. The initial trial balancing weights of two 9.5 ounce weights were inserted in balancing holes 21 and 22. From the calculations generated from the balancing program, the total required unbalance is only 11 ounces at 132° which corresponds closely to hole 16. If the initial trial weight were left in place, then a large trim balance of 15 ounces would be required near hole 6. This trial weight was later removed.

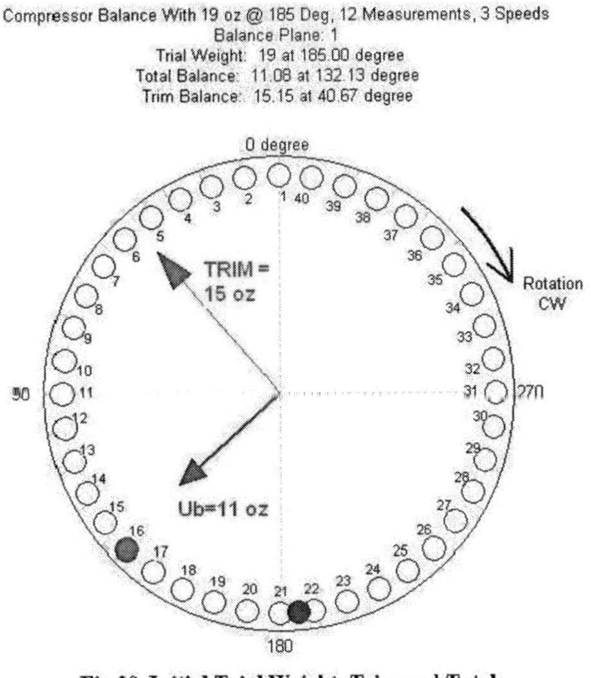

Fig 39 Initial Trial Weight, Trim and Total Computed Compressor End Balancing

It was seen that in reviewing the response of a trial weight placed at the compressor balance plane, that there was very little influence to the generator motion. Therefore when the weighting functions for the four generator probes was placed at zero, the balancing calculations were basically unchanged.

It was also determined that a fairly reasonable balance could be computed by using only the four absolute turbine exhaust and compressor inlet measurements. For example, using only the four turbine absolute measurements, the total balance was calculated to be 11.5 ounces at 124°. This compares favorably with the predicted value of 11 ounces at 132° as shown in Figure 39 using all 12 relative and absolute turbine and generator measurements.

As was reported earlier with the balancing of the 70 MW 501 gas turbines, the balancing was best accomplished by using the absolute turbine values in which the relative probe displacements and casing motions were combined.

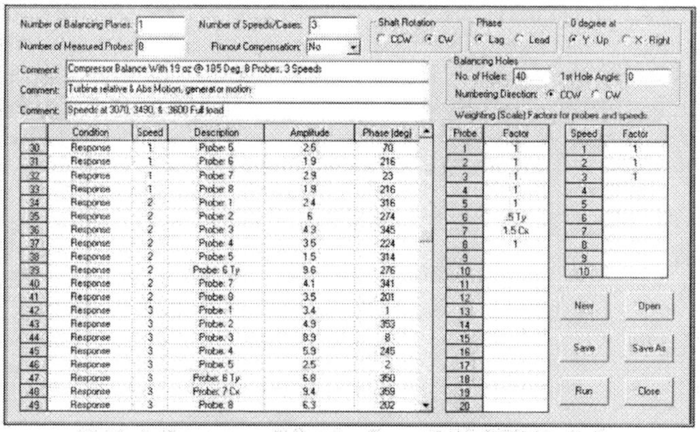

Number of Balancing Planes: 1 | Number of Speeds/Cases: 3 | Shaft Rotation: CCW / CW (selected)
Number of Measured Probes: 8 | Runout Compensation: No | Phase: Lag (selected) / Lead | 0 degree at: Y Up (selected) / X Right
Comment: Compressor Balance With 19 oz @ 185 Deg, 8 Probes, 3 Speeds
Comment: Turbine relative & Abs Motion, generator motion
Comment: Speeds at 3070, 3490, & 3600 Full load
Balancing Holes: No. of Holes: 40 | 1st Hole Angle: 0 | Numbering Direction: CCW (selected) / CW

	Condition	Speed	Description	Amplitude	Phase (deg)
30	Response	1	Probe: 5	2.5	70
31	Response	1	Probe: 6	1.9	216
32	Response	1	Probe: 7	2.9	23
33	Response	1	Probe: 8	1.9	216
34	Response	2	Probe: 1	2.4	316
35	Response	2	Probe: 2	6	274
36	Response	2	Probe: 3	4.3	345
37	Response	2	Probe: 4	3.5	224
38	Response	2	Probe: 5	1.5	314
39	Response	2	Probe: 6 Ty	9.6	276
40	Response	2	Probe: 7	4.1	341
41	Response	2	Probe: 8	3.5	201
42	Response	3	Probe: 1	3.4	1
43	Response	3	Probe: 2	4.9	353
44	Response	3	Probe: 3	8.9	8
45	Response	3	Probe: 4	5.9	245
46	Response	3	Probe: 5	2.5	2
47	Response	3	Probe: 6 Ty	6.8	350
48	Response	3	Probe: 7 Cx	9.4	359
49	Response	3	Probe: 8	6.3	202

Weighting (Scale) Factors for probes and speeds

Probe	Factor
1	1
2	1
3	1
4	1
5	1
6	.5 Ty
7	1.5 Cx
8	1
9	
10	
11	
12	
13	
14	
15	
16	
17	
18	
19	
20	

Speed	Factor
1	1
2	1
3	1
4	
5	
6	
7	
8	
9	
10	

New | Open | Save | Save As | Run | Close

Table 4 Compressor Balancing Input Table With Ty & Cx Weighting Functions Applied

Table 5 Vibration Measurements Before and After Compressor Plane Balancing Calculations With Weighting Functions Applied To Absolute Turbine Exhaust Measurement Tx and Absolute Compressor Measurement Cx

	3071 RPM			3450 RPM			3600 RPM Full Power		
(Wf)	***Initial***	***Trial***	**Final**	***Initial***	***Trial***	**Final**	***Initial***	***Trial***	**Final**
Txr	**8.1**	5.7	5.1	1.0	2.4	0.3	4.6	3.4	2.8
Tyr	2.0	1.7	1.6	5.4	6.0	3.2	4.9	4.9	3.9
Cxr	1.6	2.9	2.2	4.0	4.3	4.4	4.4	8.9	5.3
Cyr	2.1	1.4	2.3	2.9	3.5	3.3	4.8	5.9	5.1
Tx	1.1	2.5	0.5	1.5	1.5	0.9	3.7	2.5	2.3
Ty *(.5)*	3.7	1.9	2.8	**9.1**	**9.6**	5.5	7.5	6.8	5.4
Cx*(1.5)*	1.7	2.9	2.4	3.5	4.1	3.9	3.9	**9.4**	5.1
Cy	0.8	1.9	1.3	2.5	3.5	3.0	4.2	6.3	4.

Table 4 shows the application of weighting functions applied to the turbine absolute measurement Ty and the compressor measurement Cx. The above Table 5 indicates the importance of taking vibration measurements at speeds below the operating speed. For example it is seen that at 3071 RPM, the relative turbine exhaust motion Txr is 8.1 mils. At the higher speed of 3450 RPM, it is seen that the absolute turbine exhaust amplitude Ty is 9.1 mils. This is approaching shutdown values.

It is apparent that the choice of trial balancing weights of 19 oz was excessive as the trial amplitude on Ty has increased from 9.12 to 9.6 mils. In particular, it is also observed that the absolute compressor motion Cx at 3600 RPM has increased from 3.9 to 9.4 mils. When a balancing weight is applied, it is preferable that the chosen value will generally cause a reduction in overall vibration.

In the original calculations by the least squared error program, the large value of Ty encountered at 3450 RPM caused an excessive amount of unbalance to be applied to reduce this amplitude. The value of Cx at full power was also on the high side. To compensate for this, a weighting function of .5 was applied to the Ty, so as not to over balance, and a weighting factor of 1.5 was applied to Cx so that the absolute x amplitude at the compressor location at full power would be under control.

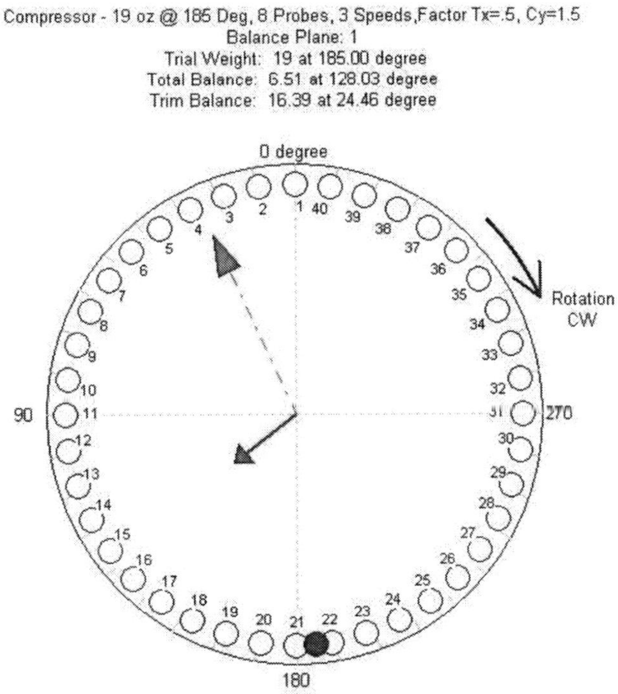

Fig. 40 Compressor Plane Balance Calculation With Tx and Cx Weighting Functions Applied

Figure 40 above shows the recomputed compressor end balancing calculations with the weighting functions applied to minimize over balancing of the turbine absolute exhaust motion Ty at 3450 RPM and to reduce the compressor absolute motion Cx at the full power level.

It is of interest to observe the final compressor balancing has been reduced to 6.5 ounces to be applied at hole 15 as compared to the earlier predicted value of 11 ounces to be applied at balance hole 16.

This represents a considerable shift from the initial trial balance of 19 ounces applied at holes 21 and 22. Such a balancing prediction can only be arrived at by the use of the computer to analyze the turbine vibrations at the lower system turbine-casing critical speed as well as the response at full power level at 3600 RPM. This predicted compressor end balancing value cannot be arrived at by simply examining the rotor polar plots.

W501D 2 Plane Balancing Calculations

Table 6 represents the input and results for a two plane balancing calculation using the initial trial weight of 19.27 oz applied at the compressor, balance plane 1 and also the results of the 7.4 oz trial weight applied at the turbine exhaust, balance plane 2.

By combining the results of these two initial unbalance trials, a combined final balance combination can be computed. These two balance corrections are then added simultaneously. For example, for the compressor inlet plane 1, the final balancing prediction is a correction weight of 16.45 oz at 70.4°. This balance would be achieved by removing the 19 oz weights applied at holes 21 and 23 and shifting the final reduced balancing weight over to hole 16 as shown in Figure 41.

The 2nd turbine exhaust balance plane correction is predicted to be 8.4 oz applied between holes 2 and 3. A trim weight of 7.6 oz could also be added to balance hole 36 as shown in Figure 42.

Table 6 2 Plane Balancing Prediction

Balance Plane 1 - Compressor Inlet, Ut = 19.27 oz@ 185° (Holes 21 & 22)
Ub1 = 16.45 oz@ 72° (Hole 16)

Balance Plane 2 - Turbine Exhaust, Ut = 7.4 oz @ 72° (Hole 9)
Ub2 = 8.4 oz @ 14° (Holes 2 & 3)

		3070 RPM			**3600 Full Power**		
Probe	**BAL Plane**	**Initial**	**Trial**	**Final**	**Initial**	**Trial**	***Final***
Tx	1	1.1	2.5	3.3	3.7	2.5	1.6
Ty	1	3.7	1.9	1.5	7.5	6.8	2.2
Cx	1	1.7	2.9	2.0	3.9	9.4	3.7
Cy	1	0.8	1.9	0.3	4.2	6.3	2.3
Tx	2	2.5	2.3	3.3	2.5	1.7	1.6
Ty	2	1.9	4.4	1.5	6.8	5.6	2.2
Cx	2	2.9	2.5	2.0	9.4	5.3	3.7
Cy	2	1.9	0.8	0.3	6.3	3.8	2.3

Figure 41 represents a vector plot of the initial trial, trim and final balance weight correction for the compressor plane 1. If the initial trial weight of 19 oz is left in the balance holes 21 and 22, then a fairly substantial trim balance of 16.5 oz would be required at balance hole 9.

It is of interest to note that the final balance correction is only 3 oz lower than the initial trial weight. The final balancing weight is to be applied to hole 16, which is only 60° away from the original guess. Although the original trial weight was high, it was of an acceptable amount and location.

The final balancing correction for the turbine balance plane 2 is shown to be around 8 oz as compared to the initial trial weight of 7.4 oz. Thus the initial turbine trial weight was of the proper magnitude and was off by approximately 60° from the final balance location. With regard to the balancing correction at the hot end of the gas turbine, it is sometimes desirable to leave the original trial weight in as it is often difficult to extract. Under these circumstances then a trim weight of 7.6 oz would be applied to balancing hole 36. This would seem to be a reasonable procedure.

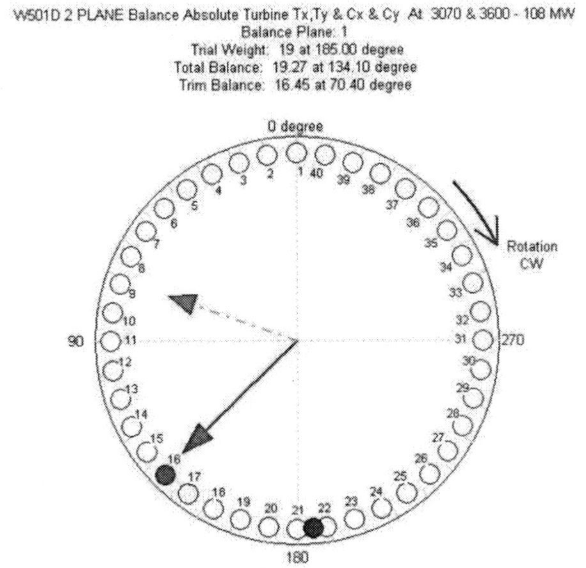

Fig. 41 Plane 1 - Compressor Initial Trial and Final Balance Correction

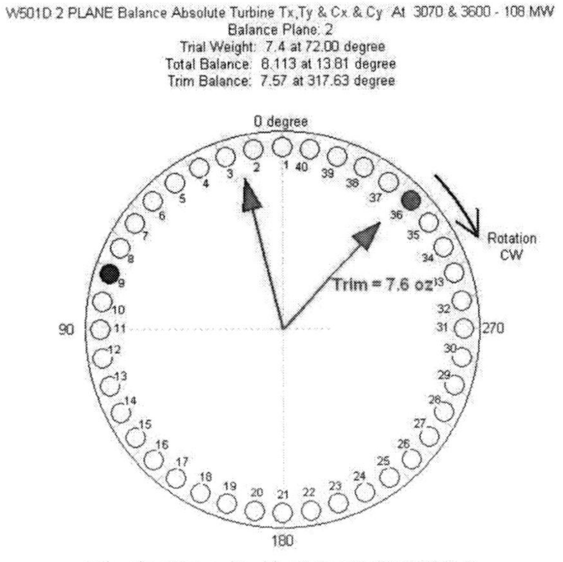

Fig. 42 Plane 2 - Turbine Initial Trial and Final Balance Correction

Discussion

The W501 series of gas turbine - generators represent a very complex system with challenging problems with respect to balancing and dynamic analysis. The situation is made more complicated by the thermal distortion of the gas turbine and generator under various load conditions. Of particular influence is the flexible support under the gas turbine exhaust tilting pad bearing. In a previous paper, it has been shown that the flexible turbine bearing support, not only reduces the effective bearing stiffness, but can also reduces significantly the turbine exhaust tilting pad bearing damping by over 90%.

In addition to the unusual behavior of the gas turbine due to the extremely flexible turbine bearing support, there is considerable turbine casing dynamics involved. The casing dynamics introduces additional modes into the operating speed range. This effect is not usually encountered so dramatically with steam turbines which have much more massive cases.

Other complications that can enter the picture is the fact that the design of the 120 deg partial arc generator bearings are over hundred years old! These partial arc bearings provide very little horizontal support and damping. In addition, the generator is predicted to be unstable at speed, exciting the generator first critical speed at 3600 RPM. This instability is also predicted by the OEM. Although not reported in practice, it is evidence of the poor damping provided by the generator bearings. One must also be careful that a generator rub does not occur as this will distort the influence coefficients and upset the balancing calculations. Hence, one must also monitor the full vibration spectrum as the rub can generate a 3x supersynchronous vibration.

Another problem that was not encountered with the 70 MW W501 gas turbines is the possibility of the occurrence of self excited subsynchronous turbine whirling. With the power level of the system increased from 70 MW to up to 208 MW, the Alford cross coupling forces are increased. There have been observed occurrences of self excited whirl instability in which the turbine 1st vertical critical speed around 860 RPM appears to be excited.

Conclusions

There have been situations with the W501 in which field balancing attempts have met with limited success. It appears that the selection of the initial compressor and turbine exhaust trial weights have been acceptable. These balancing trial weights are usually based on the polar plots of the vibration performance.

With respect to the number of planes required for balancing flexible rotors, Kellenberger had presented a paper on the N or N+2 method of balancing. The minimum requirement is N balance planes to balance N critical speeds. It was stated that N+2 balance planes are required if bearing forces are also to be minimized.

Therefore, it is impossible to have a single balance plane produce desirable results at both running speed and as well as at the additional critical speeds encountered through the entire speed range. If, for example, an aggressive single plane balancing correction is installed at full speed, full power, this balance correction runs the risk of causing excessive excitation of one or more of the lower system critical speeds. The high vibrations encountered may cause the unit to trip out and not allow the unit to reach full operating speed.

It is important to note that after two successful trial balancing weights have been applied to the turbine and compressor locations, then a computer program is required to crunch the numbers to further define a proper balance solution. It is not possible to simply use the polar plots to accurately compute the amount and particular the angular locations of the final balance corrections to be installed in the two balance locations. The system, due to the low turbine end damping and also operation near a rotor-casing mode, is very sensitive to the phase location of the balance weights.

When collecting balancing data, it is essential that the data before and after the placement of the trial weights, be taken under the same conditions of speed and power level. Due to the thermal effects in the gas turbine and also warpage of the generator at higher power levels, influence coefficients change with a change in the power level.

Recommendations

After the first two trial weights at the turbine and compressor location have been successfully installed, it is recommended that a least squared error computer program be used to compute the final balancing calculations. It should be noted that a trial balance at the coupling may also be incorporated with the other two planes to form a three plane balancing solution.
The W501 gas turbines have the unusual characteristics that the cross coupled influence coefficients are of greater magnitude then the direct coupled coefficients. This means that a balance weight applied at one end of the turbine causes a higher response at the other end. This again is due to the extreme flexibility under the turbine exhaust tilting pad bearing.
To further attempt to refine the balance by the use of polar plots only would not be productive.

One of the advantages of the least squared error method of balancing is that the influence coefficients collected from previous balancing cases may be used for future balancing situations. This approach has been reported to me to have been successfully used on several W501 units by Ed Shepard of Mechtell, the inventor of the Turbobalancer. The assistance of Mr. Shepard and the Turbobalancer was vital in the earlier successful balancing reported on the CP&L 70 MW 501 units.

There are several other recommendations concerning the modification of future W501 gas turbine-generator systems.

The generator 120° partial arc bearings should be replaced by 2 100° tilting pad or 2 fixed pad bearings with load between pads. This would add needed system damping and reduce the occurrence of horizontal generator rubs. The current generator bearing design provides very little horizontal support or damping. An increase in the generator horizontal stiffness and damping characteristics could aid in reducing some of the extreme rotor asymmetry characteristics observed.

The flexible support under the exhaust tilting pad bearing is required because of temperature considerations. However it would be of value to consider if this support could be stiffened. By stiffening up the turbine bearing support, the tilting pad effective damping will be greatly improved. This may also have the added effect of shifting the rotor casing mode to a more manageable speed. A detailed finite element analysis should be performed on the current tilting pad bearing supporting structure to review the feasibility of a design change.

With the improvement in electronic monitoring and miniature sensors, it is recommended that pressure sensors be embedded in future tilting pad and generator bearings. These bearings could monitor the rotor static and dynamic loads generated. The static and dynamic hydrodynamic bearing pressures generated may be used for both alignment and balancing. The use of static and dynamic bearing pressure measurements for alignment and balancing have been reported on earlier by both Lund and Gunter.

REFERENCES

Alford, J., 1965, ***Protecting Turbomachinery From Self Excited Rotor Whirl,*** *Journal of Engineering For Industry*, pp. 333–334

Badgley, R. H., 1974, ***Recent Developments in Multiplane–Multispeed Balancing of a Flexible Rotors in the United States***, *International Union of the Theoretical and Applied Mechanics*, Lyngby, Denmark

Bently, D. E. 1982, ***Polar Ploting Applications For Rotating Machinery***, Vibration Institute, Clarendon Hills, Ill.

Blake, M. P., 1967, ***Use Phase Measuring to Balance Rotors in Place***, Hydrocarbon Processing,

Chen, W. J. And E. J. Gunter, 2007 ***Introduction to Dynamics of Rotor–Bearing Systems***, Trafford

Darlow, M., 1989, ***Flexible Rotor Balancing By the Unified Balancing Approach***, Rensselaer Polytechnic, Troy New York

Goodman, T. P., 1964, ***A Least–Squares Method for Computing Balance Corrections***, Journal of Engineering for Industry, 86(3): 273 – 279

Gunter, E. J., Barrett, L. E., and P. E. Allaire, 1976, ***Balancing of Multi-Mass Flexible Rotors: Part I Theory And Part II Experimental Results***, Proceedings of The Fifth Turbomachinery Symposium, Texas A&M University, College Station, Texas, October

Gunter, E. J. And R. Humphris, 1986, ***Field Balancing of 70 MW Gas Turbine Generators***, Proceedings International Conference On Rotor Dynamics, JSME, Tokyo, pp. 135 – 143

Gunter, E. J. And C. Jackson, Chapter 3: ***Balancing of Rigid and Flexible Rotors, pp 3.1 - 3.117***, Handbook of Rotor Dynamics

Gunter, E. J., 2006, ***Unbalance Response and Field Balancing of an 1150–MW Turbine – Generator with Generator Bow,*** Seventh IFToMM Conference on Rotor Dynamics, Vienna September 25 – 28

Jackson, C. 1971, ***Using the Orbit to Balance***, Mechanical Engineering

Kellenberger, W., 1972, ***Should a Flexible Rotor to Be Balanced in N or N +2 Planes?***, Engineering For Industry, 94, 548 – 560

Lund, J. W., And J. Tonnesen, 1972, ***Analysis and Experiments on Multi–Plane Balancing Of Flexible Rotors***, Journal of Engineering for Industry, 94(1): 233

Palazzolo, A. B., And E. J. Gunter, 1977, ***Multi-Mass Flexible Rotor Balancing By the Least Squares Error Method***, Vibration Institute, Clarendon Hills Illinois

VIBRATION CHARACTERISTICS OF MACHINES ON RESILIENT MOUNTS

Robert J. Sayer, PE
Applied Structural Dynamics
Medina, Ohio
sayerinc@bright.net

Abstract: Machines are commonly mounted on vibration isolation systems to minimize the dynamic forces that are produced by a machine from transmitting into surrounding structural systems or to minimize the vibration from the surrounding environment from transmitting to the machine. Isolation systems typically consist of a set of soft resilient springs placed under the machine. The resilient mounting of equipment results in its structural dynamic response characteristics being much different than that of machines which are positively connected to structures and foundations. The flexibility of the isolation springs results in a nearly free-free support condition.

Key Words: isolation, resonance, shop testing of motors, structural dynamics, vibrating screens

Structural Dynamics of Free-Free Systems:

The structural dynamics of a machine on isolation springs is different than a machine supported directly upon structures or foundations. The resilient mounting condition results in the machine responding as though it is suspended in a free-free state. To illustrate the unique structural dynamic response characteristics of machines supported on a resilient mounting condition, consider a double-wide, double-inlet centrifugal fan rotor. This type of fan has a wheel mounted on a shaft, mid-span between the bearings. If the structural constraint provided to the rotor by the bearings were infinitely rigid, the natural frequency associated with the principal flexural mode of the rotor would be 19.3 Hz. The mode shape is shown in Figure 1 superimposed upon the undeformed shape (shown in green). It consists of simple flexural deformation of the shaft between the bearings. The displacement of the shaft is zero at the bearings. It is maximum at the mid-span point between the bearings. Note that this is the 1st mode of natural frequency for the shaft. The fan wheel participates only as added mass, simply going along for the ride.

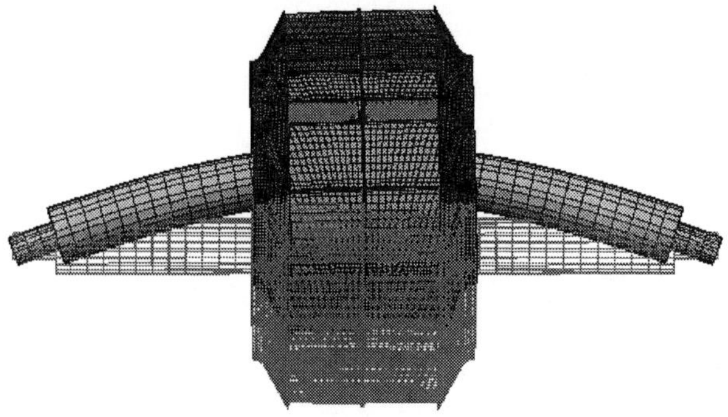

Figure 1: Rotor Mode in Rigid Bearings

Now consider the same fan rotor, this time totally unsupported, in the free-free state. Figure 2 shows the flexural mode shape superimposed upon the undeformed shape. The flexural deformation of the rotor looks almost identical to the mode shape for the infinitely rigid support condition. However, the natural frequency is 62.3 Hz, which is larger than the natural frequency of the rotor with rigid supports. This appears to be contrary to reason; if natural frequency reduces with a reduction in stiffness, why is the natural frequency of the unsupported rotor greater than the natural frequency of the rigidly constrained rotor?

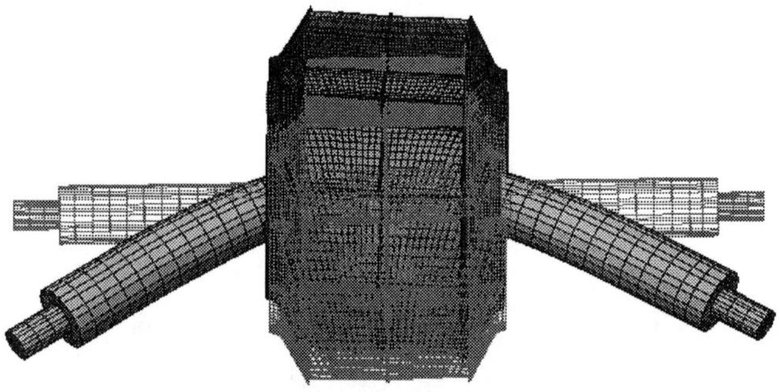

Figure 2: Rotor Flexural Mode (Free-Free)

A detailed investigation of the mode shape is required to provide an answer to this perplexing question. Even though the mode of natural frequency shown in Figure 2 is the principal flexural mode of the shaft, the displacement of the shaft at the bearing is no longer zero. It is out-of-phase with the displacement of the fan wheel at mid-span of the shaft. The zero displacement locations are somewhere between the bearing locations. This would be similar to the mode shape for a rigidly constrained rotor with bearings located at the zero displacement locations rather than the actual bearing locations. Since the flexural stiffness of a shaft is dependent upon the span between bearings, it increases as the span decreases, the effective flexural stiffness of the free-free supported shaft is greater than the actual stiffness of the rigidly supported shaft.

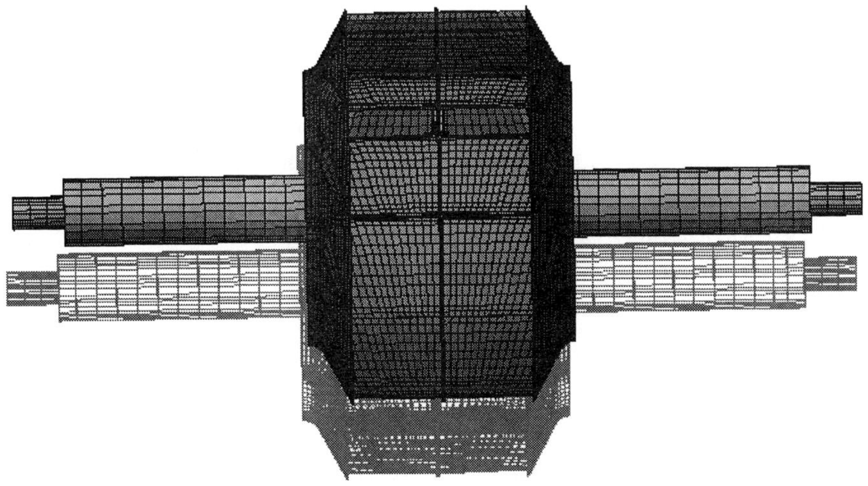

Figure 3: Rotor Rigid Body Mode

Unlike the rigidly supported rotor, the flexural shaft mode of Figure 2 is not the 1st natural frequency for the free-free state. Figure 3 shows another mode shape superimposed upon the undeformed shape (shown in green). The entire rotor participates as a rigid body; it does not have any internal deformation. The natural frequency for this mode is 0.00 Hz, which is lower than the natural frequency of the flexural mode.

The rigidly supported rotor does not contain a rigid body mode; it is actually part of the flexural mode. The principal flexural mode of the rigid support condition is a combination of the two modes for the free-free condition. Therefore, the natural frequency of the rigidly supported rotor falls somewhere between 0 Hz and 62.3 Hz, the natural frequency of the flexural mode for the free-free state. The resilient mounting condition results in a bifurcation of the two modes of natural frequency.

Although large centrifugal fans, like the fan used in the previous example, are not usually supported upon isolation springs, they are regularly installed in fluid-film bearings. The stiffness of this type of bearing is typically much less than the flexural stiffness of the rotor and structural stiffness of the supporting foundation. In many cases, it can be considered a resilient mount. Large centrifugal fan rotors on oil-film bearings generally will have both of the two bifurcated modes of natural frequency.

Fans & Pumps Supported on Isolator Bases:

Commercial air-handling units and pumps are commonly supplied with isolation bases. Isolator springs are provided between the isolation base and the floor or foundation. Machines supported upon isolation bases present a unique set of structural dynamic considerations.

These mechanical systems consist of a set of isolation springs, an isolation base, motor, fan or pump rotor and housing. The stiffness of the isolator springs relative to that of the isolation base is insignificant. The isolation base and all the attached equipment approximates an unsupported structural system operating in a free-free state. This supporting condition must be accounted for accurately when calculating the natural frequencies of an isolated system.

There are numerous different fan and pump isolation base arrangements. A discussion of each and every possible arrangement is beyond the scope of this discussion. An isolation base can have many natural frequencies. The primary modes can be classified as torsional and flexural modes. Figure 4 contains the mode shape for the natural frequency associated with the principal torsional mode of an isolation base as calculated by a numerical finite element analysis.

In the principal torsional mode, the maximum deformation occurs at each corner of the base. The motion of every corner is out-of-phase with that of each adjacent corner. Since neither the fan nor the motor is typically located at either the center of stiffness or center of mass, this mode can be sensitive to dynamic forces produced by the motor and/or the fan. The vibrational response of the isolation base can become quite severe if either the operating speed of the motor or the fan approaches the natural frequency of this mode. If excited, the vibration motion of the motor will be out-of-phase with that of the fan. The relative motion of the motor to the fan leads to the development of secondary dynamic forces in the drive belts and bearings, which will adversely affect the reliability of the equipment, by reducing bearing and belt life.

Although the isolator springs provide structural support for the weight of the structural-mechanical system, the springs do not provide support for the structural dynamics of the system. In most cases, the effect of changing the stiffness of a spring constant will result in an insignificant change to the natural frequency of the isolation base. For the example above, the natural frequency of the torsional mode was calculated at 39.9 Hz using the isolator springs that were originally specified for the design. The analysis was redone after increasing the stiffness of the isolator spring under the motor by fifty percent. The calculated natural frequency increased by less than one percent, to only 40.1 Hz.

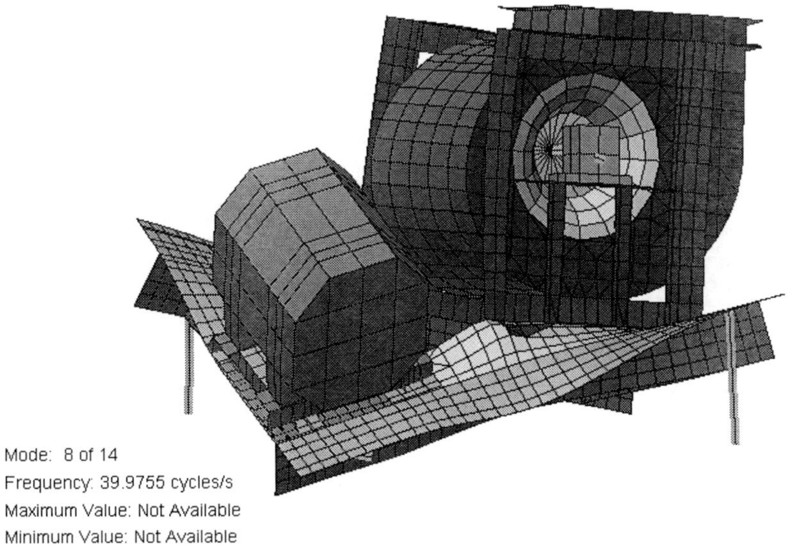

Figure 4: Torsion Mode of Fan on Isolation Base

All of the mechanical equipment supported upon the isolation base participates in the torsional mode shown in Figure 4. Not only does the fan housing add to the mass of the system, it provides stiffness that resists the torsional deformation of the base. Both the stiffness and mass characteristics of the fan must be included in any calculation for this natural frequency. Likewise, the stiffness and mass characteristics of the motor and fan rotor must be considered in the natural frequency calculation.

Vibrating Screens & Feeders:

Vibrating screens and feeders are used in many industries to convey, dewater and screen materials. This type of equipment is very common in the mining industry where material flow is generally accomplished by gravity, requiring placement of the screens in elevated towers. Directives aimed toward increasing plant efficiencies have resulted in vibrating screens becoming very large. The screens are typically placed upon flexible support structures that frame across large distances. The large distances that structures span between columns result in increased structural flexibility and reduced natural frequencies. The combination of the greater dynamic forces produced by these larger screens with the increased flexibility of the structural supporting systems has become a serious design issue.

Figure 6: Incline Screen

Figure 7: Multi-Slope Banana Screen

Figure 6 is a picture of a typical incline screen. This screen is powered by two motors, driving two eccentric shafts. It has three inclined decks. The eccentric drive shafts are located between the middle and bottom decks. The magnitude of the dynamic force produced by the eccentric drive mechanism is a function of the square of the operating speed of the screen.

Figure 7 is a picture of a typical multi-slope screen, also referred to as a banana screen. This screen has two decks. The slope of the decks varies from the feed end to the discharge end. The banana screen is driven by mechanisms located near the top of the screen, above the upper deck.

Regardless of the type of screen or drive arrangement, multiple eccentric weights can be phased in such a way as to provide circular, elliptical or linear motion. When analyzing the structural dynamic response of the screen, the isolation efficiency of the springs, and the response of the structural supporting system, it is important to identify the type of motion that a screen is designed to operate under. This defines the direction of the dynamic forces produced by the drive mechanisms. The operating speed of the screen is another important factor that must be considered.

Similar to other types of equipment on isolation springs, vibrating screens will have torsional and flexural modes. Because of the size and flexibility of these machines, it is not always possible to provide a structural design that results in all natural frequencies of the screen body being greater than its operating speed. For cases where it is possible to do so, the machine is classified as a rigid screen. For cases where all natural frequencies, except that of the principal torsional mode, are greater than operating speed, the machine is classified as a semi-rigid design. Finally, if multiple modes of natural frequency exist below the operating speed, the screen is classified as a flexible design. Even though a screen may be classified as a rigid design, it still could be subjected to resonant excitation, as illustrated by the following case study.

Case Study (Self-Excitation of Natural Frequency in a Vibrating Screen):
Interaction between the structural dynamic characteristics of a screen body with that of the flexible supporting system can lead to many problems. A vibrating screen at an aggregates plant, that was used to wash product, was experiencing high vibration in the z-direction of the side plates. Since screen side plate failure in other applications has been found to be related to z-direction vibration, this was a concern.

The x-direction is defined as the horizontal in the direction of flow of material across the screen. The y-direction is vertical. The z-direction is perpendicular to the plane of the screen body. The screen was driven by two eccentric mechanisms that where phased perfectly together using a timing belt to provide a perfect harmonic excitation force. Theoretically, the response of the screen should have been circular, with motion constrained to the x-direction and y-direction of the machine.

Based upon available sheave sizes and production requirements (i.e. screen throughput), the screen could have been operated at either 12.67 Hz (760 rpm), 13.5 Hz (810 rpm), or 14.33 Hz (860 rpm). It was initially set at the lower speed (12.67 Hz). Tests were performed to determine the natural frequencies of the screen body. A natural frequency was identified at 20.25 Hz. The ratio of this natural frequency to operating speed, at the lower operating speed, was 20.25/12.67 = 1.59. Usually this separation margin would be considered outstanding. However, the intent of this screen was to wash a product, not to size it. There were many material grades that stayed on and traveled across the top deck. It is normally considered, throughout the mining industry, that the material moving across the decks of a screen comprises a suspended mass and does not mass load the screen. However, in this case, the mass loading of the top deck was extreme enough to mass load the screen body, thus, lowering its natural frequencies.

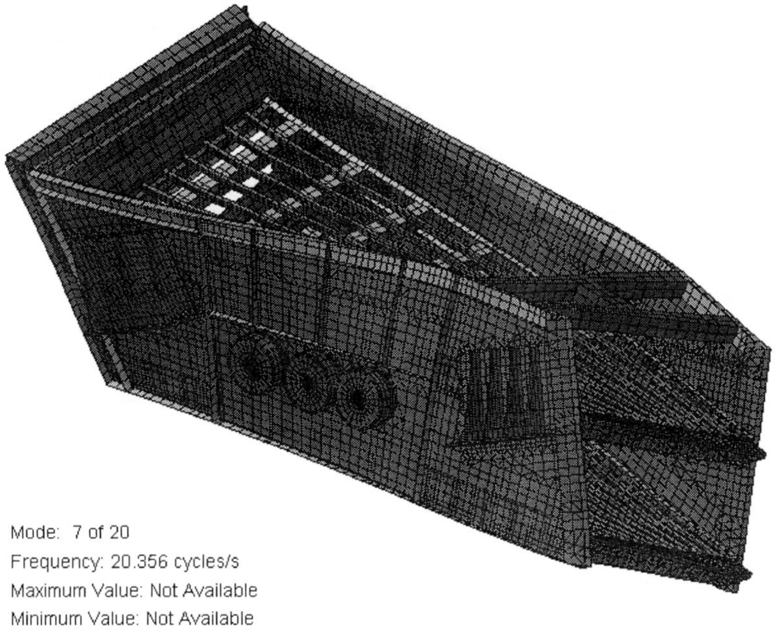

Figure 8: Torsional Mode Shape of Screen Body

At an operating speed of 760 rpm, the mass loading of the material caused the natural frequency of the torsional mode to drop to 19.0 Hz, which was almost exactly 1.5X the rotational speed of the eccentric drive. This natural frequency was associated with the principal torsional mode of the screen body (Figure 8).

There were two screens of identical design that were located side-by-side. They were loaded by a single conveyor belt through a pant-leg chute. This provided momentum to the material toward one side of the screen, instead of loading it directly down the center. The mass loading of material caused the center-of-mass to shift toward one side of the screen. This created an eccentricity between the location of the dynamic force produced by the drive and the center-of-mass, which produced a dynamic torsional force applied at the operating speed of the screen. Normally this force would not be considered significant, but its interaction with the torsional mode made it severe.

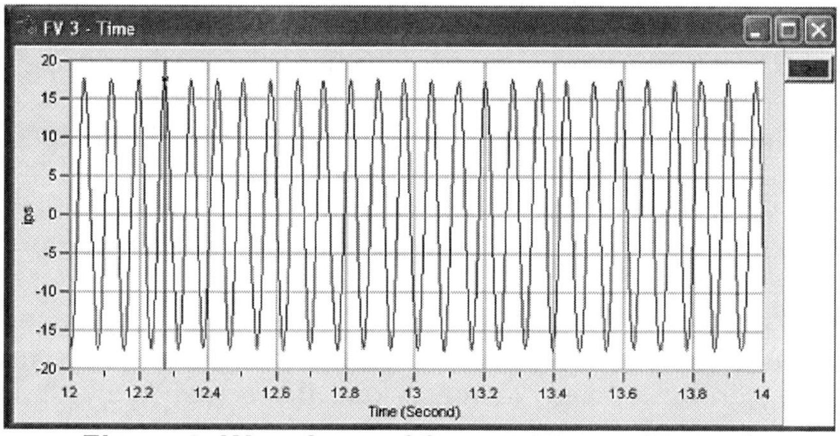

Figure 9: Waveform of Screen Throw (Empty)

The y-direction vibration of a screen should be a pure harmonic if everything is operating properly. Figure 9 is a waveform of the y-direction vibration measured when the screen was operating empty. As expected, it was a pure harmonic, with peak positive and negative motion at 17.5 ips. After material began to flow across the top deck, the waveform changed significantly to that shown in Figure 10. In this waveform, the maximum positive peak increased to around 25 ips. However, every other throw of the machine, the maximum positive peak decreased to around 10 ips. In the negative direction, the maximum motion was around 20 ips.

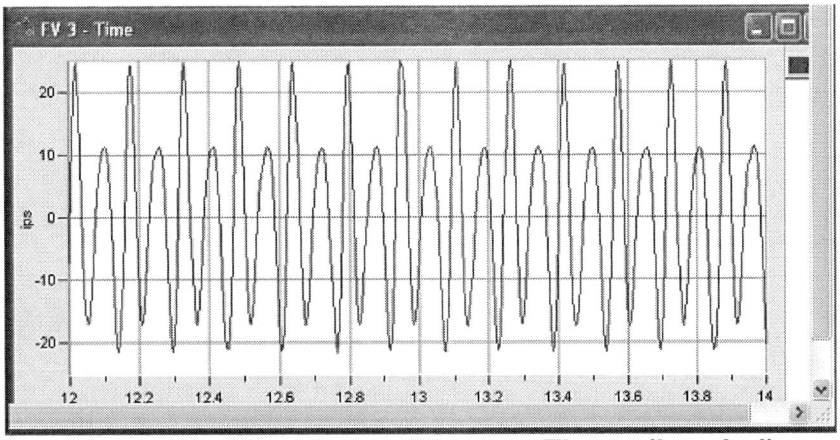

Figure 10: Waveform of Screen Throw (Loaded)

This waveform can be explained as the response of the torsional mode superimposed upon the normal rigid body response of the machine.

Figure 11 is a mathematical formulation of three different curves. The first curve (blue line) is y1 = 17.5 cos(ωt) and represents the normal rigid body motion of a screen. The second curve (red line) is y2 = 7.5cos(1.5ωt) and represents the resonant torsional response. The third curve (green line) is the addition of the other two. It matches the data very well.

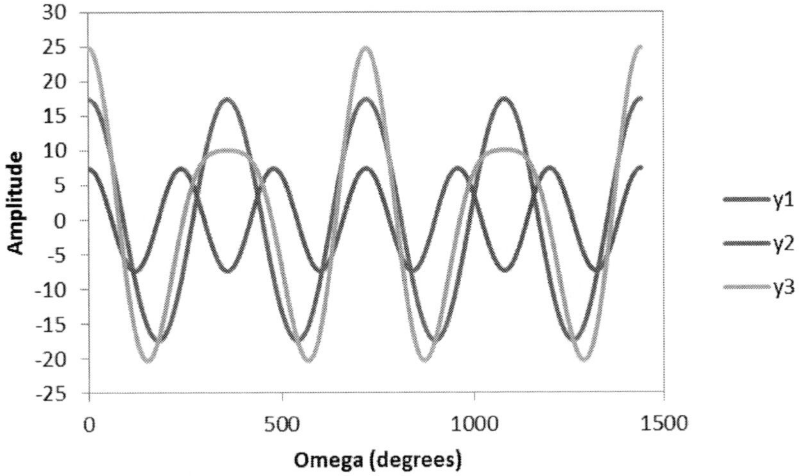

Figure 11: y3 = y1 + y2

Every other cycle of y1 is in-phase with y2. These responses are out-of-phase with each other for the cycle between. The result of this is a maximum positive peak response that varies in magnitude every second cycle. It also results in the lower of the two positive peaks being more rounded, which also matches the actual data. The FFT of the resultant response waveform y3 contains components of vibration at 1X and 1.5X operating speed.

Because the torsional mode of natural frequency was being excited, the z-direction motion of the side plate, measured at the upper feed-end corner of the screen exceeded

4.0 ips. Typically, the industry standard, acceptable z-direction vibration level of a new screen is 1.0 ips. Since this vibration far exceeded industry standard acceptance criteria, the operating speed of the screen was increased to 14.3 Hz (860 rpm) to provide separation between 1.5X the operating speed of the screen and the natural frequency of the torsional mode.

Shop Testing of Electrical Machines on Resilient Mounts:
The National Electrical Manufacturers Association (NEMA) publishes a standard [1] that includes a recommendation for the evaluation of mechanical vibration in electric motors. For the shop testing of motors, a resilient mounting condition is allowed. During this test, the motor is mounted upon a soft isolation material (mechanical springs, elastomeric pads, etc.) designed such that the natural frequency of the isolation system is less than 33% of the lowest speed that the motor will be tested at. If done properly, the resilient mount will minimize the effects of the foundation from the performance of the motor. The following argument is used to support testing the motor on a resilient mount: if the motor meets the vibration requirements of the standard on resilient mounts, where the effects of the foundation have been removed, and does not perform satisfactorily when mounted in the field, then the reason for the poor vibration performance should be the foundation upon which the motor is mounted.

Figure 12: Generator on Shop Test Bed (from Biby & Plymon [2])

To illustrate the potential problem with this argument, consider the shop testing of the generator shown in Figure 12. It was supported upon several layers of elastomeric pads, designed to meet the natural frequency requirements of the NEMA specification regarding resilient mounts. Metal sheets were provided between each layer of the elastomeric pads to insure proper distribution of the load through the pads. The vibration performance of the machine was to be evaluated at its intended operating speed of 900 rpm. Figure 13 contains a coast-down plot of the vibration levels measured during an intentional over-speed test. The maximum rotational speed of the test was 1000 rpm.

The vibration levels at 900 rpm are acceptable per the NEMA specification. There is a noticeable amplification of vibration as the machine speed coasts down through 385 rpm, due to the excitation of the natural frequency of the resilient mount, indicating that the resilient mounting fell slightly short of the NEMA requirements. The ratio of the natural frequency of the resilient mount to the operating speed of the machine (385/900 = 0.428) was slightly greater than the 0.333 specified by NEMA.

The main concern over this test is the fact that the vibration levels increase dramatically as the rotation speed was increased to 1000 rpm, indicating the presence of another natural frequency, most probably associated with the twisting mode of the stator frame. According to the data, it can be concluded that this natural frequency is somewhat larger than 1000 rpm. Also, it can be concluded that, as supported, this natural frequency does not have any significant effect on the vibration characteristics of the generator at 900 rpm. The problem with this statement is that the generator will not be supported in the field as tested in the shop.

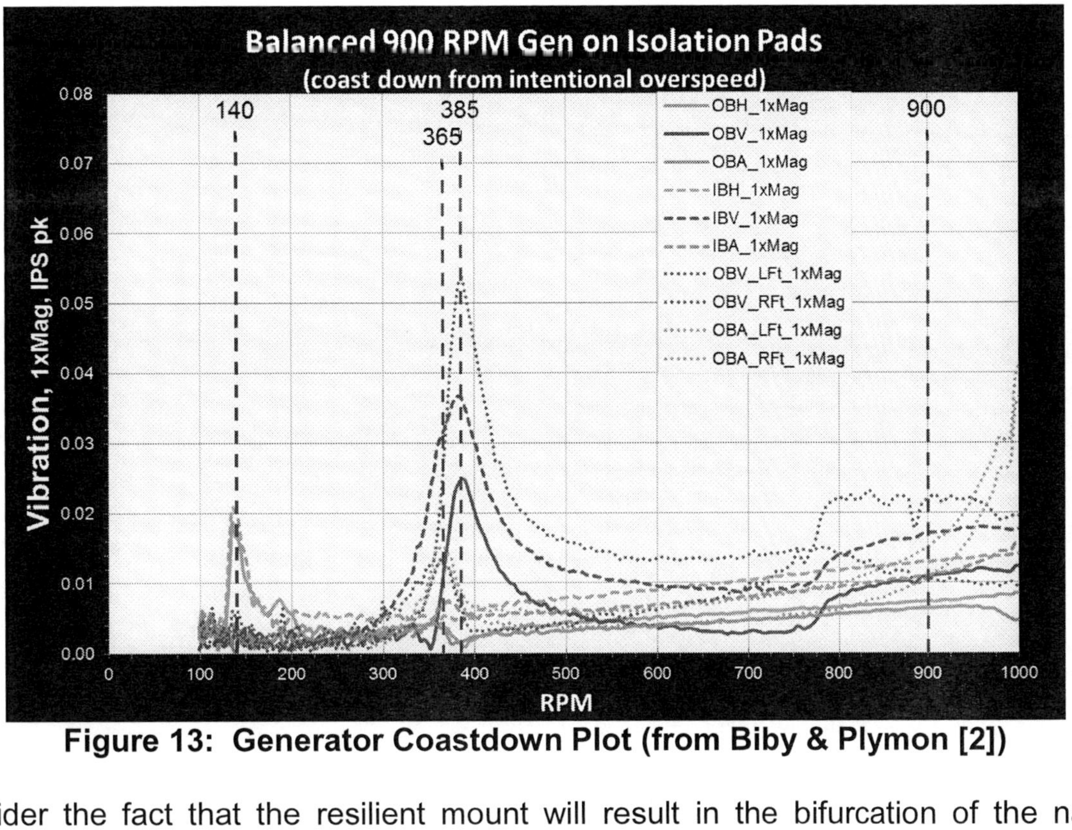

Figure 13: Generator Coastdown Plot (from Biby & Plymon [2])

Consider the fact that the resilient mount will result in the bifurcation of the natural frequencies of the generator frame. It will have a nearly rigid body mode (which is the 385 rpm isolator mode of natural frequency) and torsional and flexural modes. The natural frequencies of these torsional and flexural modes will be greater on the resilient mounts than a rigid or semi-rigid foundation. If the natural frequency of the machine frame is slightly greater than the operating speed when shop tested on resilient mounts, then it could be very close to operating speed when supported in the field.

References:

[1] NEMA MG- 1-2011, *Motors and Generators*, Part 7, Section 1 Mechanical Vibration - Measurement, Evaluations and Limits, National Electrical Manufacturers Association, Rosslyn, VA, 2011

[2] Brian Biby and Allen Plymon, *Acceptance of Motor Testing II: Mounting Techniques*, Power Point Presentation, 39th Annual Training Conference of the Vibration Institute, Indianapolis, IN, July 2015.

CAVITATION AND FOULING INDUCED VIBRATIONS IN LARGE STEAM TURBINES

S. Murugan

Rotating Equipment Engineer, Dubai Petroleum, Dubai, UAE
smuruganp2001@yahoo.com

Abstract: High noise was observed in one of the 40 MW capacity steam turbines. Vibration analysis and condition assessment were carried out, on emergency call. A drastic increase in case vibration level of 15mm/sec-pk was noticed at the turbine exhaust end. Normal level is 3mm/sec-pk. Impact energy too were high at 47Gs compared to normal value of 15Gs. Spectral analysis showed series of multiples and sidebands with significantly higher amplitudes, which are usually indicators of severe damage or imminent damage to components. But the multiples and side bands were non-synchronous. These are usually seen in rolling element bearings. Also, data from proximity probes had an increase in vibration amplitude but were at lower levels. Having journal bearings, we had anticipated heavy cavitation effects at final stages of turbine, coupled with heavy fouling. So, we decided to decrease the load and observe the changes. Significant reduction in both the vibration and noise levels were seen for just 4% reduction in load. Therefore, we planned and carried out an on-line wash in a month's time. Vibration and noise levels came back to normal. This paper outlines how vibration analysis helped minimise damage to components and minimise maintenance work and downtime with an optimal performance output.

Key Words: Cavitation effects; diaphragms; erosion; fouling; on-line wash.

Introduction: This is a 12 stage multistage turbine, Figure 1, with 1st and 3rd Curtis stages. It is the extraction and final exhaust condensing stages of the steam turbine that are of concern. The rated capacity of the turbine is 40MW at a Speed of 3800rpm. The turbine uses very high pressure steam of 100barg, 500Deg-C at the flow rate of 650Tons/hr.

During routine walk through, the plant operators observed high noise from the turbine drive end bearing, in odd hours. It is the exhaust end of the turbine. The noise is not just flow noise but more likely a mechanical impact sound. So, we were called in the night on urgent basis, for condition assessment of the turbine. Vibration data were collected and analysis was carried out, on the spot. Vibration analysis revealed nothing mechanical, rather the vibration was cavitation and fouling related. It gave the much required time, in the middle of the night, to make appropriate decisions amicably. Next morning, a decision was taken to reduce the load and watch for changes. The vibration value was reduced to ¼ of the maximum amplitude measured, for just 4% reduction of the load. It was the combination of load, deposition and steam quality that led the machine to exhibit and operate under dangerous level of vibration and noise. Prolonged operation of turbine at this condition would lead to erosion and pitting damage to costly rotor and diaphragm.

Figure 1: Steam Turbine Arrangement

Noise was so high and was of an impacting nature that was heard from exhaust end bearing housing. Similar noise appeared throughout the machine extending up to the steam inlet end. So, vibration data were collected and analysed on the spot, to ascertain exact nature of the vibration. Extensive vibration analysis carried out on both the offline casing vibration data and an online proximity probe vibrations, in order to get a clear picture and confirmation of the problem.

Turbine Case Vibration Data Analysis: Overall vibration amplitude levels collected from turbine exhaust case were very high at 15 mm/sec pk from a normal level of 3mm/sec pk. The vibration spectrum showed a series of multiples extending up to 60 times running speed with series of sidebands. They were misleading to the suspicion of impending problem and or imminent danger to the machine. Also, waveform data showed very high impact energy levels as high as 48 G-s pk compared to normal level of 15 G-s. This had further added to the suspicion of impending damage to the mechanical components.

Careful analysis of captured spectrum, Figure 2, revealed the multiple frequencies were nonsynchronous as are commonly seen in rolling element bearings. The fundamental frequency was 4.69 times the running speed. The sidebands are 0.69 times running speed. Having the turbine mounted on tilting bad journal bearings, the conclusion of fouling related vibration components was made as opposed to mechanical condition related danger of damage to components which are usually shaft speed related and are of synchronous vibrations.

The pattern of vibration was no way related to any resonance in the system - rotor, case, or piping and supports.

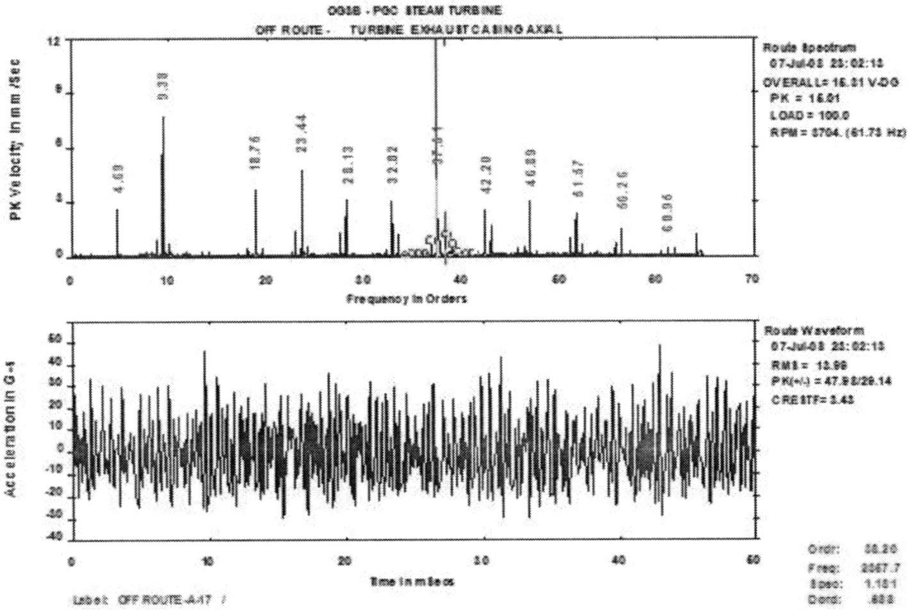

Figure 2 Turbine Case Vibrations

Bearing Case Vibration Data Analysis: Vibration data collected from bearing housing too had similar signatures, Figure 3, with higher amplitude levels of 7.4 mm/sec pk compared to a normal level of 2mm/sec pk. This led to the confirmation of actual problem within the machine itself rather transferring from external sources such as piping, inlet/extraction valves, and driven equipment or nearby machines. Though vibration components were of nonsynchronous ones, there was no other choice than to believe that it was originating from rotor. This is due to the interaction between rotor and stator masked by hard deposits and heavy fouling. Thus affecting steam turbine flow path engineering and design.

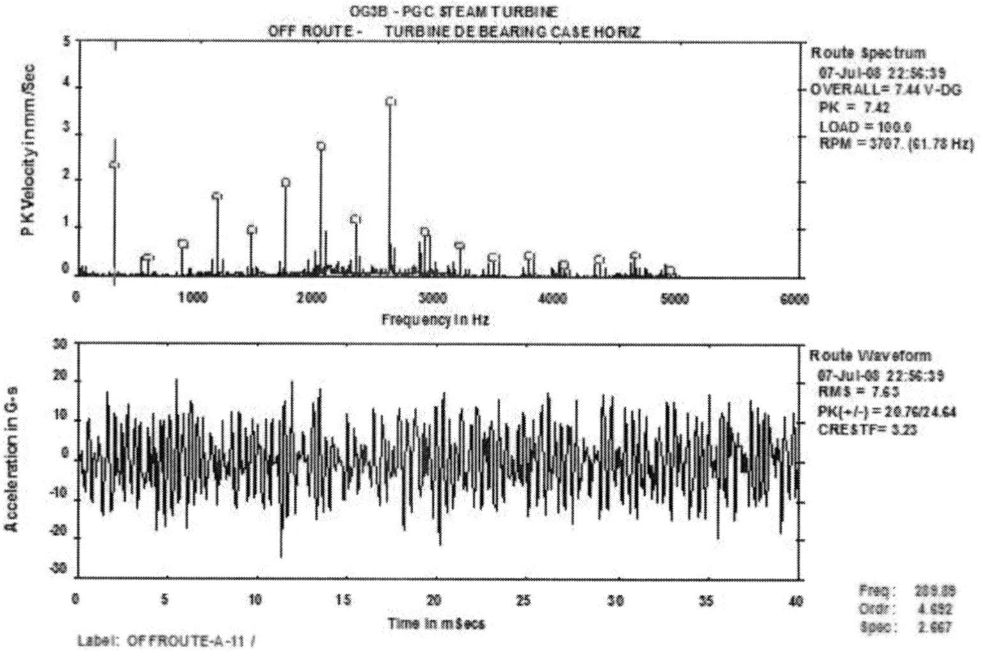

Figure 3: Turbine Bearing Case Vibration

Shaft Vibration Data Analysis: Shaft vibration data from installed proximity probes had significant levels of overall vibration amplitudes increase, Figure 4a, at the time of fouling induced vibrations. The spectrum plot too had similar non-synchronous component at 4.63xorder, Figure 4b, and its multiples that are comparable in amplitudes to that of 1x of the running speed components. These non-synchronous components extended up to the compressor drive end bearings, Figure 4c, sounding the seriousness of such incidents and confirming it as originating from steam turbine rotor as the compressor bearing housing/case are independent to that of turbine.

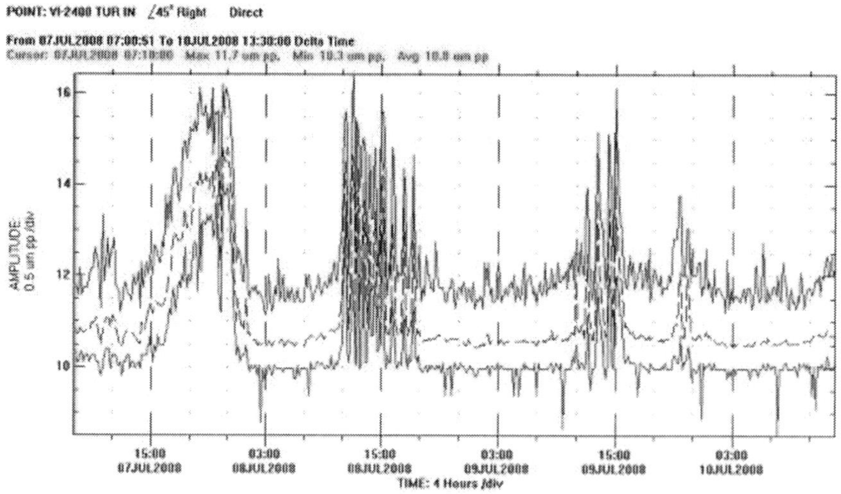

Figure 4a: Turbine Shaft Overall Vibration Trend

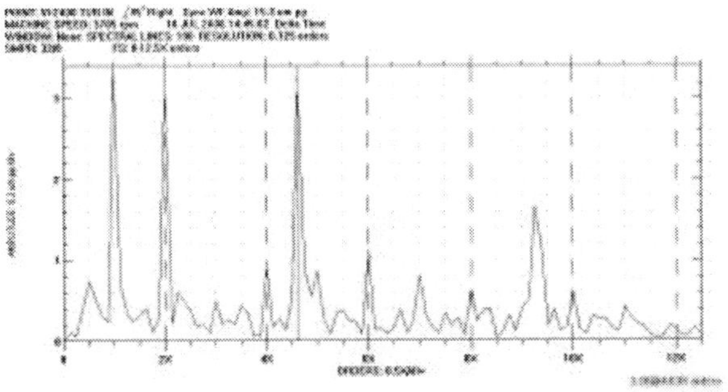

Figure 4b Turbine shaft vibration

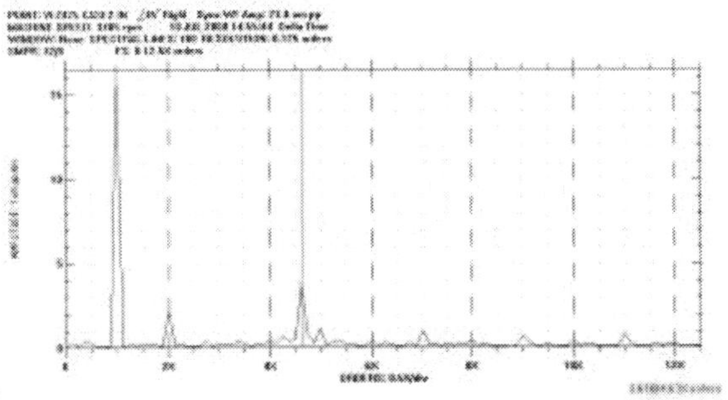

Figure 4c Compressor DE vibration

Action Taken: Based on above analysis of both case and shaft vibration data, it was concluded that there was no means to predict / foresee any damage or imminent damage to components to shut down the machine immediately. The machine was monitored, until morning. A decision was made to reduce the load and watch for changes. The result was positive, as expected. Recorded high noise and vibration levels were subdued and reduced significantly to normal and acceptable levels.

Overall case vibration amplitude came down to 3.3 mm/sec pk from a whopping 15.3mm/sec pk., Figure 5a. It was about 5 fold reduction in vibration for just 5% reduction in load. Similar reduction was seen in impact energy levels from waveform data. That was reduced from as high as 48 G-s to just 15 G-s. Turbine bearing vibrations were reduced from 7.4mm/sec pk to 2 mm/sec pk. Shaft vibration levels improved from 16 microns to 10microns. The 4.63x non-synchronous component and its multiples disappeared, Figure 5b completely.

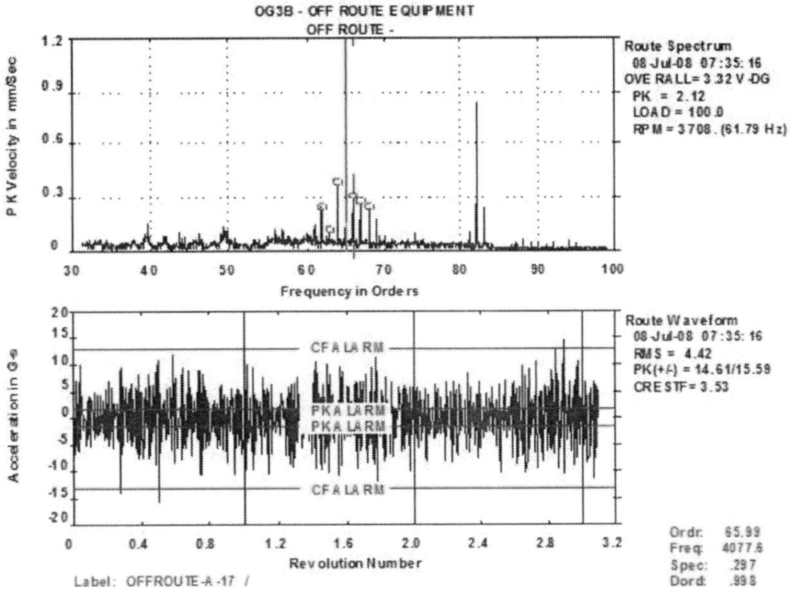

Figure 5a: Turbine Case Vibration after Load Reduction By 4%

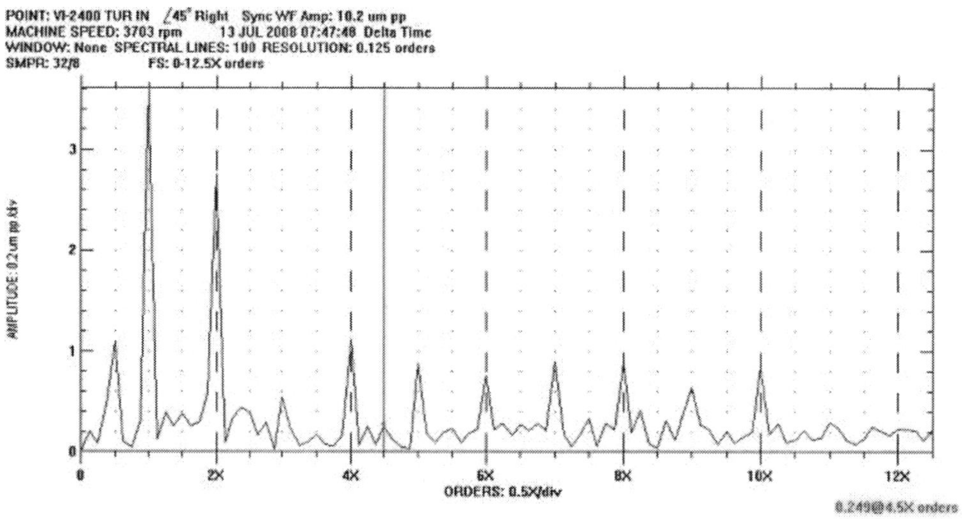

Figure 5b: Turbine Shaft Vibration-Nonsynchronous Component Disappeared

On-line Wash: Though vibration levels had come down to acceptable level, to maintain at this level, operation at lower loads was necessary. So, on-line wash was planned and carried out in a month, to improve performance and elimination of high vibration and cavitation effects at higher loads. During online wash, it was noticed that there was sudden change in vibration amplitude, Figure 6, and returned back to normal within minutes. This was suggesting that considerable amount of fouling & scaling getting detached and washed away by the flowing stream. The same indications of heavy fouling & scaling have been evidenced from pre-wash vibration data showing 66x and 82x running speed components with 1x running speed sidebands, Figure 5a.

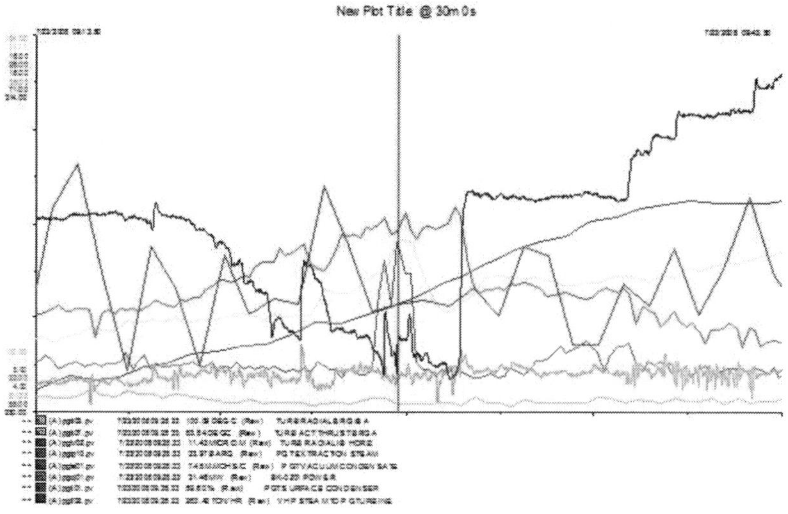

Figure 6: On-line Wash Trend Showing Unusual Vibration Amplitude

Internal Findings: While on-line wash is effective in cleaning up fouling caused by water soluble contaminants such as salts, the deposition caused by presence of non-water soluble contaminants such as silica in the steam doesn't. With all the above analysis, observation and action taken, we had anticipated possible hard deposits and heavy fouling present in the turbine. The same had been included to turnaround scope which was planned within a year, to open up the turbine for inspection and cleaning. As anticipated, very hard and heavy deposits built up were found in all the rotor blades, Figure 7a & 7b, and diaphragm vanes and nozzles, especially on the lower half sets. Final stages of diaphragm and shrouds had erosion and pitting due to cavitation effects in steam turbine. Out-of-service cleaning was carried out using aluminium oxide grit blasting.

Figure 7a & 7b Deposits and Fouling in turbine rotor blades and diaphragm vanes

Cavitation and Its Effects: The cavitation effects on blades resulted from converting the high velocity steam flow to mechanical energy, especially in the final stage blades in the wet steam section. The wet steam was curdled easily during running, in high velocity motion, with many directional changes. The tiny water particles / droplets thus formed get blasted off due to the centrifugal action by the high speed rotating blades and resulting high velocity flow stream. This is called cavitation effects in steam turbines. The same can be seen from severe erosion and pitting in final 3 stages, 10th, 11th & 12th stages, of turbine diaphragms, Figure 8a & 8b, resulting in very high rotor tip to diaphragm shroud radial clearance of as high as 4 mm as against maximum allowable design clearance of 1.5mm.

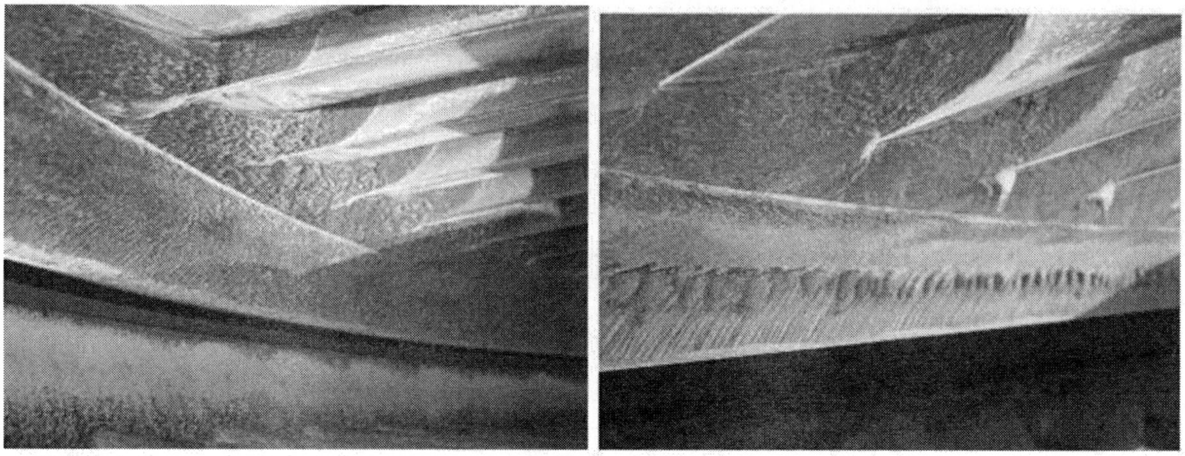

Figure 8a & 8b Erosion and pitting in final stages diaphragms

Turbine Deposition, Fouling and Erosion [1]: The development of modern, high-efficiency steam turbines has led to an increase in deposition, erosion, and corrosion problems. Close tolerances in the turbines, the use of high-strength steels, and impure steam all contribute to these conditions.

Boiler water solids are carried over with steam from deposits in non-return valves, super heaters, and turbine stop and control valves. Carryover can contaminate process streams and affect product quality. Superheated steam turbines are particularly prone to damage by carryover. Sticking of governor and stop valves due to deposits can cause turbine over speed and catastrophic damage. Solid particles in steam can erode turbine parts, while deposition on turbine blades can reduce efficiency and capacity. Losses of 5% in turbine efficiency and 20% in turbine capacity have occurred due to deposition. When large slugs of boiler water carry over with steam, the resulting thermal and mechanical shock can cause severe damage
Adherent deposits form in the steam passage, distorting the original shape of turbine nozzles and blades. These deposits, often rough or uneven at the surface, increase resistance to the flow of steam. Distortion of steam passages alters steam velocities and pressure drops, reducing the capacity and efficiency of the turbine.

Silica can vaporize into the steam at operating pressures as low as 400 psig. This has caused deposition problems in numerous turbines. As deposits accumulate on turbine blades, stage pressures increase. Figure 9 shows the effect of gradual deposit build-up of stage pressure, during a 15-month period. Deposits caused by the use of contaminated water lower steam effectiveness. In a fouled condition, this 30-MW turbine lost over 5% of its generating capacity.

Erosion of turbine blades results in rough, uneven surfaces that alter steam flow paths. This reduces turbine efficiency and can also limit capacity. Erosion at the high-pressure end of a turbine is usually caused by solid particles (usually iron oxide) present in the steam. Erosion of intermediate and low-pressure blades is usually caused by water in the steam. Operation below design inlet steam temperature or at low load can cause condensation in these stages, leading to erosion problems.

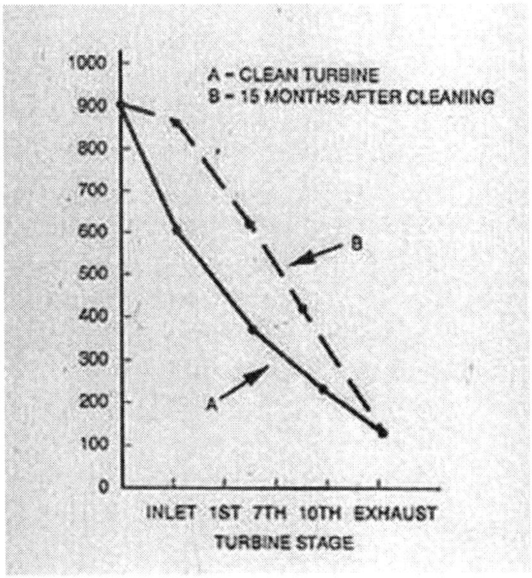

Figure 9: Fouled 30-MW turbine, Increasing Stage Pressures and Decreasing Capacity by Over 5% during a 15-month Period [1].

Conclusion: While hard deposits and heavy fouling had caused unusual vibration behaviour at elevated amplitude levels, the quality of the steam with increased load played a role in developing devastating cavitation effects at final stages resulting in noise and severe impacts. Steam purity, though controlled, too had a role in deposition and fouling of the turbine stages. Ultimately, it was affecting the steam flow path engineering and design. Continuous operation under this condition would have caused severe erosion and pitting damage to costly components such as the diaphragms and rotor with efficiency and capacity loss of 5 to 20%. This is evident when just 4% reduction in load helped bring down both the vibration and noise levels from very high levels of 15mm/sec pk to 3mm/sec pk and from 48 G-s pk to 15G-s pk. Therefore, vibration analysis played an important role in many aspects.

1. Continued plant operations and production and organized planned outage for rectification.
2. Helped minimise damage to components by marginally bringing down load on time, by just 4%. Prolonged operation of turbine otherwise would have led to severe erosion and pitting damage, to costly rotor and diaphragms.
3. Avoided downtime and maintenance work for unnecessary inspections, with optimal performance output.
4. Helped plan and carry out remedial actions on time, develop and prepare work scopes for major turnaround and performance improvement.

Reference:

1. Handbook of Industrial Water Treatment, http://www.gewater.com/handbook/index.jsp, Boiler Water Systems chapter 16 & 18

Biography:

Chartered Engineer having over 25 years of experience in the field of rotating equipment maintenance, troubleshooting, Condition monitoring & Vibration analysis. Currently practicing as a Rotating Equipment engineer at Dubai Petroleum, Dubai. Holds Equivalent Degree in Mechanical engineering from The Institution of Engineers (India). A certified category-III Vibration analyst from Vibration Institute and Certified Maintenance and Reliability professional (CMRP) from SMRP. Member of Vibration Institute, SMRP and the Institution of Engineers, India.

CASE STUDY: TROUBLESHOOTING HIGH VIBRATION ISSUE FOR NEWLY COMISSIONED MOTOR PUMP SYSTEM IN A PETROCHEMICAL COMPLEX

Rishikesh Kumar
Rishikesh.kumar@sadara.com, Vibration Engineer

Abdulhadi Alhazmi
Abdulhadi.Alhazmi@sadara.com, Vibration Technologist

Michael Huston
Michael.Huston@sadara.com, PPM Supervisor
Reliability Department, Sadara Chemicals, Jubail, Saudi Arabia

Abstract:
Vibration Analysis has been a powerful tool for a long time in troubleshooting and fault diagnosis of rotating equipment in industry. Application of this diagnostic methodology brings a lot of value during start-ups and commissioning of equipment when vibration domain is interfaced with design information of rotating pieces to identify inherent flaws. In this paper, a case study has been presented wherein successful troubleshooting through vibration analysis identified and mitigated a chronic design issue during both solo and coupled run of equipment. In the solo run, the motor experienced high vibrations; unacceptable by ISO standards. Through spectrum, natural frequency tests and waterfall plots, the root cause of the problem was located and through epoxy grouting, the vibration problem was corrected. However, when this motor was coupled with the pump, the system now had high vibration issues on the pumps and piping. Through in-depth spectrum analysis being interfaced with design aspects of the pump, a serious design flaw was located and after certain modifications, the entire system has been running smoothly at site with acceptable vibration levels. This sort of troubleshooting and diagnosis helps to locate the root cause of problems, enhancing the long term reliability of equipment and directly impacting productivity and profitability of the organization.

Keywords: Commissioning; Design; Epoxy grouting; Reliability; Start-ups; Vibration

Introduction

The Start-up phase in an organization poses many challenges as unexpected and unforeseen issues / problems may originate due to improper installations or wrong design considerations for the desired operational requirements. Reliability bath curve (Figure 1) also represents commissioning / startup phase as the Infant mortality phase represents a challenging period in which initial failure frequencies match the peak frequencies of the wear out phase. High technical expertise of personnel with diverse diagnostic approaches remains key for the successful start-up of plant machines for readiness to get into the production phase. The situation is very demanding as per project timelines and locating and fixing flaws at an early stage is appreciated to be reliable, effective, efficient and productive. It involves a lot of coordination and collaboration among different stakeholders viz. vendors, EPC's, OEM's and plant personnel to develop a strategic plan to execute test runs and agree on a resolution plan to eliminate or mitigate the source of the problem. Authors of this paper have presented a case study delineating high vibration issues during the solo run of two identical motors at certain speeds. When these motors were coupled to the pump, high vibration levels, along with higher noise levels, were noted on bearing housings and along pipelines, posing a threat for its operation from reliability and safety perspective. Through various tests and detailed investigations, root cause of the issues were identified to resolve the vibration problems.

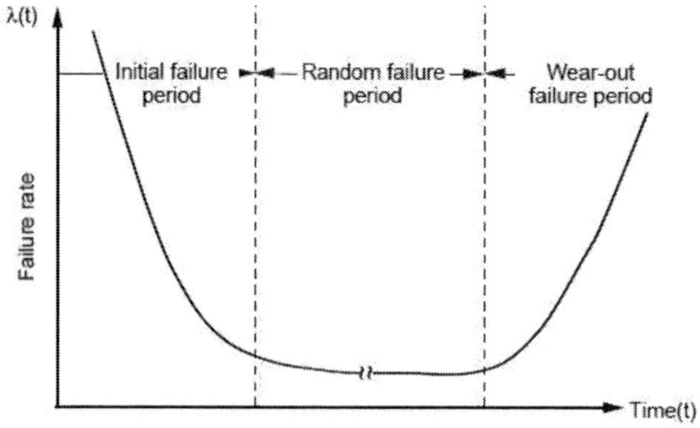

Figure 1: Reliability Bathtub Curve

Case –Study

1. Solo Run Test

Two identical motors A & B, rated at 400 H.P for Recycle Solvent Pumps, exhibited high vibration issues during solo run tests at the site. These motors were run through Variable Frequency Drive to meet operational requirements. Higher vibration levels were noted in speed ranges of 2500-3000 rpm. Spectrum (Figure 2) indicating dominant 1X peak at different speeds, provided symptoms of unbalance of rotor, possible operational issues like misalignment, looseness, soft foot and distortion of foundations. Operational flaws were not located and the unbalance of the rotor seemed to be the potential source of the problem. These motors were sent to the OEM's factory for shop testing to locate the source of the problem. At the shop, these motors were tested on a rigid steel base and had low vibration levels (<1.5 mm/sec Pk) at different speed ranges up to 3000 rpm. When the foundation bolts were loosened; there was no significant change in vibration levels. With tests performed at OEM's shop, it was concluded there were no inherent flaws in the motor.

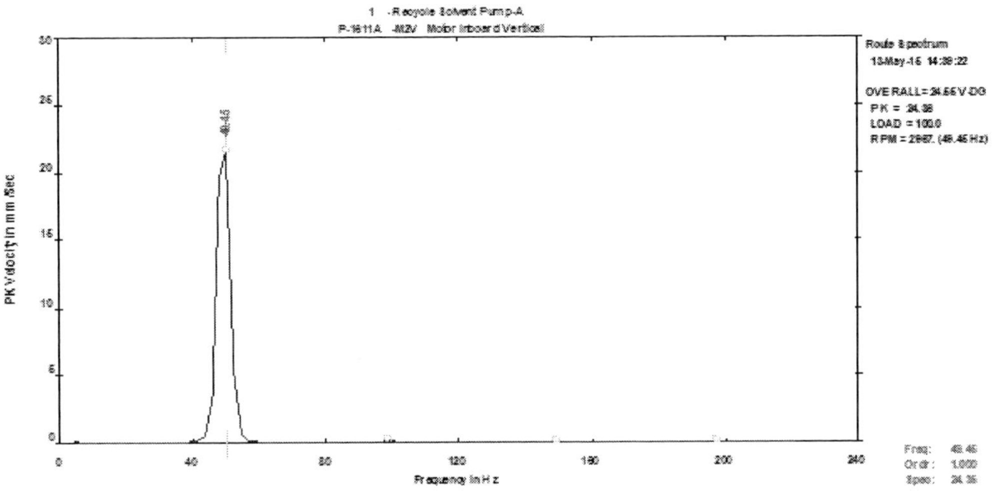

Figure 2: Spectrum for Motor A [DE Vert]

The motors were shipped back to the site and another solo run was performed; once again high vibration levels were noted on both motors; A exhibited higher levels (24.5 mm/sec Pk) than B (6.74 mm/sec Pk) around 2500 to 3000 rpm. All vibration measurements mentioned in this paper are expressed in mm/sec Pk. Following tests were conducted on these motors to locate root cause of problem:

(a) Natural Frequency test on bearing housings of motor.
(b) Natural Frequency test on foundation frame / skid of motor.
(c) Vibration measurement at different speeds / load conditions.
(d) Waterfall Plot.

Impact tests were performed on the Motor DE and NDE bearing housings to locate its natural frequencies. Through analysis of Impact test spectrum (Figure 3(a)), motor A's natural frequency at DE Vertical was located around 50 Hz. For motor B, the natural frequency (Figure 3(b)) at DE Vertical was 48.60 Hz.
A hollow sound was noted when the skid structure / foundation bed at site was struck with a hammer at different locations. This had given an indication of the presence of air pockets and improper structural integrity of the entire mounting structure. Impact tests were performed on reference points as per Figure 4 on the motor foundation structure to locate its natural frequencies. The natural frequencies of foundation structure for both motors A and B at labelled points were noted around 350-400 Hz (Figure 5).

Vibration measurements on the bearing housings for motors A and B were performed at different speeds and load conditions (Table 1.0). Through VFD, the motor was run through different speeds with 2970 rpm as reference for 100% load and vibration data was recorded through a CSI 2140 Emerson analyzer. Higher vibration levels of 24.5 mm/sec Pk was observed on DE Vertical for motor A, whereas for motor B, vibration levels of 6.74 mm/sec Pk were noted around 3000 rpm.

The waterfall plot is representative of vibration amplitudes at different speeds at different time intervals. The waterfall plot is generated from a series of spectrum plots taken over a period of time or changing load conditions. The Waterfall plot is useful in determining changes to various frequency components. These plots (Figure 6(a,b)) were captured through the Emerson Analyzer once the machine was switched off - as it coasted to a stop.

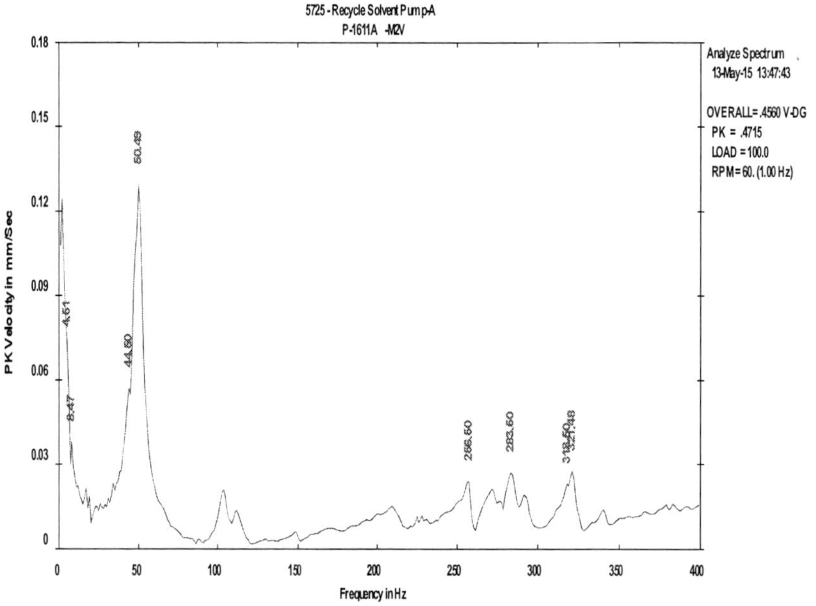

Figure 3(a): Natural Frequency test Motor A [DE Vert]

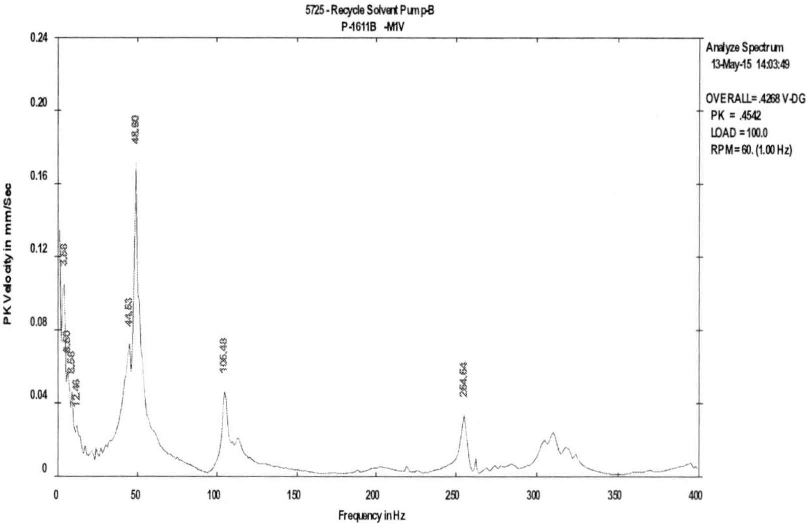

Figure 3(b): Natural Frequency test Motor B [DE Vert]

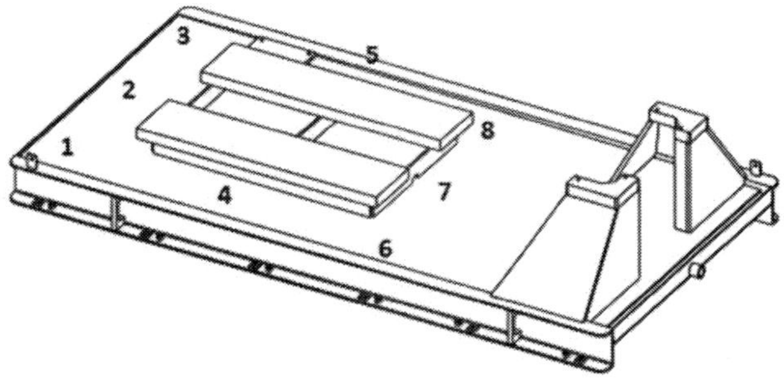

Figure 4: Foundation reference points for Impact test

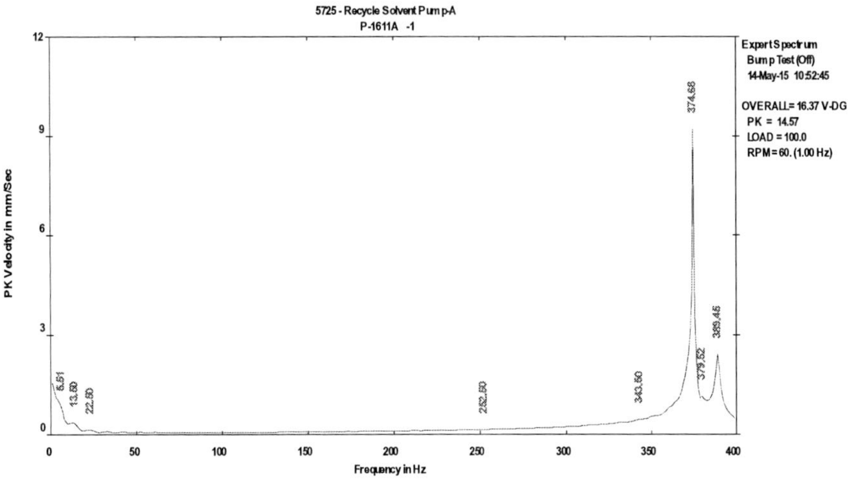

Figure 5 (a): Impact Test [Motor A –Point 1]

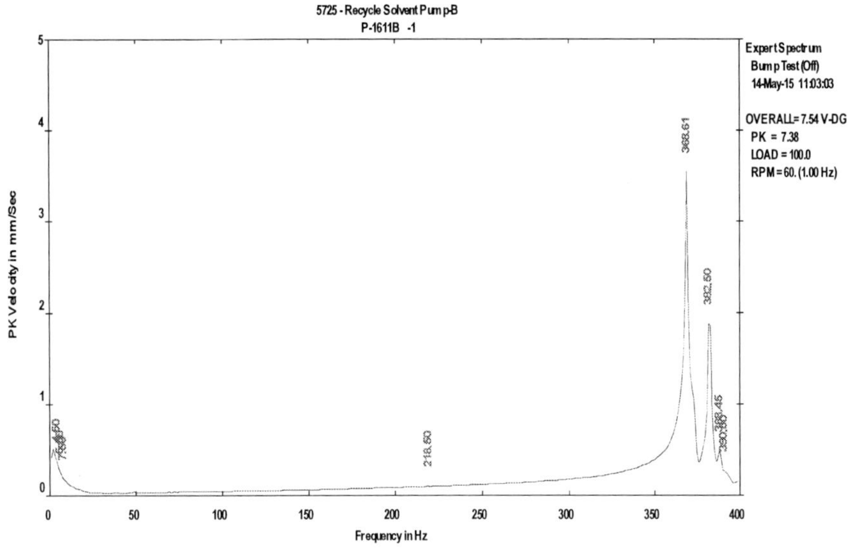

Figure 5 (b): Impact Test [Motor B –Point 1]

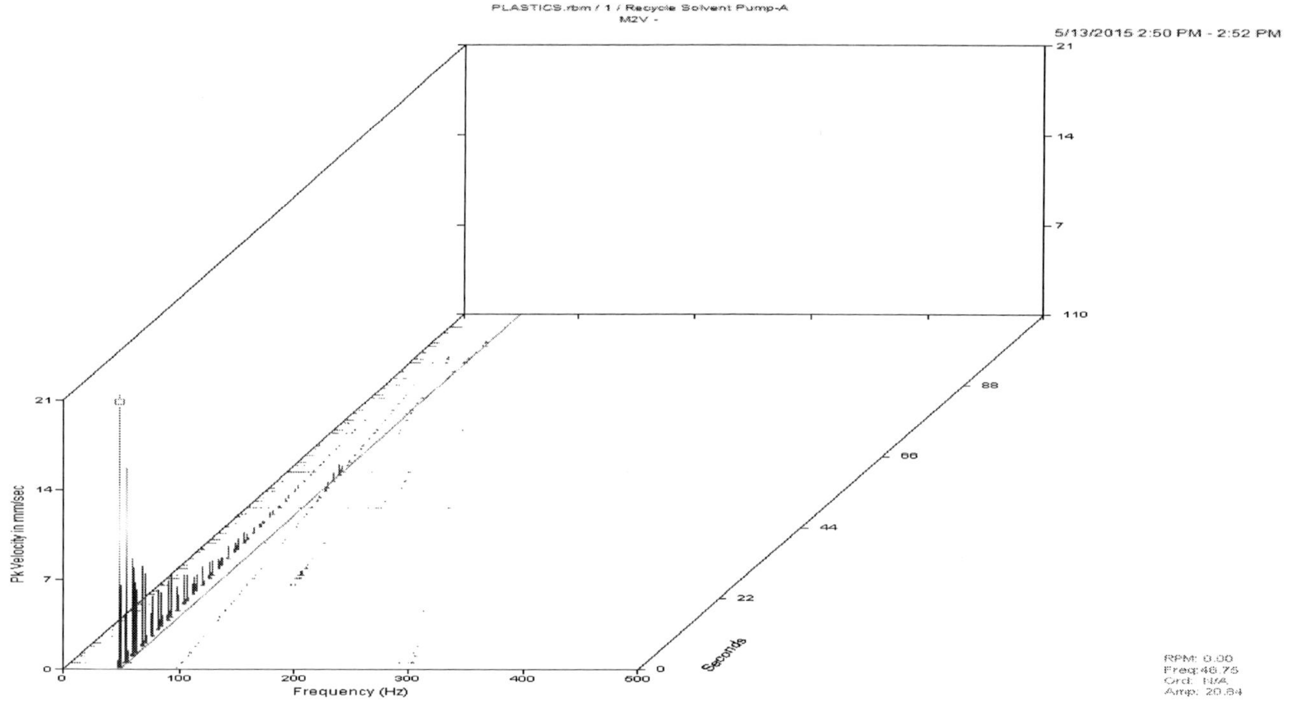

Figure 6 (a) : Waterfall Plot - Motor A[DE Vert]

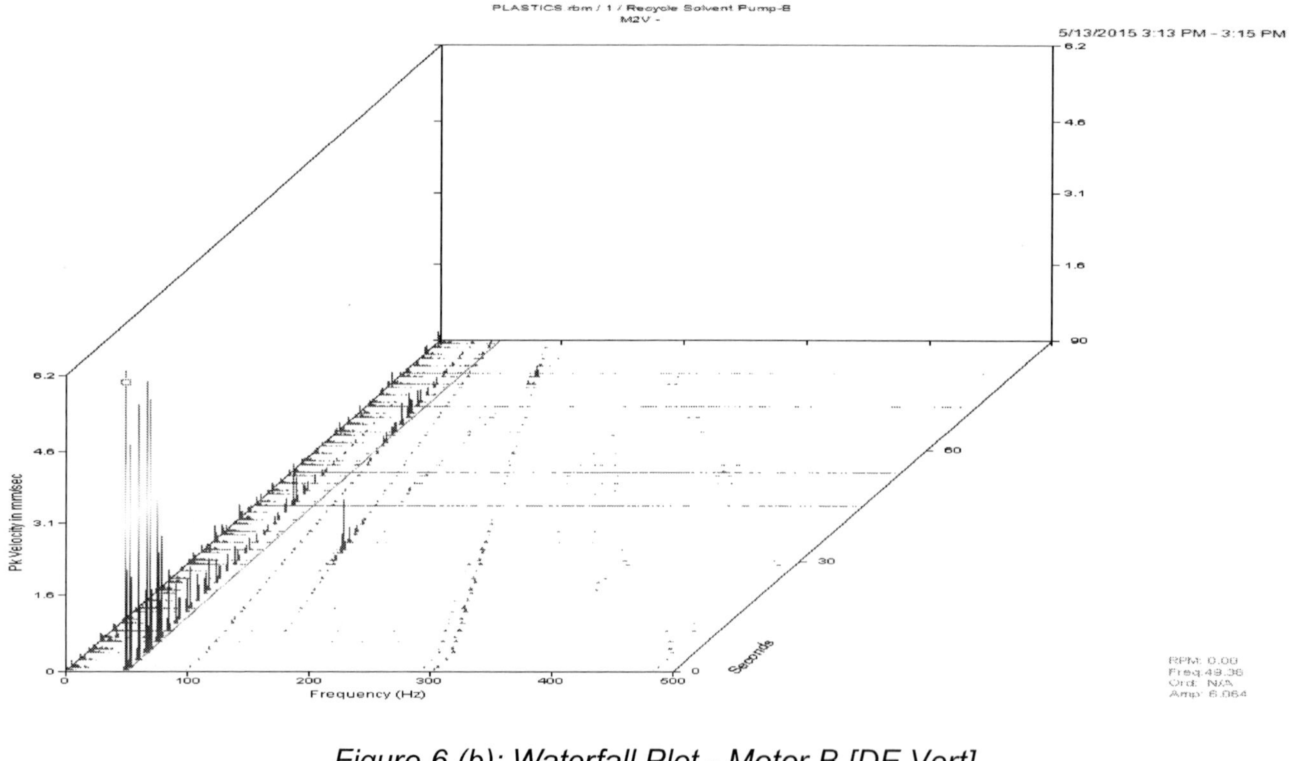

Figure 6 (b): Waterfall Plot - Motor B [DE Vert]

Results and Discussion:

1. Bearing Housing natural frequency for motor A is 50.49 Hz, which is coinciding with the running speed of 3000 rpm of the motor. This motor is operating in resonant condition. The Hollow sound from the mounting frame structure provided an indication of poor structural integrity. The waterfall plot indicated high vibration amplitudes around 3000 rpm and afterwards vibration amplitude levels dropped and followed the speed curve.
2. Bearing Housing natural frequency for motor B is 48.60 Hz, which is in close proximity of the running speed of 3000 rpm. The waterfall plot shows an amplitude peaking for longer duration for this motor as compared to A. This motor is operating near to resonant condition. Operating motor in speed ranges of 2800 rpm to 3000 rpm is detrimental and needs an appropriate correction for reliable and safe operation.
3. Natural frequencies obtained around 360-400 Hz at reference points 1-8 on mounting frame foundations for motors A and B were not in range of the running speed regime.
4. High Vibration levels were observed in the vertical direction, indicating low stiffness levels of the entire system in this direction.
5. High amplitude peaking during natural frequency test on foundation skid frame of motor A was observed as compared to B. This confirms motor B's structure has better damping characteristics than motor A. This is the reason for higher vibration levels on motor A's bearing housings as compared to motor B.

Remedial Action:

It was concluded motor A was operating in resonant condition while motor B was operating near to a resonant condition. It is desired to have 20-25% margin on bearing housing natural frequencies to running speed to avoid resonant condition and for safe and reliable operation. The stakeholders of the project decided to stiffen the baseplate through epoxy grouting. Poor structural integrity (Figure 7) of the vertical cross member of the foundation structure was located and through scale measurement, the volume of epoxy for grouting was calculated as 50 mm thick X 3 m X 1.6 m. A hole was drilled through the foundation frame structure and epoxy was poured through it. After grouting, the hole was closed and bearing housings were tested for natural frequencies. New noted natural frequency was around 300 Hz; grouting action has drastically shifted the natural frequency values (Figure 7) of the bearing housings from 50 Hz to 300 Hz indicating a very high degree of stiffening achieved through this process. After epoxy grouting, motors A and B had exhibited very low vibration levels (Table 2(a,b)) below 2 mm/sec Pk at various running speeds / load conditions. With reduced vibration levels throughout the entire operating speed regime to meet operational requirements, the entire system was safe and reliable for further operation. The successful implementation of epoxy grouting on two foundation frames / skid structures has given confidence to plant personnel and this learning was leveraged on 6 foundation structures across the site to improve structural integrity.

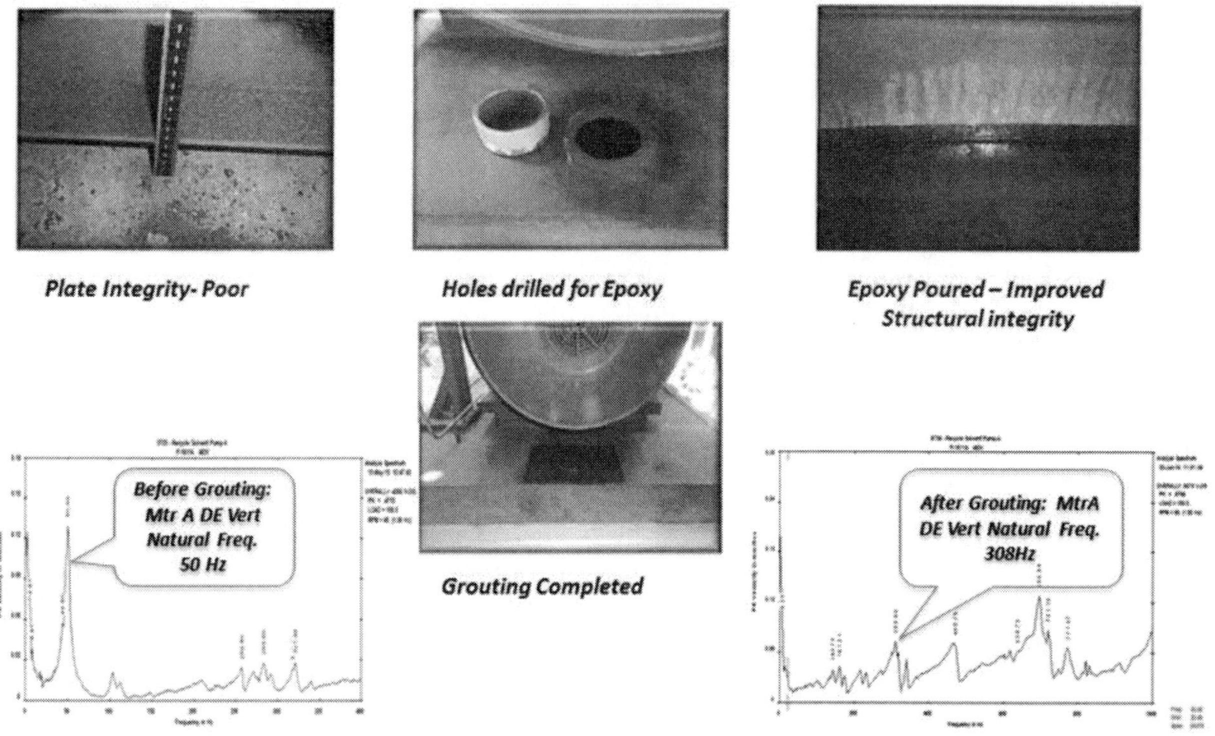

Figure 7: Structural Integrity- Grouting Process

Measurement Points	Measurement Point Description	**Motor A Vibration data at different Load*/ Speeds**					Units
	Load*	25%	50%	75%	85%	100%	
M1H	Motor Outboard Horizontal	0.7	1.9	1.2	*3.1*	***11.5***	mm/Sec
M1P	Motor Outboard Horz Peakvue	0.2	0.6	0.3	0.5	0.3	G-s
M1V	Motor Outboard Vertical	0.7	1.3	1.4	2.7	***17.4***	mm/Sec
M1A	Motor Outboard Axial	0.2	0.2	0.2	0.4	***5.3***	mm/Sec
M2H	Motor Inboard Horizontal	1	2	1.2	*3*	***13.6***	mm/Sec
M2P	Motor Inboard Horz Peakvue	0.4	0.4	0.4	0.4	0.2	G-s
M2V	Motor Inboard Vertical	0.5	1	1.3	*3.1*	***24.5***	mm/Sec
M2A	Motor Inboard Axial	0.6	0.6	0.9	0.9	***5.5***	mm/Sec

Table 1(a): Motor A Vibration Data at different Loads/ Speed (2970 rpm as reference for 100% Load)*

Measurement Points	*Measurement Point Description*	***Motor B Vibration data at different Load */ Speeds***			
		25%	50%	75%	100%
M1H	Motor Outboard Horizontal	0.5	0.8	0.7	1.6
M1P	Motor Outboard Horz Peakvue	0.3	0.6	0.6	0.3
M1V	Motor Outboard Vertical	0.6	0.7	0.6	***3.4***
M1A	Motor Outboard Axial	0.1	0.2	0.3	1.9
M2H	Motor Inboard Horizontal	0.6	0.8	0.8	***2.1***
M2P	Motor Inboard Horz Peakvue	0.4	0.4	0.6	0.3
M2V	Motor Inboard Vertical	0.6	0.8	0.8	***6.7***
M2A	Motor Inboard Axial	0.6	0.6	1	***2.2***

Table 1(b): Motor B Vibration Data at different Load/ Speeds

Motor -A					
Measurement Points	*Measurement point description*	*Before Grouting*		*After Grouting*	
		13/05/2105	25/05/2105	3/6/2015	3/6/2015
		100% (2970rpm)	90% (2510 rpm)	100%	90%
M1H	Motor Outboard Horizontal	***11.4***	***3.0***	0.9	0.9
M1V	Motor Outboard Vertical	***17.4***	***2.7***	0.7	0.7
M1A	Motor Outboard Axial	***5.3***	0.4	0.5	0.4
M2H	Motor Inboard Horizontal	***13.6***	***2.9***	0.9	0.9
M2V	Motor Inboard Vertical	***24.5***	***3.1***	0.6	0.7
M2A	Motor Inboard Axial	***5.5***	0.7	0.6	0.6

Table 2(a): Motor A Vibration Data – Before & After Grouting process

Motor -B					
Measurement Points	*Measurement point description*	*Before Grouting*		*After Grouting*	
		13/05/2105	25/05/2105	3/6/2015	3/6/2015

	*Load**	100% (2970 rpm)	90% (2510 rpm)	100%	90%
M1H	Motor Outboard Horizontal	1.6	1.0	0.9	0.9
M1V	Motor Outboard Vertical	***3.3***	0.9	0.7	0.9
M1A	Motor Outboard Axial	1.9	0.4	0.5	0.5
M2H	Motor Inboard Horizontal	***2.1***	1.4	0.9	1.0
M2V	Motor Inboard Vertical	***6.7***	1.1	0.9	0.9
M2A	Motor Inboard Axial	***2.2***	0.6	0.8	0.8

Table 2(b): Motor B Vibration Data – Before & After Grouting process

2. *Mechanical Test Run*

After the successful solo run test, motors A and B were coupled with recycle solvent pumps. Technical specifications of the centrifugal pump included 4 vanes, an impeller diameter of 483 mm, flow rate of approximately 200.00 m^3/hr, and NPSH over 4 meters. This pump is intended to transfer solvents to meet production requirements. After basic site checks, including proper alignment, the mechanical test run was conducted and vibration levels were recorded. High vibration levels on pump bearing housings (6.5 mm/sec Pk) and 40 mm/sec Pk on suction piping were accompanied with high noise levels all across layout. Vibration spectrum (Figure 8) indicated presence of Vane pass frequencies around 4X and its harmonics / super harmonics of running speed at 15X and 16X all along pipelines, with X being the running speed of the system around 2510 rpm. Running the pump at these higher vibration level posed risk for accelerated wear of wear rings, bearings and cracking of pipelines.

Various tests were conducted at site by running the pump at different speeds, noting flow rates and vibration levels. An effort was made to run the pump near to its best efficiency point BEP and follow the pump curve. In all tests, high vibration levels of 6.5-10.8 mm/sec Pk were noted with spectrum indicating the presence of dominant peaks around harmonics and super harmonics of the vane pass frequencies / running speed on the bearing housings and on different measurement points across the pipelines. Pipe strain, soft foot, alignment, coupling and seal positioning were checked and adjusted, but it had not brought down vibration levels of system.

Detailed investigation was carried out to locate the source of the problem; it was noted the pump was designed using a maximum sized impeller of 483 mm, i.e 19 inches. The clearance between the impeller tip to volute casing was near to an inch. It has been normal practice to select the pump with maximum impeller diameter; the theory being by minimizing the clearance gap between the impeller tip and casing tongue hydraulic efficiency is maximized. For this reason, many hydraulic engineers design [1] the impellers with as little as 5% clearance values. This may provide good mechanical performance at the designed flow, fluid angles leaving the impeller blades at off design flows are mismatched with the tongue angle. This results in the development of significant pressure pulsations. These pressure pulsations [2] impact the impeller, causing a radial shaft deflection at a frequency equal to a blade passing frequency. This motion also occurs at seal faces resulting in a translating motion of the faces, rather than a pure rotary motion for which the faces were designed. Translating motion will cause fluid migration across the faces and if solids (wear / external particles) are present, can pull solids between seal faces. This may lead to entire seizure of pump.

Providing greater opportunity for the fluid to adjust to the tongue angle after it leaves the impeller vanes can mitigate this effect. This can be done either by reducing impeller diameter or pump speed. The smaller the impeller diameter, the more reliable the pump is, but design consideration needs to optimize reliability, efficiency, and cost for given operational parameters. Normally OEM's use maximum impeller diameter to enhance hydraulic, mechanical efficiency and lower NPSH to get the bids, posing threats of high vibration and noise issues at operational stages.

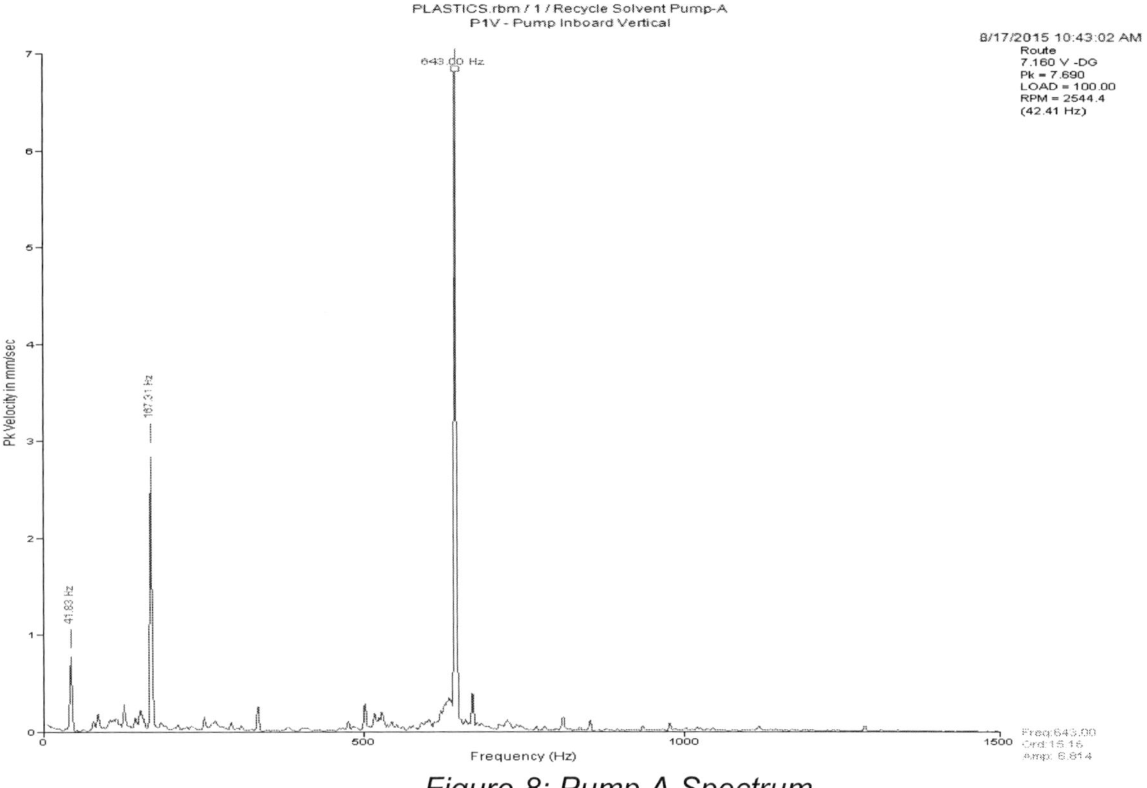

Figure 8: Pump A Spectrum

Results and Discussion:

1. Spectrum analysis (Figure 8) indicated dominant peaks around Vane pass frequencies and its harmonics (4 X Vane pass), super harmonics (17 X running speed) all across pipelines. This indicated the system is operating under strong pressure pulsations.
2. Clearance between the impeller tip to volute casing was around an inch, which is around 5.2% of the impeller diameter (19 inches). Increasing cut water clearance can bring down the vibration levels around vane pass frequencies and its harmonics.
3. The pressure pulsations in narrow clearance zone is the source of the problem and as it transverses across pipelines, it produces lot of noise. Noise is generated due to pressure pulsations. It impinges on the wall of the pipelines resulting to vibratory forces increasing the probability of fatigue failure of pipelines.

Remedial Action:

Impeller diameter was trimmed from 19 inches to 17.2 inches to increase the cut water clearance. Trimming action was accompanied by adjusting pump parameters as per pump affinity laws to meet the operational requirements. Photographs of trimmed impellers are shown in Figure 9. Pump affinity law states that for a given pump with a constant speed, the capacity (Q) will be directly proportional to the impeller diameter (D), Head (H) will be directly proportional to the square of the impeller diameter and the required power (P) will be directly proportional to the cube of the impeller diameter. These proportionalities give rise to following relationships:

$$\frac{Q1}{Q2}=\frac{D1}{D2} \quad \frac{H1}{H2}=\left(\frac{D1}{D2}\right)^2 \quad \frac{P1}{P2}=\left(\frac{D1}{D2}\right)^3$$

Affinity laws can be used for determining pump performance parameters as per changes in impeller diameter:

$$\frac{Q1}{Q2}=\frac{N1}{N2}\left(\frac{D1^3}{D2^3}\right) \quad \frac{H1}{H2}=\left(\frac{N1}{N2}\right)^2\left(\frac{D1}{D2}\right)^2 \quad \frac{P1}{P2}=\left(\frac{N1}{N2}\right)^3\left(\frac{D1}{D2}\right)^5$$

Where N represents speed of the pump while subscripts 1 and 2 denote values before and after changed parameters. Trimming from 19 inches to 17.2 inches to maintain a flow rate of 200.0 m^3/hr motor speed was increased to 2800 rpm from 2500 rpm; Head was increased to 256.02 m from 247.08 m and NPSH increased to 5.35 m from 4.35 m. Trimming action has slightly sacrificed pump efficiency from 70.5% to 69%. Pump performance curves for original and trimmed impellers depicting pump parameters is shown in Figure 10. After trimming impellers, the balancing job was performed in the shop as per ISO balancing grade G1.0. After installation of trimmed impellers, a test run was conducted and as per Table 3(a,b) very less vibration levels were noted on bearing housings and measured levels on piping were below 12 mm/sec Pk.

Vibration levels reduced significantly around the vane pass frequencies (4X) and its harmonics (16X), as shown in the comparative plot (Figure11) before and after trimming action of impellers. Similar pumps with impellers having poor cutwater clearances were located across site and trimming action was performed. It has been implemented on 7 impellers, including one spare impeller.

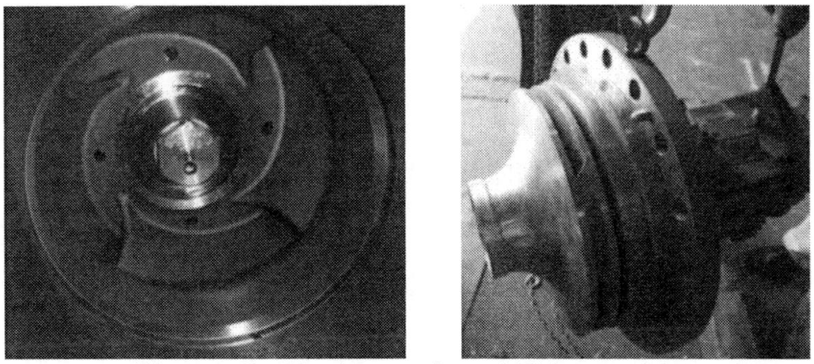

Figure 9: Photographs of Trimmed Impellers

Conclusion:

Multidisciplinary diagnosis through integrating vibration spectrum data with design, operational and process parameters of rotating equipment remains key for successful troubleshooting / diagnosis of chronic issues. In complex cases, the resolution of vibration problems may need alteration in the design parameters to meet operational requirements. Reengineered design to meet reliability / production targets remains a challenge for an organization as it involves a confluence of different technical domain expertise and collaboration of stakeholders from different departments to find an optimum solution.

References

1. R. Barry Erickson, Eugene P. Sabini, "Hydraulic selection to minimize the unscheduled maintenance portion of life cycle cost."
2. Bloch, H.P and Geitner, F.K.,"An Introduction to Machinery Reliability Assessment", 2nd ed., Gulf Publishing Co. Houston, TX (1994).

3(a) Recycle Solvent Pump-A					
Measurement Points	*Measurement point description*	*19Inches Impeller*		*17.2 Inches Impeller*	
		6/9/2015	*15/09/15*	*25/10/15*	*26/10/15*
	Speed /Flow Rate	*2520 rpm*	*2200 rpm*	*2800 rpm 200 m^3/hr*	*2800rpm 200m^3/hr*
M1H	Motor Outboard Horizontal	1.3	0.8	1.3	1.5
M1V	Motor Outboard Vertical	1.1	0.8	1.0	1.2
M1A	Motor Outboard Axial	0.5	0.5	0.4	0.4
M2H	Motor Inboard Horizontal	1.6	1.0	1.6	1.8
M2V	Motor Inboard Vertical	1.0	0.7	1.0	0.8

M2A	Motor Inboard Axial	0.6	0.4	0.3	0.3
P1H	Pump Inboard Horizontal	2.2	1.9	0.1	1.4
P1V	Pump Inboard Vertical	***6.5***	***6.5***	1.8	1.8
P1A	Pump Inboard Axial	2.4	2.0	1.8	1.4
P2H	Pump Outboard Horizontal	2.9	1.3	0.9	1.1
P2V	Pump Outboard Vertical	1.5	3.9	1.0	1.3
P2A	Pump Outboard Axial	3.5	1.8	1.6	1.6

3(b) Recycle Solvent Pump-B					
Measurement Points	*Measurement point description*	*19 Inches Impeller*		*17.2 Inches Impeller*	
		21/09/15	*21/09/15*	*11/11/2015*	*12/11/2015*
	Speed /Flow Rate	2400 rpm 215 m^3/hr	2500 rpm 205 m^3/ hr	2800 rpm 200 m^3/hr	2800 rpm 200 m^3/hr
M1H	Motor Outboard Horizontal	0.5	0.5	0.5	0.5
M1V	Motor Outboard Vertical	1.1	1.2	0.7	1.0
M1A	Motor Outboard Axial	0.4	0.4	0.3	0.3
M2H	Motor Inboard Horizontal	0.7	0.7	0.6	0.7
M2V	Motor Inboard Vertical	0.9	1.1	0.6	0.8
M2A	Motor Inboard Axial	0.5	0.4	0.4	0.3
P1H	Pump Inboard Horizontal	4.2	4.3	1.8	1.6
P1V	Pump Inboard Vertical	4.0	***7.8***	3.9	3.3
P1A	Pump Inboard Axial	***10.8***	***5.5***	2.8	2.3
P2H	Pump Outboard Horizontal	3.8	***4.8***	1.6	1.1
P2V	Pump Outboard Vertical	1.7	2.0	1.5	1.2
P2A	Pump Outboard Axial	4.2	4.1	1.7	1.0

Table 3 (a,b): Pump A/B Vibration Data – Before & After Trimming Impellers

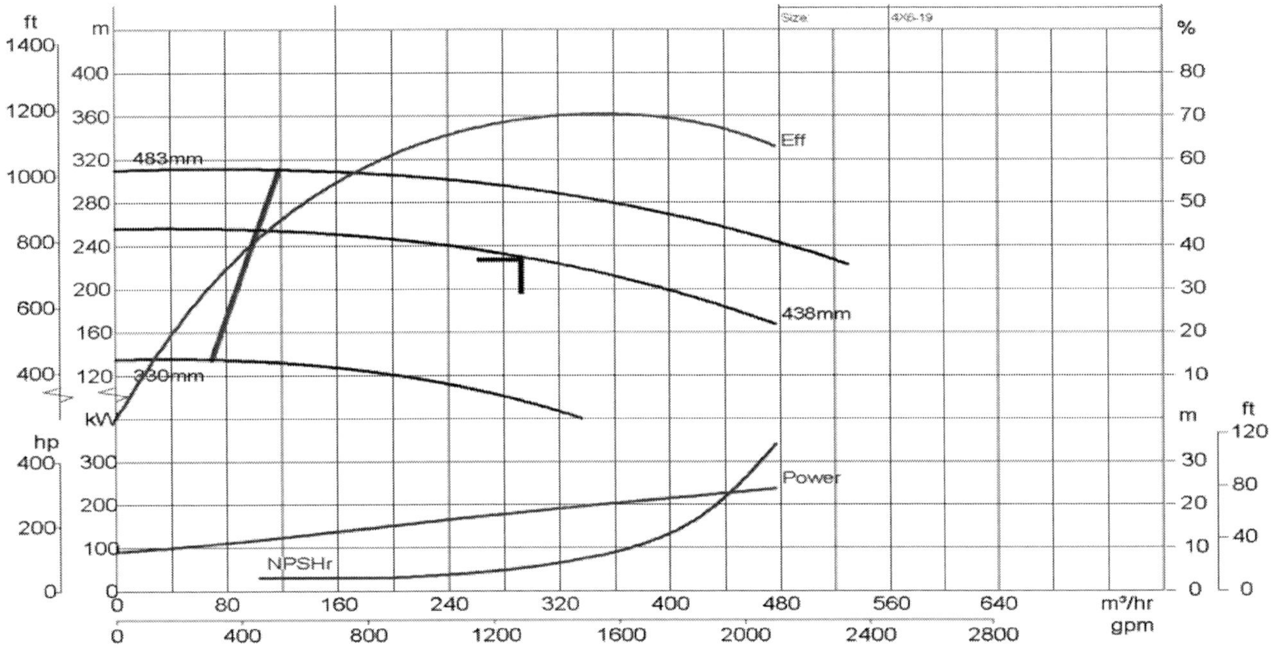

Figure 10: Pump Performance Curves

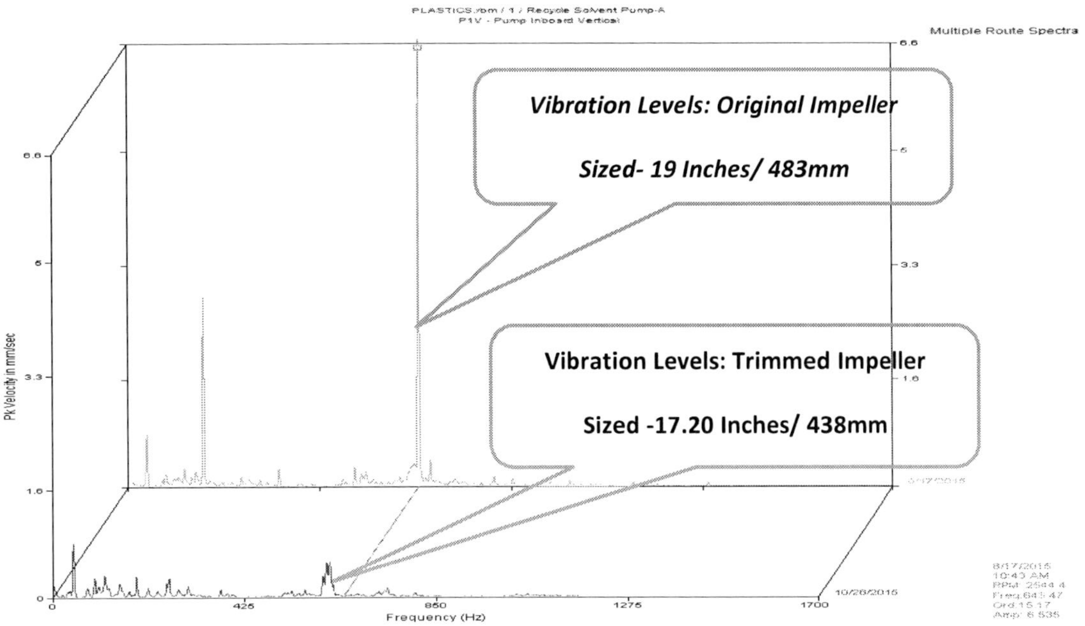

Figure 11: Comparative Vibration levels

Note: All Vibration measurements represented in mm/sec Pk

BIOGRAPHY:

Rishihesh Kumar works for Sadara Chemicals. He received his Bachelor's of Mechanical Engineering, and a Masters of Technology at ISM, India. His expertise is in noise, vibration, tribology, rotordynamics, and failure analysis. He has been working in the industrial field for 13 years. His areas of interest are tribodynamics, human vibration, and multidisciplinary diagnosis. He is a member of ASME, TIS and the Vibration Institute.

PIPING VIBRATION PROBLEM SOLUTIONS

Gyorgy Szasz, Ph.D.
Stress Engineering Services, Inc
3314 Richland Ave
Metairie, LA 70002
gszasz@stress.com

Abstract

Piping vibration related failures at industrial facilities have the ability to present safety, environmental and economic risks. This paper focuses on mitigation approaches for various types of piping vibration problems after the problem has been identified and classified.

Key Words: piping; vibration; mitigation; lateral; shell; mode; broadband; narrowband;

Introduction

Potential piping vibration issues are supposed to be taken into account at the plant or unit design phase, but surprises can still occur during startups and various other operating modes. Once a plant is successfully up and running, the miles and miles of piping that are installed are generally not monitored for vibration. It is indeed rare for a piping vibration issue to surface without some operational parameter or physical configuration change taking place. Flow rate, for example, is one of the most common parameters to influence vibration.

If a piping vibration issue does emerge, it is typically found in one of two ways:

(i) the vibration is visible and is spotted by operators,
(ii) a failure occurs which draws attention to it.

Once the problem has been spotted, the next phase is discovery, when relevant data is gathered which can include vibration, strain, dynamic pressure as well as process parameters such as flow rate, fluid density, temperature and pressure. Properly obtained, processed and analyzed vibration data is essential to success in solving piping vibration problems. Methods for evaluating the data are described in [1-5], therefore this paper focuses on mitigation strategies. The mitigation strategy depends heavily on the type of piping vibration.

Some examples of piping vibration types include:

1. Very low frequency lateral vibration excited by flow turbulence
2. Cantilevered small bore branch vibration excited by main line vibration
3. Low frequency vibration due to mechanical excitation
4. Low frequency vibration excited by pulsation
5. Broadband vibration due to control valve cavitation
6. High frequency tone due to liquid flow through a partially open control valve
7. High frequency vibration excited by compressor vane/lobe pass frequency pulsation
8. High frequency broadband acoustic excitation induced vibration

This paper is intended to be a high level review of the types of remedies that can be applied to these types of piping vibration problems. A brief discussion of each case is presented on the following pages that are based on first hand field experience with these types of issues.

1. <u>Very low frequency lateral vibration excited by flow turbulence.</u>

This type of piping vibration frequently occurs with sparsely supported large bore lines carrying hot liquid or gas (feed water, crude oil, steam, etc...). The vibration is usually visible as the piping vibrates at one or more of its lower natural frequencies due to broadband excitation from turbulent flow inside the piping. While some turbulent noise is generated in straight runs of pipe, the majority of turbulent noise typically comes from other elements (elbows, tees, reducers, valves, orifice plates, etc…), which can create a significant amount of turbulence depending on the complexity of their geometry and the amount of restriction they present. Reducing this type of vibration can be accomplished via reducing turbulence in the piping by smoothing the flow (Figure 1) reducing the flow rate, raising the piping natural frequencies by adding rigid supports, or adding damping to the piping via viscous dampers (Figure 2). If rigid supports are added, the pipe flexibility must be re-evaluated to ensure that the piping is not overstressed due to restricting thermal growth (for pipe conveying hot or cold fluids).

A similar phenomenon can also occur for long unsupported sections of small bore piping and branch connections excited by the main line vibrations. For small bore piping the most convenient solution is to reduce the length of the unsupported spans, making sure that enough flexibility is maintained to accommodate thermal movements if applicable.

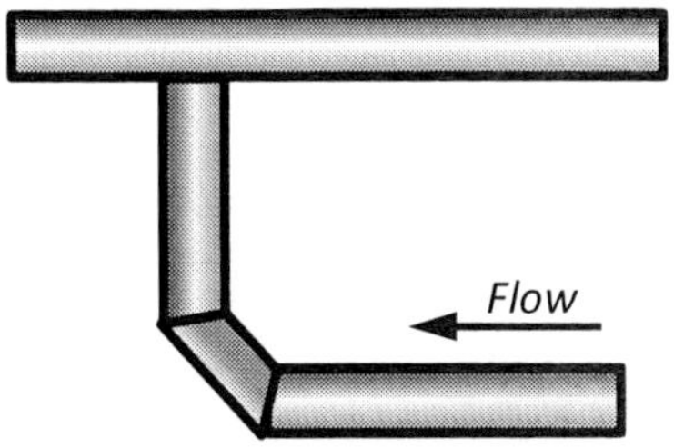

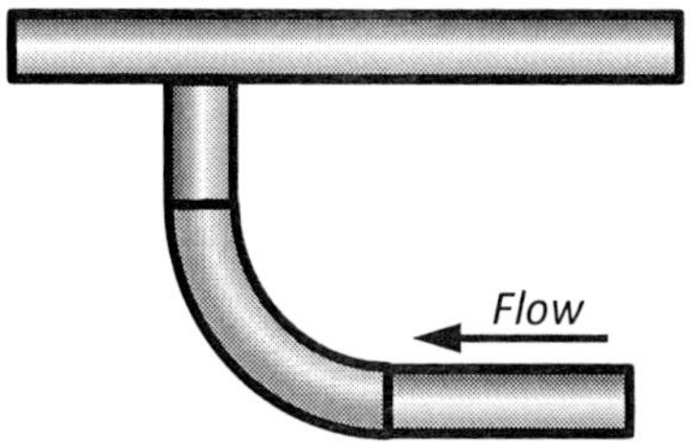

Figure 1: Flow smoothing via swept bend elbow

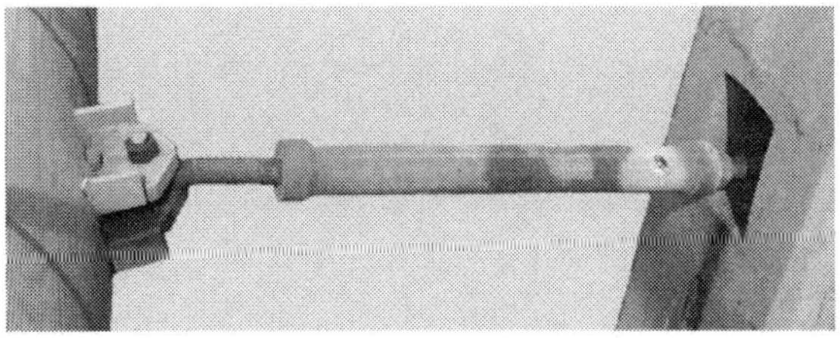
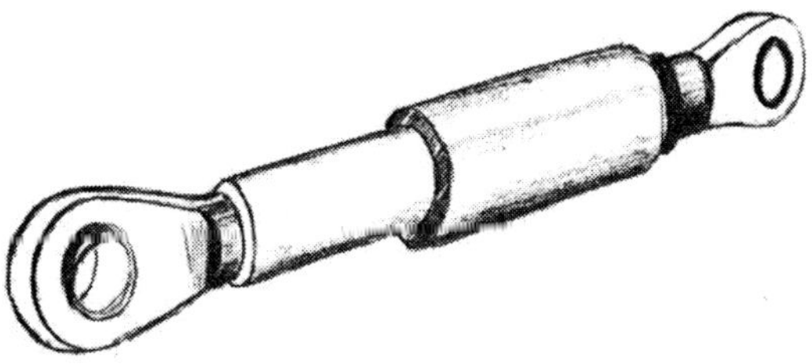

Figure 2: Rigid piping support (left) and viscous damper (right)

2. <u>Cantilevered small bore branch vibration excited by main line vibration</u>

Cantilever mounted small bore pipes with a concentrated mass on the end (such as a valve or instrument) vibrate at high amplitudes when excited near their natural frequencies. Raising the natural frequency of the cantilevered component is often a successful remedy to the problem regardless of whether the excitation is broadband or narrow band. Raising the natural frequency can usually be accomplished by adding bracing or reducing the cantilevered mass (Figure 3). Cantilevered mass can be reduced by, for example, switching to a different type valve. Any bracing applied has to provide a stiffening effect in two directions (parent pipe axial, and parent pipe perpendicular) which can be accomplished with a large enough section modulus for the brace or multi angle bracing. Note that the goal is to reduce relative motion, so small bore piping should be braced to the parent pipe rather than external structures. If the length of the cantilever makes bracing difficult, the side branch design may have to be altered (Figure 4).

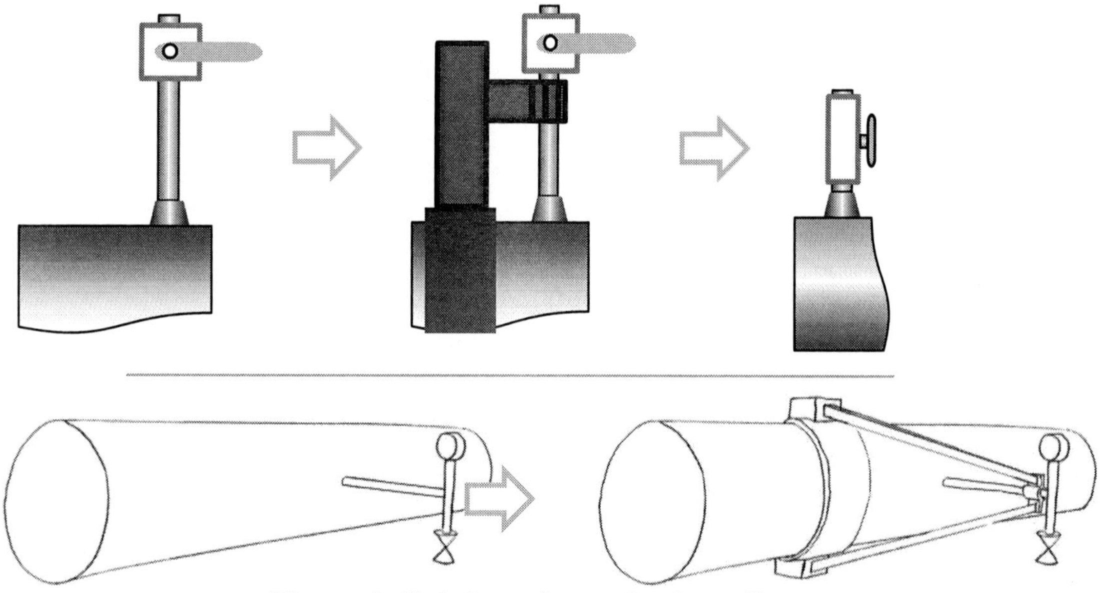

Figure 3: Solutions for a short cantilever

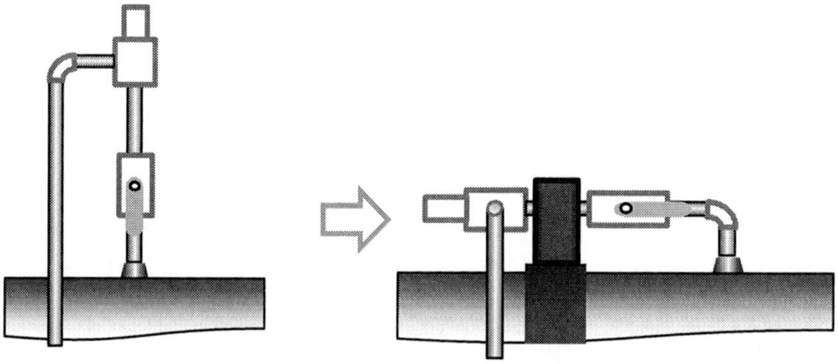

Figure 4: Solution for a long cantilever

3. Low frequency vibration due to mechanical excitation

Piping vibration driven by mechanical sources such as pumps or compressors will occur at the running speed and/or its harmonics. This type of forcing is typically not an issue unless the piping has a resonance near a forcing frequency. In those cases tuning the piping out of resonance should always be the primary goal, as it is not good practice to have piping natural frequencies near running speed and/or its harmonics. If tuning is not feasible or does not yield sufficient results, the excitation forces should be reduced by achieving better balance, shaft alignment, etc., depending on the fault that is the source of the excitation.

In less typical cases, it is possible to encounter excessive forced vibration. An example for such a case can be the suction bottle of a reciprocating compressor connected to two cylinders with a long straight section of the suction line attached directly to it. If the pistons in the cylinders are not driven in phase, the bottle will be exposed to a cyclic torque which can cause it to slightly rotate. As the bottle rotates, any unsupported long straight sections can reach fairly high vibration amplitudes even without resonance as shown on the example in Figure 5. The vibration levels shown in the figure are in terms of inches per second rms and a phase angle. These values were measured in the horizontal direction, perpendicular to the crank shaft and parallel to the cylinder throw direction.

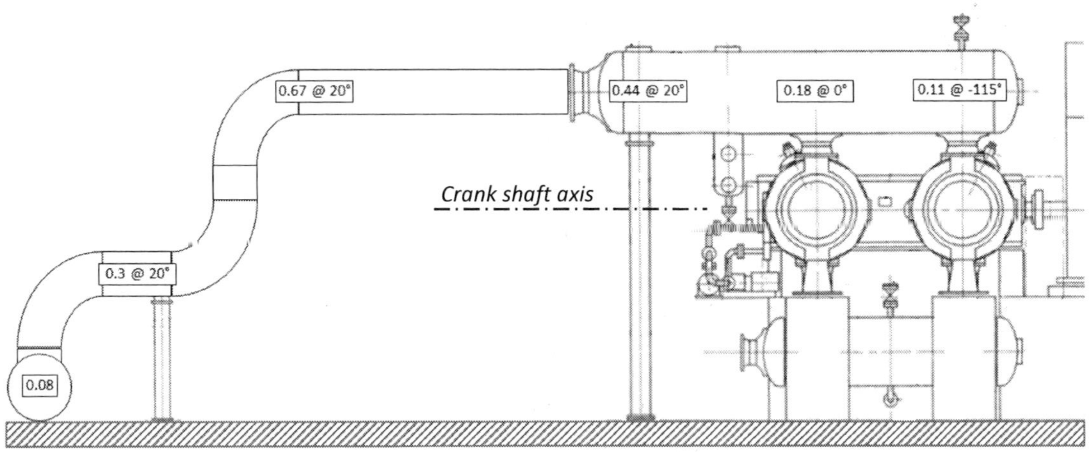

Figure 5: Forced vibration example

4. Low frequency vibration excited by pulsation

Low frequency pressure pulsations in main line piping have the potential to cause elevated vibrations. The source of the pulsations can be flow induced, such as vortex shedding at a branch connection or reciprocating machinery. If the frequency of those pulsations match the natural frequency of an acoustic volume such as a dead leg, they are likely to be amplified. The amplified pulsations often couple with piping natural frequencies which makes the vibration amplitude elevated. Eliminating the source of the pulsations is generally not feasible or not practical; therefore the best way of solving this type of vibration problem is to prevent acoustic amplification. In the case of dead leg resonance, moving the acoustic boundary (for example a closed valve) closer to the main line can accomplish this goal (Figure 6).

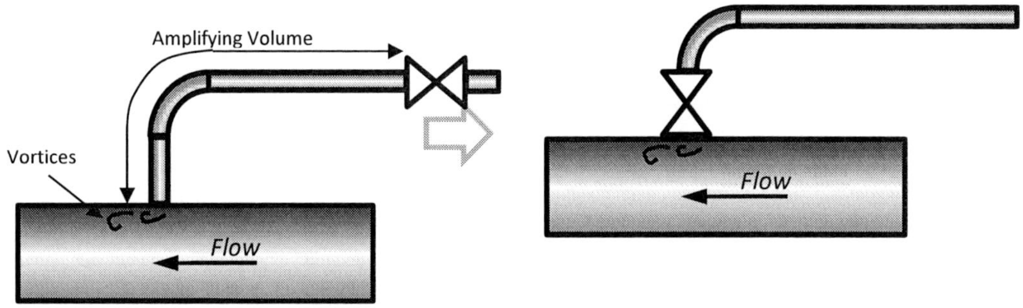

Figure 6: Eliminating dead leg resonance

For the case of reciprocating machines that generate excessive pulsations typically an acoustic study needs to be performed. Whether or not the pulsations are excessive for a given reciprocating machine can be determined using guidance from API 618 [6] or API 674 [7]. If limiting the amplification of the pulsations is not feasible or successful, bracing of the piping may be attempted, however, there are typically a large number of excitation frequency harmonics and it can be challenging to achieve adequate separation from all natural frequencies (Figure 7).

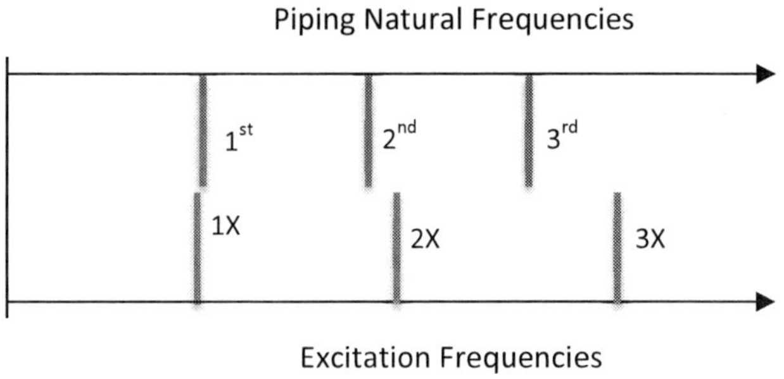

Figure 7: Reciprocating machine frequency alignment

5. Broadband vibration due to control valve cavitation

It is common to throttle flow using control valves. Based on the design of the valve, the physical properties of the liquid, and the pressure differential across the valve, the pressure in the valve may drop below the vapor pressure of the fluid, resulting in cavitation. The collapse of the cavitation bubbles as the pressure increases downstream of the valve can generate noise. This type of excitation is capable of inducing significant broadband low to high frequency lateral and shell mode vibrations in main lines, and even higher amplitude narrowband lateral vibrations of small bore branch connections due to resonance. Bracing the branch connections gets progressively harder with increasing excitation frequencies. A better solution is to distribute the pressure drop, which results in a decreased pressure drop on the individual components (Figure 8) and can avoid cavitation. Another option is to consider a multi stage valve which accomplishes the same task within one component. Figure 9shows how a basic single stage valve can have a large local pressure drop that can reach the vapor pressure of the transported liquid, resulting in cavitation, while a multi stage valve can prevent this from occuring. In the figure, P_1 is the inlet pressure, P_2 is the outlet pressure, P_V is the vapor pressure while P_{VC} is the pressure at the point of highest velocity in the valve which is referred to as the *vena contracta*.

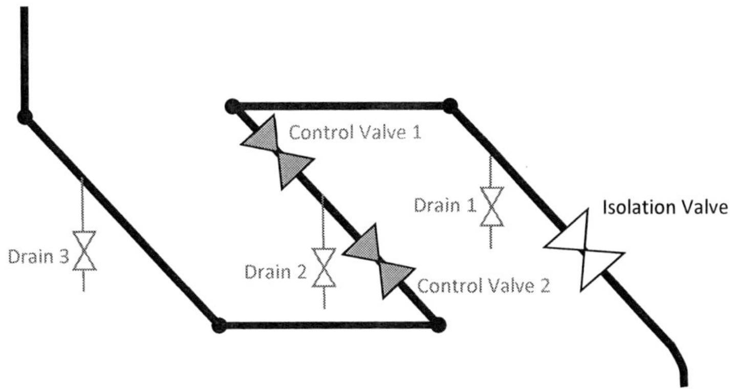

Figure 8: Expanded loop to accommodate additional control valve

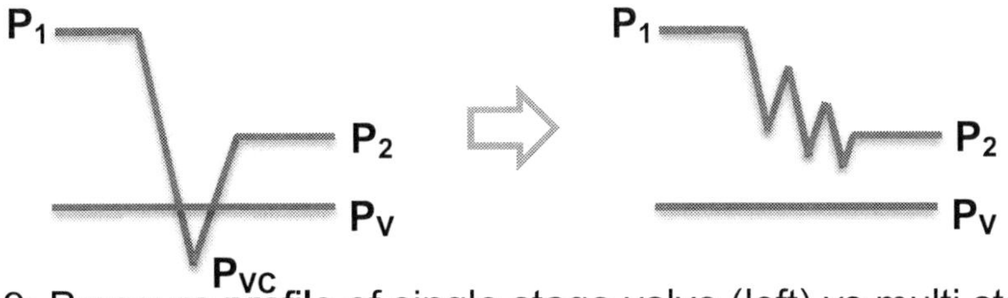

Figure 9: Pressure profile of single stage valve (left) vs multi stage valve (right)

6. High frequency tone due to liquid flow through a partially open control valve

In some cases partially open control valves can emit a very high frequency (1000 - 5000 Hz) tone that depends on the flow rate through the valve more so than the pressure drop. This high frequency excitation is usually audible and is capable of causing shell mode vibration of the main line as well as lateral vibration of branch piping. Stiffening the shell wall via strategically placed stiffening collars can help reduce the shell mode vibration. Bracing of the branch lines can be an option, though the attachment point of the brace has to be carefully selected because the high frequency excitation can excite higher modes and the brace could end up being connected to a node. The brace may also shift the side branch into another mode instead of eliminating the resonance. In addition the brace itself may go into resonance or come loose unless it's welded, since vibration at these frequencies is capable of loosening bolted connections.

This problem is typically caused by vortex shedding near the valve seat creating pressure pulsations which are then amplified by the piping acoustics and ultimately couple with the mechanical natural frequencies of the piping. If the acoustic chamber responsible for the amplification is not easy to locate or modify, the best solution to this type of problem may be to replace the valve with one that has a low noise trim, which will impede vortex shedding at the valve seat.

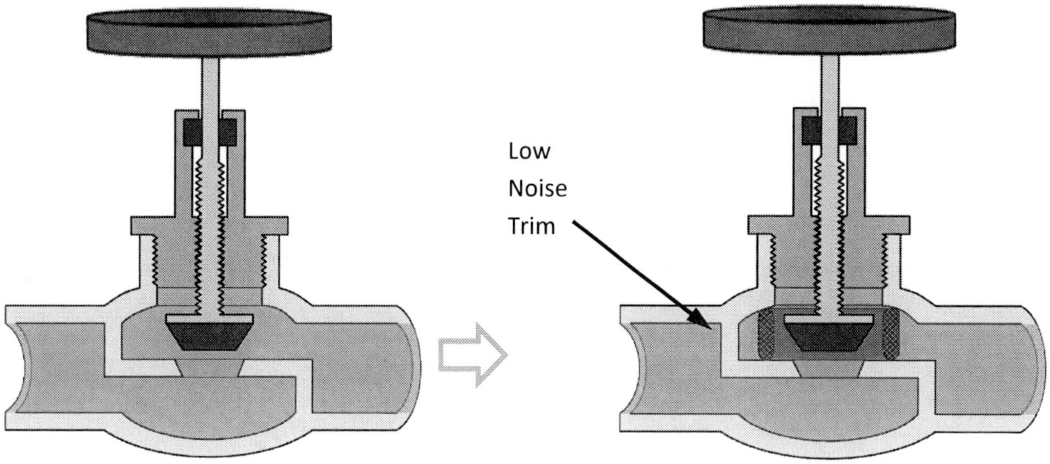

Figure 10: Basic globe valve (left) vs low noise trim valve (right)

7. High frequency vibration excited by compressor vane/lobe pass frequency pulsation

This type of mechanism produces multiple excitation frequencies since the integer multiples of vane or lobe pass frequencies are typically present in the vibration spectrum. While the fundamental frequency is typically less than 1000 Hz, its multiples can easily exceed that value. If any of those frequencies coincide with acoustic and mechanical natural frequencies, the effects, of this mechanism are very similar to those of the tonal excitation from a valve. If shell mode vibration develops, it will typically be dominated by a distinct mode which can be identified by feel around the circumference of the piping.

In severe cases, the vibration can be difficult to measure as magnetic and stinger probe mounts will not work. An adhesive mount will likely work but can be difficult or not possible to apply while the machine is running. Measuring the vibration online may therefore require an optical instrument such as a laser vibrometer.

Unlike in the case of a valve, the excitation cannot be completely removed; therefore the best solution in this case is to detune the acoustic or shell mode mechanical natural frequencies. Detuning can be accomplished by for example installing a thicker wall or larger diameter pipe. Tuning the acoustic and mechanical responses of the piping should be guided by finite element modeling as it is difficult to predict resonant frequencies of complex geometries using hand calculations. Implementing such tuning efforts will also involve a shutdown of the associated machinery. Typically the only practical solution that can be applied online is a stiffening clamp.

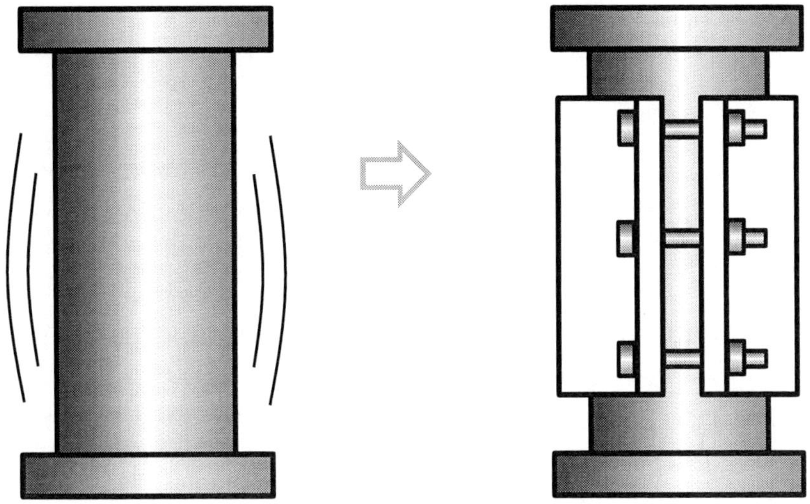

Figure 11: Piping spool piece (left) with stiffening clamp applied (right)

8. <u>High frequency broadband acoustic excitation induced vibration.</u>

This type of excitation can occur with high speed flow going through one or more orifice plates or valves, and possibly forming a jet inside the piping components. The piping will likely exhibit shell mode vibration in more than one distinct mode which may make it difficult to ascertain which modes are active without a series of measurements around the circumference and down the length of the piping. Small bore attachments may go into resonance at their natural frequencies resulting in cracks, and can even break off.

The intensity of this type of excitation is characterized by the sound power level (SPL) it generates; therefore it is best mitigated by reducing the SPL level. One way of achieving that is to reduce the amount of pressure drop at any given location, which could be accomplished by distributing the pressure drop across multiple flow elements/devices (for example orifice plates) instead of one (Figure 12). If the source is a valve, it may be possible to fit it with a low noise trim, which should reduce the SPL in line with data supplied by the valve manufacturer. Another option is to reduce the flow rate, by for example adding a parallel line.

If reducing the excitation level is not feasible, increasing the pipe wall thickness can be considered, and small bore connection should be moved as far away from the noise source as possible to take advantage of SPL attenuation with distance.

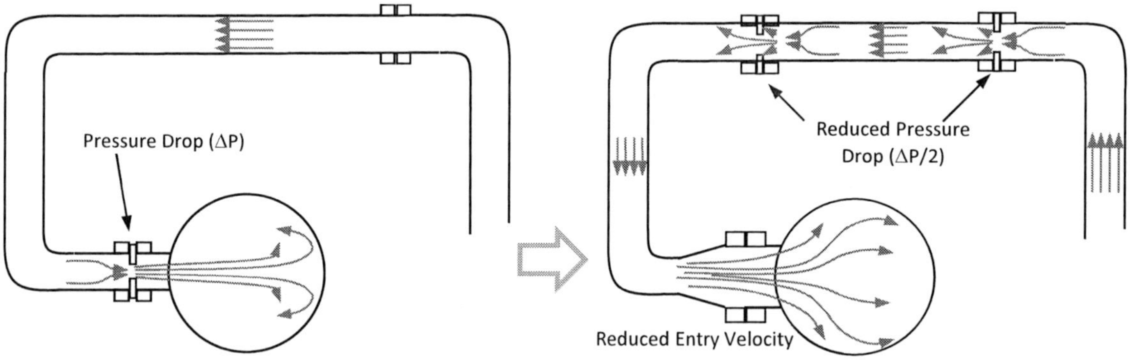

Figure 12: Concept for distributing pressure drops and slowing down flow

Conclusion

Several types of typical industrial piping vibration problems and solution approaches have been discussed in the previous sections. Table I is an attempt to provide an at-a-glance reference for the issues visited. It is not intended to be a complete list of all possible problems and remedies, but rather it is a summary of some firsthand observations in the field while dealing with piping vibration issues. The table is organized into two sections: excitation types and piping response types. The table is formatted this way because a given excitation type can elicit more than one type of response from the piping; therefore trying to capture all possible combinations would make the table too complex. For a specific problem, both the solutions related to the excitation type and the response type(s) will be applicable. In general, reducing the eliminating or reducing the excitation forces is the preferred approach when feasible. In the case of resonance, however, separating the forcing and response frequencies is a valid primary goal.

Table I: Piping vibration and solution type matrix

	Solution Strategy																
Problem	Apply Flow Smoothing	Balance and/or Align Machine	Change Machine Speed	Tune Acosutic Volumes	Eliminate dead leg	Distribute Pressure Differential	Use Multi Stage Valve	Use Low Noise Trim Valve	Reduce Flow Rate	Add Support to Main Line	Add Viscous Damper	Use Larger Diameter Pipe	Brace Side Branch	Relocate/Eliminate Side Branch	Use Thicker Pipe	Add Stiffening Clamp/Collar	Make Piping More fatigue Resistant
Very Low Frequency Turbulent Excitation	O								O								
Low Frequency Mechanical Excitation		O	O														
Low Frequency Pulsation - Machinery Induced			O	O													
Low Frequency Pulsation - Flow Induced				O	O				O								
Broadband Excitaton by Cavitating Control Valve						O	O		O								
High Frequency Pulsation				O				O	O								
High Frequency Acoustic Excitation						O	O	O	O								
Low Frequency Main Line Vibration										O	O	O					O
Side Branch Resonance												O	O	O	O		O
Main Line Shell Mode Vibration															O	O	O

Clearly there is one mitigation technique that is applicable to all piping vibration problems, and that is making the piping more fatigue resistant. Typically this refers to the locations where there are joints in the piping. Most joints will be welded while some of the small bore piping may have threaded connections. Threads will inherently present stress concentration effects and should be avoided to the extent possible. Grinding weld toes, caps and roots smooth (where feasible) can help minimize stress concentration. Socket welds can be made with a 2:1 leg ratio (toward the branch) for enhanced fatigue resistance. Stub-in type branch connections should be avoided, weldolets are an improvement but forged welding tees and sweepolets are preferred. Using thicker wall piping usually helps in this regard as well.

References:

1. Szasz, G. "Vibration Related Small Bore Piping Connection Failures," Proceedings of the Vibration Institute National Technical Training Symposium and 38th Annual Meeting, San Antonio, Texas, USA, June 11–13, 2014.
2. Caillaud, S., Pons, Y., Moussou, P., Gaudin, M., "A 12 mm/s RMS Screening Vibration Velocity for Pipes Using ANSI-OM3 Standard and Regulatory Design Rules," Proceedings of the ASME Pressure Vessels and Piping Conference, Denver, Colorado, USA, July 17–21, 2005.
3. Energy Institute, "Guidelines for the Avoidance of Vibration Induced Fatigue Failure in Process Pipework," 2nd Edition, Energy Institute, London, 2008.
4. Wachel, J.C., Morton, Scott J., Atkins Kenneth E., "Piping Vibration Analysis"; Proceedings of the 19th Turbomachinery Symposium; Turbomachinery Laboratory, pp.119-134; Texas A&M University, College Station TX, 1990.
5. ASME OM-SG-2009, "Standards and Guides for Operation and Maintenance of Nuclear Power Plants – Part 3, Requirements for Pre-operational and Initial Start-up Vibration Testing of Nuclear Power Plant Piping Systems," American Society of Mechanical Engineers, New York, NY. 2009.
6. "Reciprocating Compressors for Petroleum, Chemical, and Gas_Industry Services." API Standard 618, Fifth edition, December 2007
7. "Positive Displacement Pumps – Reciprocating" API Standard 674, Second edition, June 1995

KAYBOB REVISITED: WHAT WE HAVE LEARNED ABOUT COMPRESSOR STABILITY FROM SELF-EXCITED WHIRLING

Edgar J. Gunter, Ph.D.
RODYN ***Vibration Analysis, Inc.***
Professor Emeritus, Mechanical and Aerospace Engineering
Former Director, Rotor Dynamics Laboratory, Univ. of Virginia
Fellow ASME
DrGunter@aol.com

Brian K. Weaver, Ph.D.
Research Associate, Mechanical and Aerospace Engineering
Rotating Machinery and Controls Laboratory, Univ. of Virginia
Member ASME & STLE
bkw3q@virginia.edu

Abstract

The Kaybob compressor failure of 1971 was an excellent historic example of rotordynamic instability and the design factors that affect this phenomenon. In the case of Kaybob, the use of poorly designed bearings produced unstable whirling in both the low and high pressure compressors. This required over five months of vibration troubleshooting and redesign along with over 100 million modern U.S. dollars in total costs and lost revenue. The problems began with the use of inadequate five-pad tilting pad journal bearings which produced large ratios of bearing stiffness to shaft stiffness that contributed heavily to the instability. Subsequent designs included attempts at using bearing asymmetry and squeeze film dampers to improve the stability of the machine, however these solutions produced little improvement. The ultimate solution, a redesign of the compressor to a shorter, stiffer machine, resulted in the significant costs incurred as a result of these failed bearing designs. Another reason for the costly nature of this project was a lack of accurate bearing and rotordynamic analysis tools that could have identified the instability during the design process and aided in the redesign of the fluid film bearings supporting the machine. Since the 1970s, a wide range of bearing and rotordynamic analysis tools have been developed that accurately predict the performance of fluid film bearings and rotordynamic systems via the use of practical lubrication theory, finite element analysis, and simple relationships derived as a result of the physical insights gained from using these methods.

In this paper, the history of the Kaybob compressor failure is discussed in detail including a discussion of the ineffective bearing designs that were considered. Modern bearing and rotordynamic analysis tools are then employed to study both designs that were considered along with new designs for the bearings that could have ultimately restored stability to the machine. These designs include four-pad, load-between-pad bearings and squeeze film dampers with a central groove. Simple relationships based on the physics of the system are also used to show how the bearings could be tuned to produce optimum bearing stiffness and damping of the rotor vibration, producing insights which can inform the designer as they perform more comprehensive analyses of these systems.

Introduction

The rotordynamic stability of high-speed compressors and turbines has been a critical element of their design for decades. With the ever-increasing demand for greater output through an increased number of stages, higher speeds, and higher pressures, these machines are continuously being pushed to the design limits of the previous generation of machines. As they are being pushed to these new extremes of operation the designs must also adapt to accommodate the consequences of these extremes including increased rotor flexibility and an increased likelihood of subsynchronous whirl. This whirl can be excited by a number of sources in turbomachinery including seals and even the bearings supporting the rotor. The stability of that whirl can then be reduced by destabilizing cross-coupled stiffness forces from the bearings, interstage seals, balance pistons, and Alford-type aerodynamic cross-coupled forces around blades.

The increasing use of tilting pad journal bearings in the 1960s and 70s brought about new opportunities for compressor design as these bearings eliminated the self-exciting oil whirl commonly found in fixed geometry bearings. They also essentially eliminated the destabilizing cross-coupled stiffness forces produced by these bearings. However, despite the fact that a number of important papers were being published on these topics by Lund (1964) and others, it wasn't until 1974 that Lund published his landmark paper on damped eigenvalue solutions for stability analysis. It was because of this lack of accurate design and analysis tools that a significant number of high-speed machines were not designed appropriately for the conditions in which they were intended to operate.

The Kaybob compressor failure of 1971 was an historic case of rotordynamic instability from self-excited, subsynchronous whirl. The many months of troubleshooting that followed this failure served as a valuable lesson to the team involved but even to this day this classic example serves as an important motivator for understanding the underlying principles that govern the design of these machines, many of which that were developed after the incident and troubleshooting took place. With the advent of advanced bearing and rotordynamic analysis tools, designers now have the ability to fine tune their designs to avoid these failures, however it is critically important that the underlying principles and physical insights gained from the analysis of these machines are properly considered as even modern bearing technologies can contribute to stability problems when not implemented correctly.

In this study, modern bearing and rotordynamic analysis tools are applied to the Kaybob compressor instability case to highlight the effects of the design choices made with this machine and how they contributed to its ultimate failure. By combining useful design principles with these analysis tools it also shown not only why the costly solution to this failure worked at the time, but also how a number of alternative and much simpler design changes to the support structure could have restored stability to the machine including the use of properly designed tilting pad journal bearings and squeeze film dampers. It is the goal of this study to provide practical information and design approaches that designers can then use to avoid costly failures like the one experienced in Kaybob.

Stage **Case History - Kaybob Plant, Alberta**

The case of the Kaybob compressor instability began in November of 1971, a detailed summary of which is provided by Kenneth Smith of Cooper-Bessemer (13). Three natural gas re-injection trains were being commissioned at the Kaybob South Beaverhill Lake plant in Fox Creek, Alberta operated by Chevron-Standard.

These duplicate trains (Figure 1) consist of a gas turbine drive, a speed-increasing gearbox, a 9-stage low pressure compressor, and a 5-stage high pressure compressor. Both compressors have back-to-back impeller arrangements with centrally located balance pistons to control thrust and are supported by tilting-pad thrust and radial bearings.

The tilting pad journal bearings were designed as five-pad, load-on-pad bearings, a design that has been commonly employed in compressors over the years. Floating oil film ring seals were employed in both compressor casings. The trains were designed for a rated speed of 10,200 rpm with a maximum continuous operating speed of 11,400 rpm.

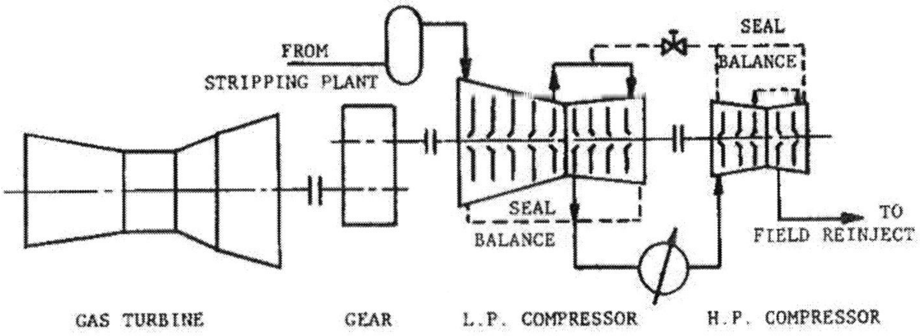

Figure 1 Kaybob Compressor Train Schematic
Smith (13)

As site testing began, a significant vibration problem became apparent in the low pressure compressor. A first train was operated at speeds up to 10,000 rpm with inlet pressures of 700-750 psi and discharge pressures of 1,300-1,500 psi. Under these conditions a significant vibration was present at a frequency of approximately 4,250 cpm, producing 3 to 4 mil peak-to-peak amplitudes in one of the bearings which had a 6 mil clearance.

A second train was then tested to rule out possible unexplained sources of resonance from the first train. This machine was ran up to its maximum speed of 11,440 rpm, followed by increases to the inlet and discharge pressures to 1,120 psi and 3,300 psi.

At this point the machine became highly unstable, resulting in a violent 5,100 cpm whirl of 9+ mils peak-to-peak amplitude. Figure 2 shows the 20-second time lapse of the instability onset in the thru-drive end bearing along with the orbit shape. Smith (13) notes that "the rotor assembly and internal labyrinths were severely damaged as a result of this violent, high-energy motion" and that the "vibration could be heard and felt in the control room some 50 feet away from the unit."

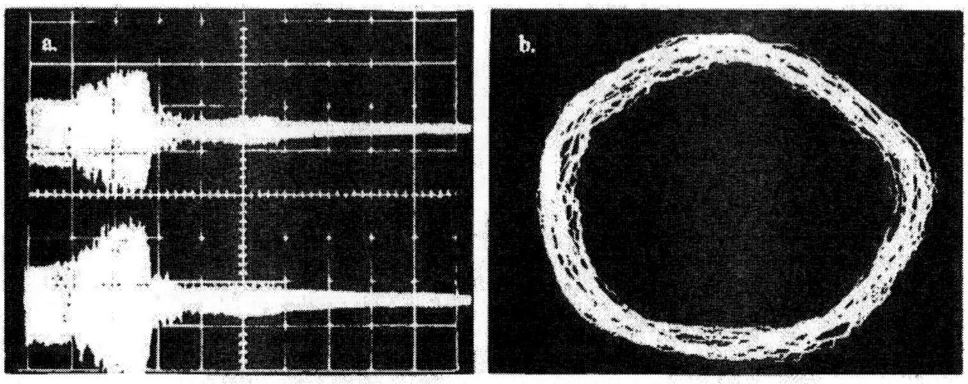

Figure 2. (a) Time-lapse and (b) orbit of the instability onset. ***From Smith (13).***

The five months that followed included a thorough investigation into the causes of this instability as well as tests of various potential solutions to the problem including investigations into both seals and journal bearings. The seal investigation involved the consideration of four different seal configurations to determine whether seal lockup or effective damping had any significant effects on the instability. It was concluded, though, that the seals were in fact floating and their configuration overall had little effect on the problem.

A number of design modifications were considered for the tilting pad journal bearings as well. These modifications included increasing the specific load of the bearing, introducing asymmetry to the bearing design, different pad-load configurations, offset pads, variable preloads, the introduction of squeeze film dampers, and combinations thereof. Two of the configurations tested are shown in Figure 3. Overall it was found that all of these bearing design-based solutions produced marginal improvements in rotordynamic stability. Additional analysis and discussion of these and other bearing designs is presented in the following sections.

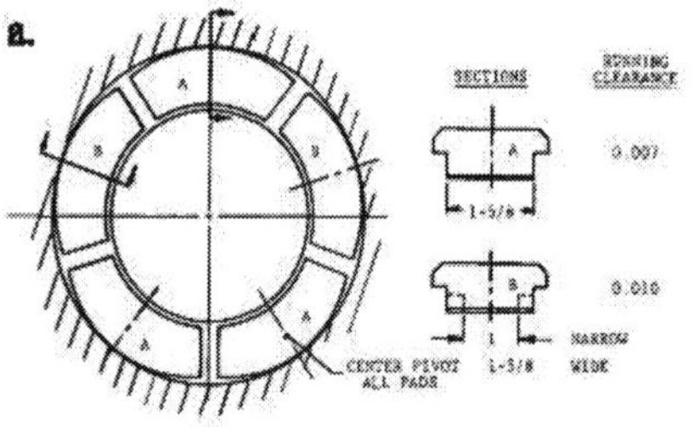

Figure 3(a) Asymmetric Tilt Pad Bearing
Smith(13)

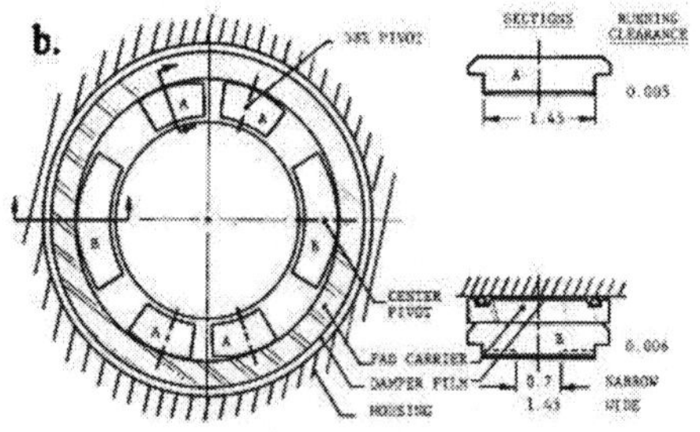

Figure 3(b) Squeeze Film Damper Design
Smith (13)

Note that in Fig. 3a, the side pads ,B, are reduced in width in an attempt to incorporate some bearing asymmetry. This modification had a negligible influence on stability. In Fig. 3b, the A pads are reduced in arc length and a squeeze film damper was added without a central groove. This design results in a damper with excessive damping and stiffness.

The ultimate solution to this vibration problem was a redesign of the rotor to a shorter, stiffer configuration with increased diameters beneath the impellers. This stiffer shaft would thereby accommodate the stiff bearing designs that were used and that will be demonstrated in the following sections. Figure 4 shows the progression of the shaft redesign process. Overall, these significant changes resulted in a redesign of most of the primary machine components, a process which would cost the plant over 100 million modern U.S. dollars.

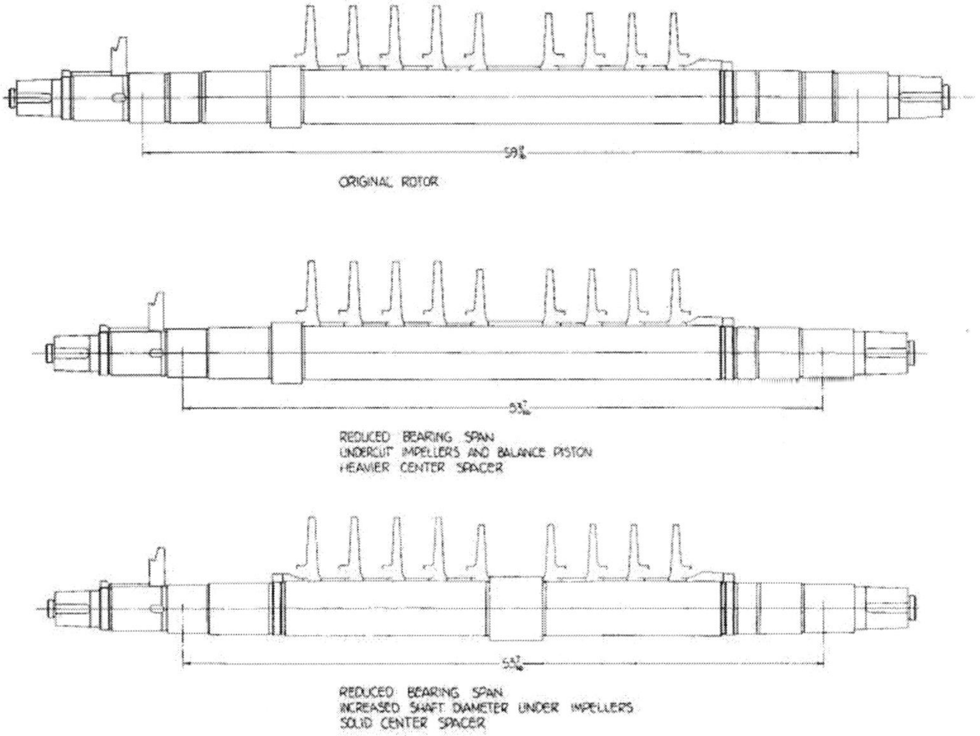

Figure 4 Redesign of the Compressor Rotor. ***From Smith (13)***

In the original design of the Kaybob rotor, the bearing span was 59.688 inches. A modal analysis of the rotor indicated that the shaft first critical speed modal stiffness was approximately 115,000 Lb/In. This fundamental shaft modal stiffness is extremely important for the proper tuning of the bearings to the shaft. From previous research studies of the optimum bearing stiffness, is has been determined that the optimum bearing stiffness is approximately one half the rotor shaft stiffness. When the bearing stiffness greatly exceeds this fundamental design value, the first mode damping is drastically reduced causing the rotor to be very susceptible to self excited whirl effects.

The fundamental stiffness of a uniform team is given by the following equation:

Kshaft − 48 EI/L^3 Lb/In (1)

In the final redesign as shown in the lower rotor of Figure 4, the bearing span has been reduced to 53.438 inches. In addition to the reduced bearing span, the center section has also been increased. This then leads to an improved shaft modal stiffness of 200,000 Lb/In.

Rotordynamic Characteristics

In revisiting the design and troubleshooting of the Kaybob compressor, the first step was to establish the overall rotordynamic characteristics of the system. A finite element model of the compressor was first developed in the Dyrobes software suite and is presented in Figure 1. The rotor spans 78 inches in total length with a bearing span of 60 inches. The model consists of 25 elements and 26 nodes with 9 disks representing the mass and inertial properties of the 9 compressor stages. The bearings are represented as linear stiffness and damping coefficients at nodes 5 and 23.

After building the shaft model, the analysis began with an assessment of the rotor critical speeds in the operating range of the compressor (Figure 6). Two modes were found in the operating range:
A first bending mode at approximately 4,600 RPM (Figure 7) which proved to be the unstable mode for the machine, and a well-damped conical mode at approximately 8,500 rpm. Overlaying the critical speed map is the vertical stiffness K_{yy} of the original 5-pad load-on-pad tilting pad bearings supporting the rotor. Its position on the curve for the first bending mode provides an early indication of a relatively stiff support structure for the compressor.

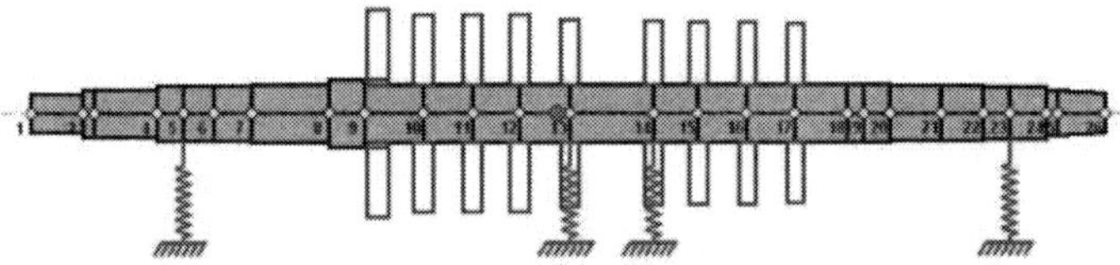

Figure 5 Kaybob 9 Stage Compressor Model With Aerodynamic Cross Coupling at St. 13 & 14

Figure 5 represents the 9 stage Kaybob compressor supported in the five pad load-on-pad tilting pad bearings. In order to evaluate the stability characteristics of this compressor, a nominal amount of aerodynamic cross coupling of 25,000 Lb/In cross coupling effects were assumed acting at stations 13 and 14. These cross coupling effects are known as the Alford effect *(1)* and can occur in all gas turbines and high pressure compressors.
These effects are often not observed on a test stand when the unit is operated under low pressure ratios. When operated in the field under high pressure conditions, the Alford effects become apparent resulting in self excited whirling which can then lead to damage of seals and labyrinths.

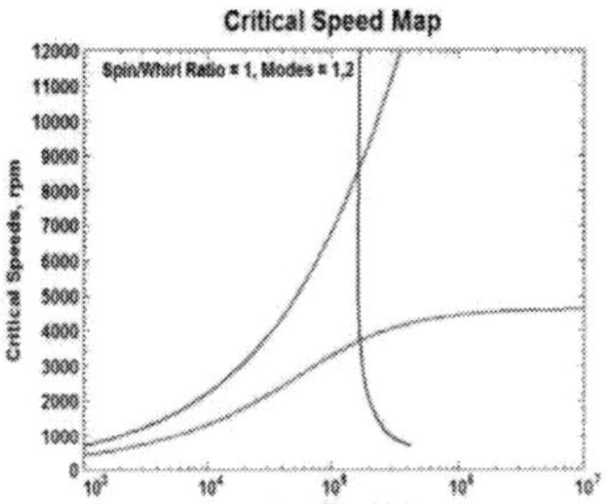

Figure 6 Compressor Critical Speed Map

Figure 6 represents the critical speed map of the 9 stage compressor.

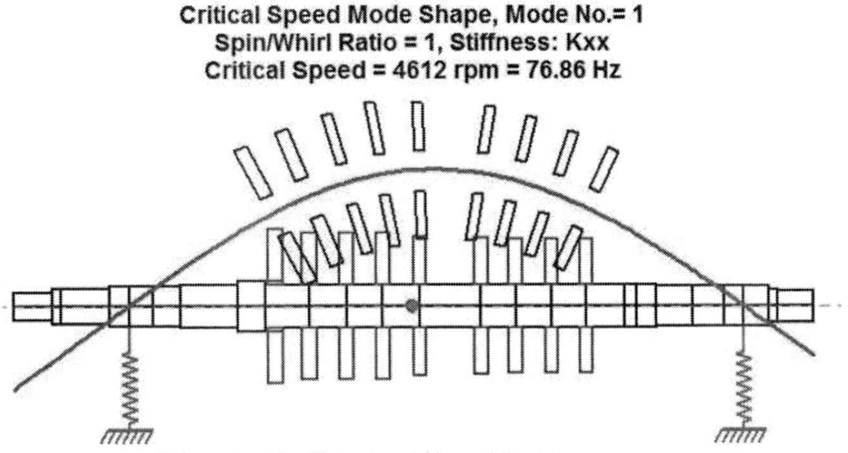

Figure 7 First Bending Mode of the Compressor

The shaft modal stresses and stiffness were also assessed for the first bending mode assuming rigid bearings with a stiffness of 10^8 lb/in (Figure 8). Peak stresses of 1,709 psi were predicted along with a shaft modal stiffness of 115,850 lb/in. This magnitude of stiffness will prove to be very important to the stability of the machine as it is compared to the stiffness of the bearings in the following sections. It was presented by Barrett, Gunter & Allaire (3) that the optimum ratio of twice the bearing stiffness to the shaft modal stiffness for maximizing modal stability is one.

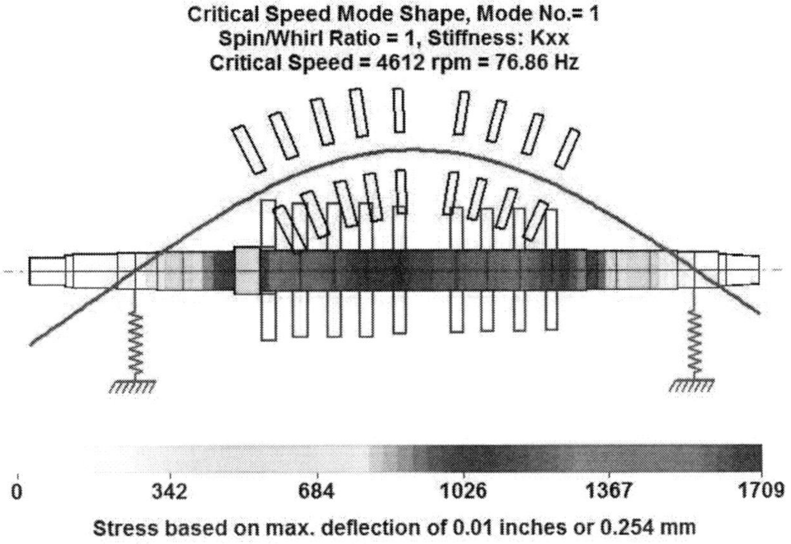

Figure 8 Shaft Modal Stresses of the First Bending Mode

To further support this important relationship in rotor dynamics, a plot was created showing the amplification factors of the first bending mode under various sets of bearing stiffness and damping (Figure 9). It is shown that the amplification factors are reduced as the bearing stiffness approaches this ideal ratio. The plot also highlights the importance of designing the bearings in such a manner that underdamped and overdamped conditions can also be avoided through proper design. The following sections on various bearing designs for the compressor will further show how various design variables affect these bearing properties and the stability of the compressor.

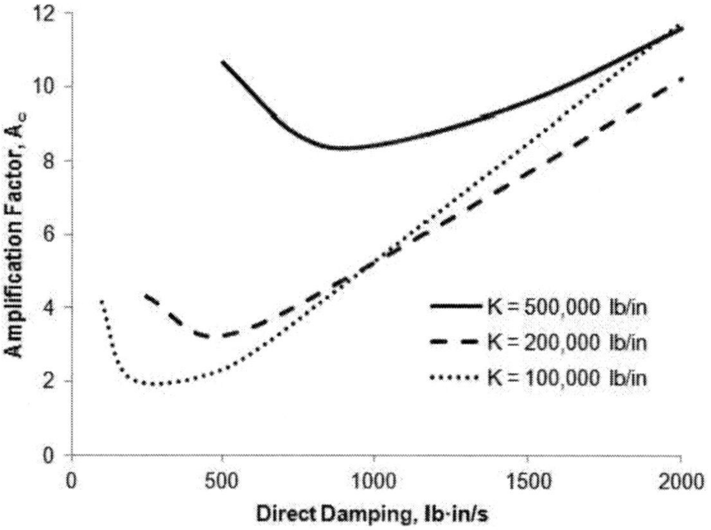

Figure 9 Amplification Factors of the First Bending Mode

Figure 9 shows the importance of tuning the bearing stiffness to the rotor shaft stiffness. For example with the original rotor design, it is seen that with a vertical bearing stiffness of 100,000 Lb/In, the optimum damping is about 250 Lb-Sec/In with an amplification factor of two. As the bearing stiffness increases, the required optimum damping increases to 500 Lb-Sec/In and the minimum amplification factor is over three.

For the case of very high stiffness bearing of 500,000 Lb/In, it is seen that the optimum bearing damping is over 800 Lb-Sec/In and the amplification factor is now over eight. Therefore it is very apparent that it is critical to select the bearing stiffness in relationship to the predicted first modal shaft stiffness. This concept will often lead to the requirement of an extended width bearing in order to have a reduced vertical bearing stiffness in order to properly tune the bearings to the fundamental shaft stiffness.

Original Bearing Design

The analysis of the Kaybob compressor continued with the development of various bearing models to better understand how the design of the bearings would impact the stability of the machine. The original design of the compressor bearings was a 5-pad load-on-pad configuration with an L/D of approximately 0.4, depicted in the Dyrobes BePerf program shown in Figure 10. This model utilizes a two-dimensional Reynolds equation solver that for the current analysis neglects thermal effects and deformation for simplicity. All bearing analyses were performed at the maximum continuous operating speed of 11,400 rpm.

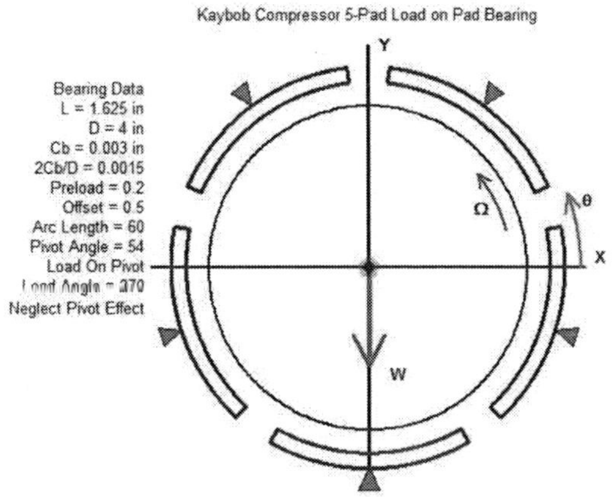

Figure 10 Original Compressor Bearing Design

The results of this initial bearing analysis are shown in Figure 11. It was found that for this configuration that the tight clearance, preload, and load-on-pad configuration together produced a relatively stiff bearing, resulting in a vertical bearing stiffness of 169,500 lb/in . Recall from the previous section that the ideal ratio of twice the bearing stiffness to the shaft modal stiffness is unity. When comparing twice the vertical stiffness of the original compressor bearing to the calculated shaft modal stiffness, this results in a ratio of over 2.9. These results alone are indicative of an overly stiff support structure for the compressor that could contribute to an unstable whirl motion.

The next step in analyzing the original rotor-bearing system was to perform a stability analysis of the compressor with the original bearings. This damped eigenvalue analysis was performed in Dyrobes Rotor and the results are presented in Figure 12. It is shown that the poor choices made in the bearing design process resulted in an unstable forward bending mode in the operating range. The combination of high stiffness and damping in the bearings resulted in less shaft motion at the bearings and therefore a greater distribution of the system energy into the shaft, resulting in the unstable mode. A critical speed analysis using the bearing vertical stiffness reveals the relative energy distribution in the shaft and bearings (Figure 13).

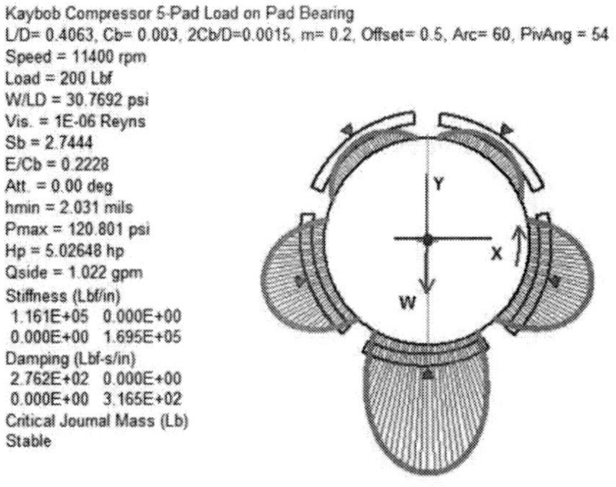

Figure 11 Original Compressor Bearing Analysis Results

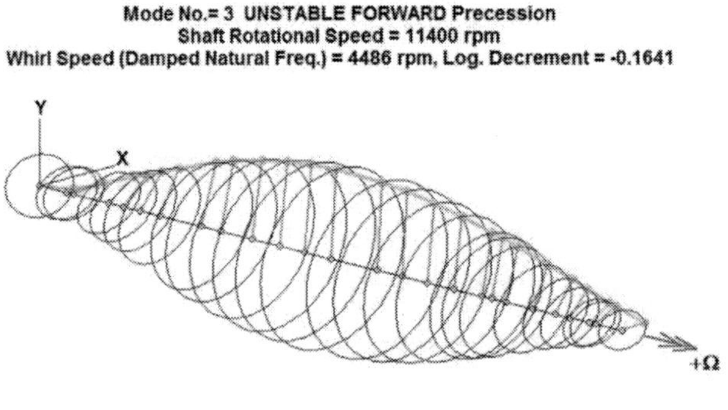

Figure 12 Stability Results for the Original Bearing Design

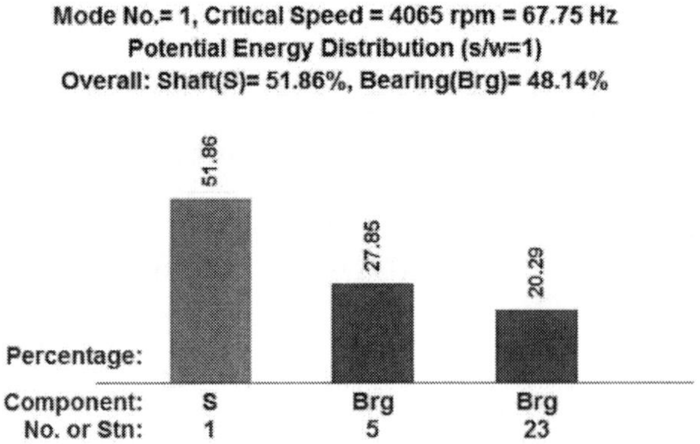

Figure 13 First Mode Shaft and Bearings Energy Distribution With the Original Bearing Design

A Lesson in Bearing Asymmetry

A second bearing design considered when troubleshooting the Kaybob compressor was an asymmetric bearing. Smith (13) cited two papers by Gunter (5,6) as evidence that introducing bearing asymmetry could lead to an improvement in rotor stability. However, this evidence was misinterpreted as asymmetry has little effect in cases of high bearing stiffness and large quantities of aerodynamic cross-coupling (Gunter (8)). This is demonstrated in the below analysis of an asymmetric bearing design similar to that considered by the team at Cooper-Bessemer.

The model below (Figure 14) shows how the bearing was clocked to a load-between-pad position. The bearing clearance was also increased in the upper-left and upper-right pads from a 3.5 mil radial clearance to 5 mils. These two pads were also narrowed from an axial length of 1.625" to a length of 1", though this modification is not captured by the model. This proves to have little effect, however, as shown in the results (Figure 15). The pads with the increased clearance show no participation in controlling the shaft motion, resulting in predicted vertical stiffness values that are still quite large when compared to the shaft stiffness. A significant drop in damping is also produced as a result of this design.

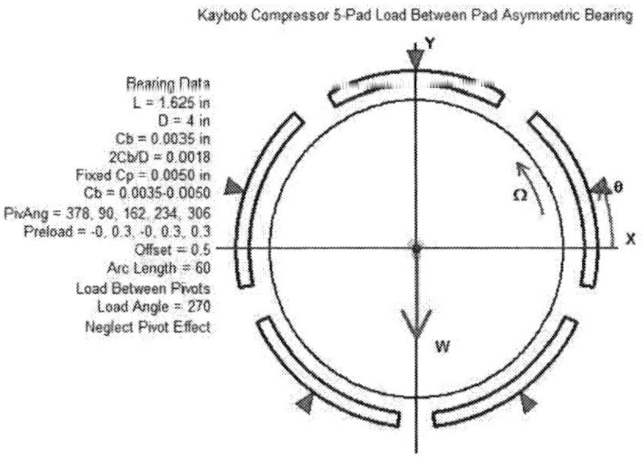

Figure 14 Asymmetric Bearing Model

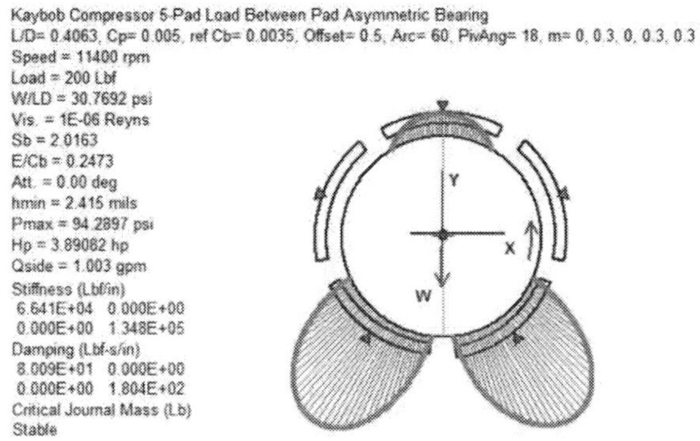

Figure 15 Results for the Asymmetric Bearing

A stability analysis of the bearing-rotor system (Figure 16) in fact shows a decrease in the log decrement of the first bending mode when compared to the original bearing design. This is due to the fact that not only is the bearing stiffness still too large, but now the system is also likely underdamped with this bearing design.

The team at Cooper-Bessemer tried multiple bearing designs that utilized bearing asymmetry, but with such high levels of bearing stiffness and aerodynamic cross-coupling they were unable to achieve any significant increases in modal stability.

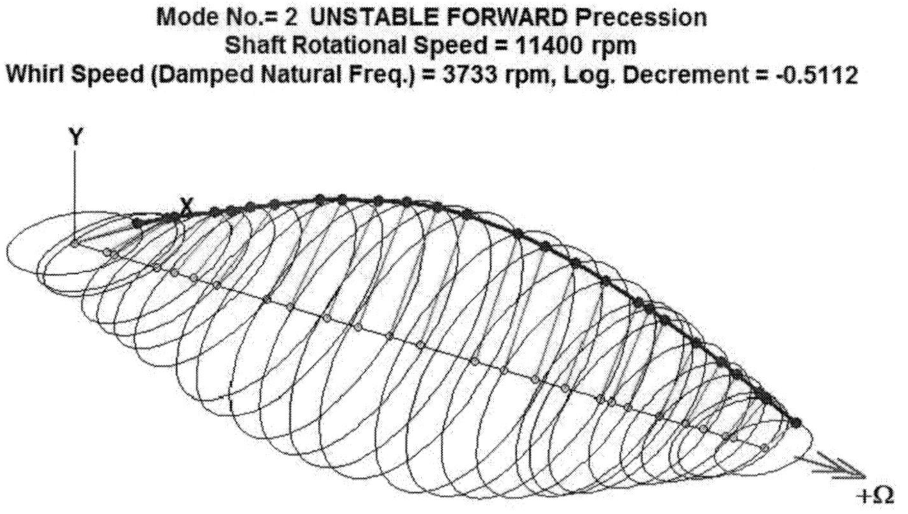

Figure 16 Stability Results for the Asymmetric Bearing

Figure 16 represents the damped eigenvalue analysis of the nine stage Kaybob compressor with the aerodynamic cross coupling included at stations 13 and 14. It has also been shown that in addition to turbines and compressors developing aerodynamic cross coupling, a center span balance piston can also create aerodynamic cross coupling effects. Therefore small values of aerodynamic cross coupling of 25,000 Lb/In were placed at the center stations 13 and 14 at the balance piston location.

It is seen that at the operating speed of 11,400 RPM, there now occurs a self excited whirl component at 3733 CPM. The log decrement shows a value of -0.51 which represents a highly unstable system. Noncontact probes are often placed at the bearings to observe the rotor motion. The motion observed at the bearing locations is often quite small as compared to the large orbital motion occurring at the rotor center due to the added whirl component. This large motion occurring at the center causes extensive rubbing and seals to be extensively damaged.

4-Pad Bearing Design

A design not considered by the team at Cooper-Bessemer was a 4-pad tilting pad bearing. These bearings, particularly when oriented in a load-between-pad position, are known to be beneficial from a rotordynamic standpoint due to their ability to more evenly distribute stiffness forces in the vertical and horizontal directions. To see how this design might impact the Kaybob compressor, an initial 4-pad load-between-pad bearing design was analyzed with other design parameters that were kept constant from the original 5-pad bearing including the axial length, bearing clearance, and preload as shown in Figure 17.

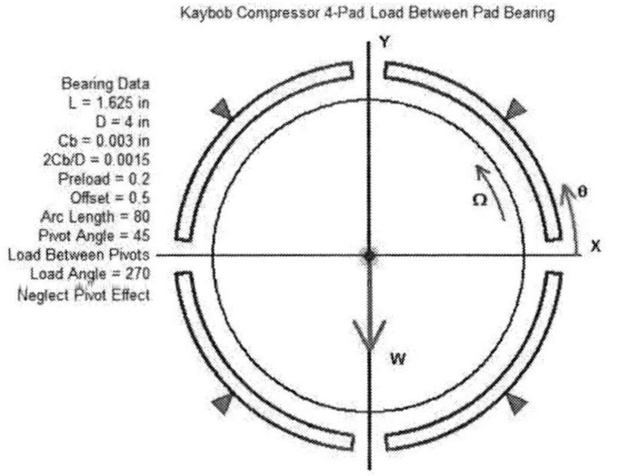

Figure 17 Initial Four Pad Bearing Design

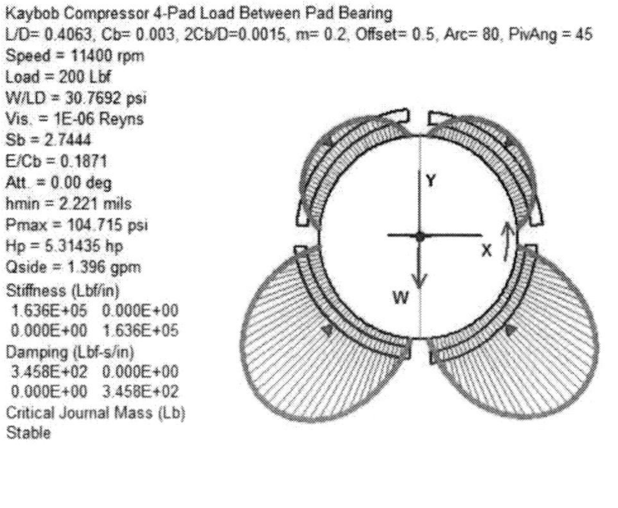

Figure 18 Results of the Four Pad Bearing Analysis

The results of the bearing analysis (Figure 18) show that the 4-pad design does produce equal values of vertical and horizontal stiffness. When compared to the original 5-pad bearing, The vertical stiffness is slightly lower along with an increase in direct damping. However, in general, this design is still very stiff for this machine.

A stability analysis (Figure 19) reveals a reduction in the log decrement of the first bending mode when compared to the original design, likely due to increases in damping and horizontal stiffness preventing the bearings from absorbing that shaft vibrational energy. These results indicate that more drastic changes to the bearing design are necessary in order to restore stability to the system.

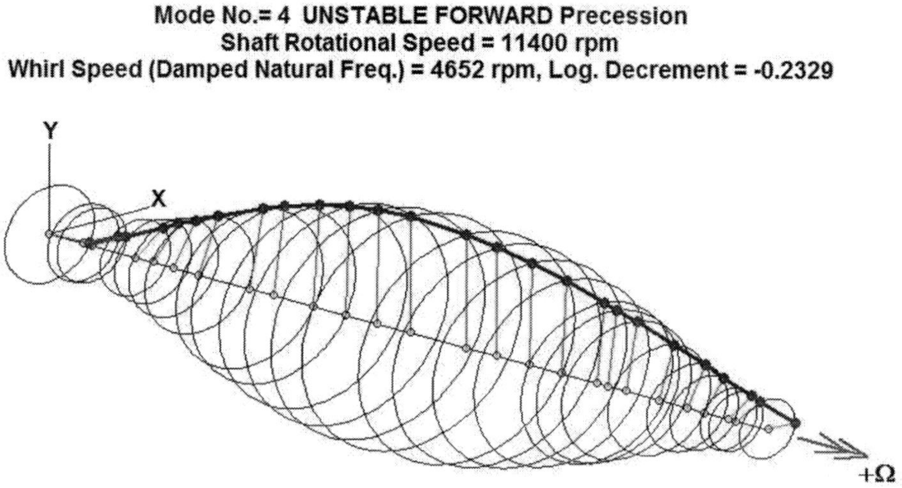

Figure 19 Stability Results for a Four Pad Bearing Design

The design for the four pad bearing is superior to the five pad bearing in that the log decrement has improved. However the dynamic analysis of the compressor with the four pad design as shown above is still unstable. The system has been improved over the 5 pad design with a reduction of the log decrement from -0.51 for the 5 pad case as compared to -0.233 with the 4 pad design.

In order to improve the rotor dynamics with the four pad bearings, it will be necessary to further reduce the vertical bearing stiffness in order to bring the values more in line with the fundamental shaft stiffness value. This requires a longer pad with a reduced bearing preload.

The paradox of this type of design is that the noncontact probes monitoring the bearing motion indicate higher amplitudes of motion, although the shaft center motion has been reduced. This phenomena led one chief engineer of a compressor company to remove the redesigned four pad bearings and replace them with the original five pad design because of the increased motion observed at the bearings. Total rotor failure occurred shortly after the five pad bearings were installed!

Improved 4-Pad Bearing

In an effort to determine what bearing design choices were necessary in order to restore stability to Kaybob compressor, a number of modifications were made to the previously analyzed 4-pad bearing. Because it has been revealed that all of the designs considered thus far have produced bearings that are overly stiff, a number of design parameters must be considered to reduce the stiffness of the bearing while maintaining damping so that the bearings can better absorb the energy of the first bending mode.

Design parameters including the bearing clearance, pad preload, load orientation, and lubricant properties have all been shown by many authors (Nicholas and Kirk 1979 and 1982, Bhat et al. 1982, Weaver et al. 2015) to affect the bearing stiffness. Therefore, for this design, it was decided to increase the bearing clearance from 3 mils radial to 6 mils and reduce the pad preload from 0.2 to 0.1 while maintaining the load-between-pad orientation to reduce the stiffness of the bearing. In order to maintain adequate damping the bearing axial length was also increased from 1.625" to 4", resulting in an increase in the L/D ratio from 0.4 to 1.

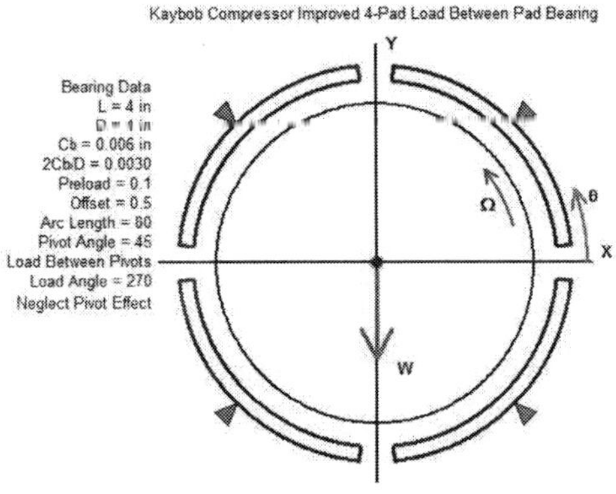

Figure 20 Improved Four Pad Bearing Design

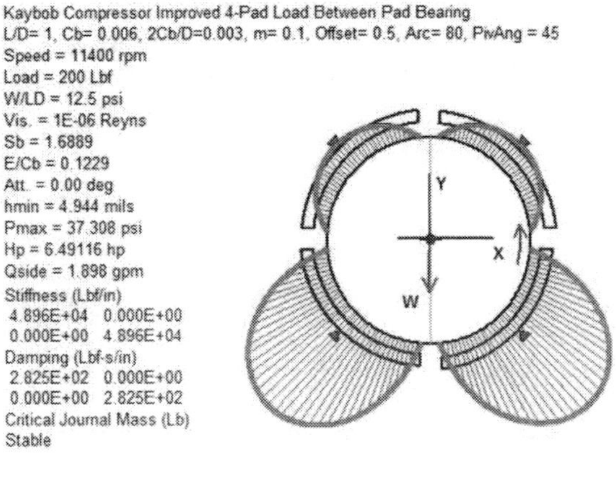

Figure 21 Analysis of the Improved Four Pad Bearing

The results from the bearing analysis show a significant change in bearing performance when compared to the previous designs (Figure 21). The bearing stiffness has now been reduced to 48,960 lb/in. When comparing twice the bearing stiffness to the shaft modal stiffness this now results in a ratio of 0.85, much closer to the desired value of unity. The increase in bearing axial length also results in reasonable levels of damping. A stability analysis of this system shows a drastic increase in the log decrement of the first bending mode (Figure 22), resulting in a fully stable mode for the compressor. An energy distribution plot (Figure 23) shows how the improved bearing design does a much better job of allowing the bearings to absorb the vibrational energy, resulting in an increase of 32% of the energy being transmitted to the bearings when compared to the original 5-pad design.

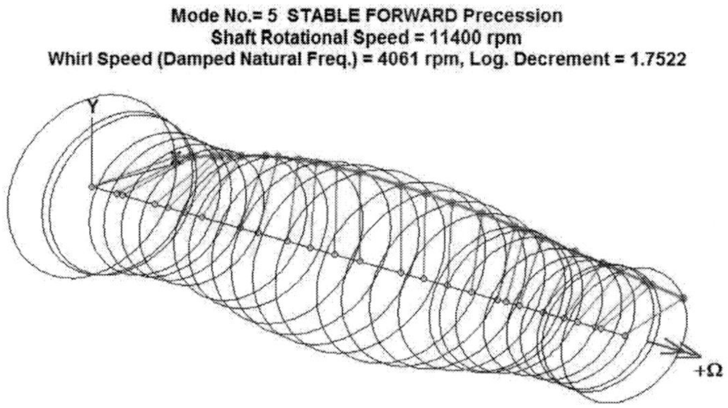

Figure 22 Stability Analysis of the Improved Four Pad Bearing

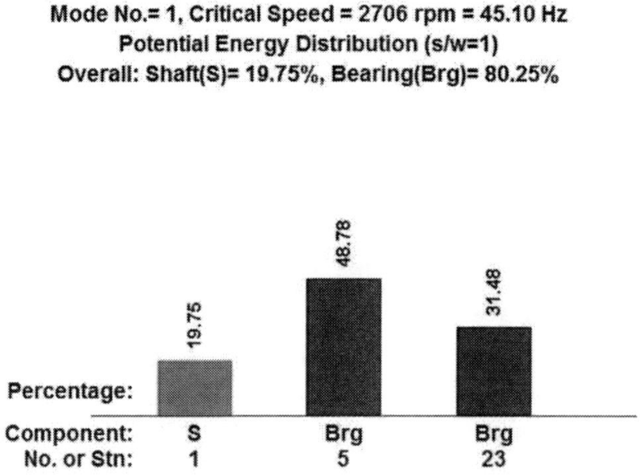

Figure 23 Energy Distribution in the Shaft and Bearings for the Improved Four Pad Bearing Design

Squeeze Film Dampers

Another solution considered by the team at Cooper-Bessemer was a squeeze film damper. Smith (1975) describes the overall design as a long damper surrounding a number of different bearing designs that were considered including a unique design that involved six tilting pads with asymmetries in the pad clearance, axial length, and arc length. However, none of these designs produced an ultimately stable solution for the machine and the use of a long squeeze film damper was likely a key factor in this.

To further make this point, the below analysis using the Dyrobes built-in squeeze film damper tool was performed on two damper designs surrounding the original 5-pad bearing (Figure 24): a long damper design and a damper with a central groove of 0.15 inches in axial length. The non-grooved damper is treated as a long bearing, but because an added groove to the damper results in a different pressure distribution within the damper the grooved damper is modeled as a short bearing.

With all other inputs equal it can be seen that the groove has a significant effect on the predicted stiffness and damping of the damper. The longer damper design - similar to the one used by Cooper-Bessemer - produces very high levels of stiffness and damping while the grooved design produces much more reasonable values.

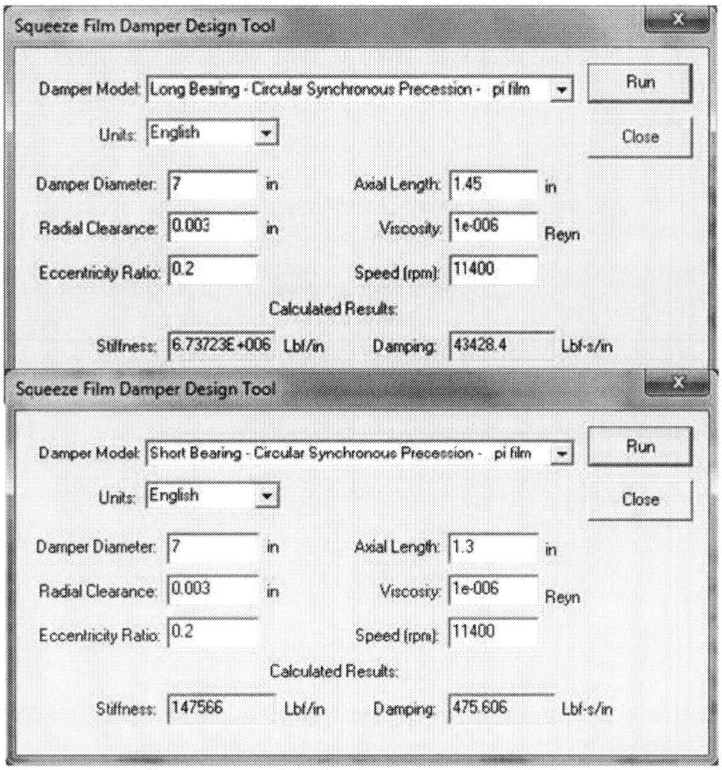

Figure 24 Non-Grooved and Grooved Squeeze Film Damper Analysis

Stability analyses of these two support scenarios also reveal the significant effect of the central groove (Figures 25 and 26). It is shown that the non-grooved damper produces only a very slight increase in the log decrement of the unstable mode when compared to the original bearing design.

The grooved damper, however, provides a substantial increase in the log decrement, resulting in a marginally stable system. Had the team at Cooper-Bessemer considered adding a central groove to their squeeze film damper design they may have actually avoided the costly rotor redesign discussed in the next section.

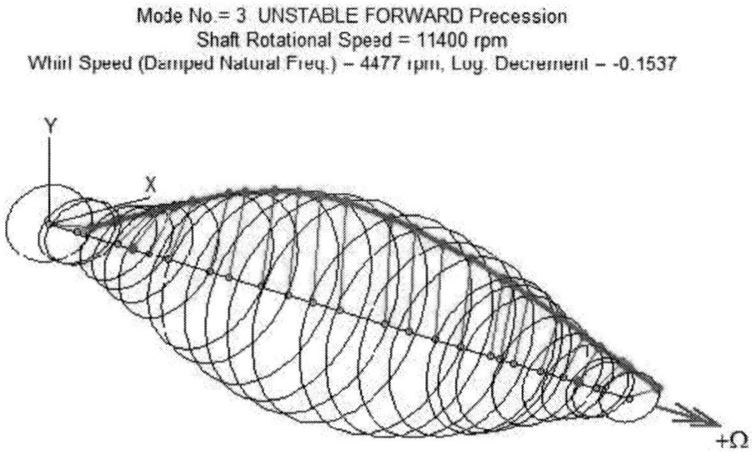

Figure 25 Stability Analysis of The Non-Grooved Squeeze Film Damper Supported System

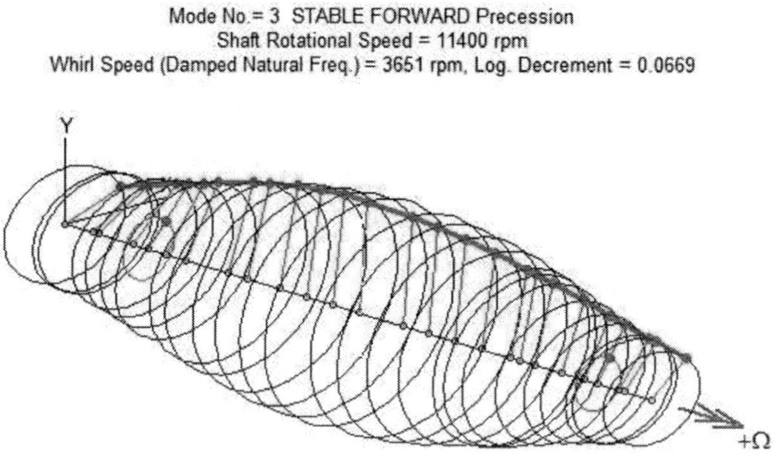

Figure 26 Stability Analysis of the Grooved Squeeze Film Damper Supported System

Rotor Redesign

To produce a stable operating environment the team at Cooper-Bessemer ultimately had to redesign the rotor. This involved multiple design iterations that resulted in a final shaft design with a bearing span reduced from about 60 inches to approximately 53 inches, modified shrink fits, an integral center seal section, and an increased shaft diameter under the impellers (Figure 4). This design, along with either a squeeze film damper or a 5-pad tilting pad bearing with no preload, was enough to restore the full operational capabilities of the compressor.

To show how these shaft design modifications made such a difference in the performance of the machine a shaft model was developed to analyze with the original bearing design and the original design with a long squeeze film damper (Figure 27). A critical speed analysis assuming rigid bearings with a stiffness of 10^8 lb/in predicted shaft stresses at the first bending mode of 2,371 psi (Figure 28), an increase of 39% when compared to the original rotor.

The shaft modal stiffness was calculated to be 201,780 lb/in, a 74% increase when compared to the original rotor. This significant increase in shaft modal stiffness made the new rotor design much more suited to the high-stiffness bearings utilized.

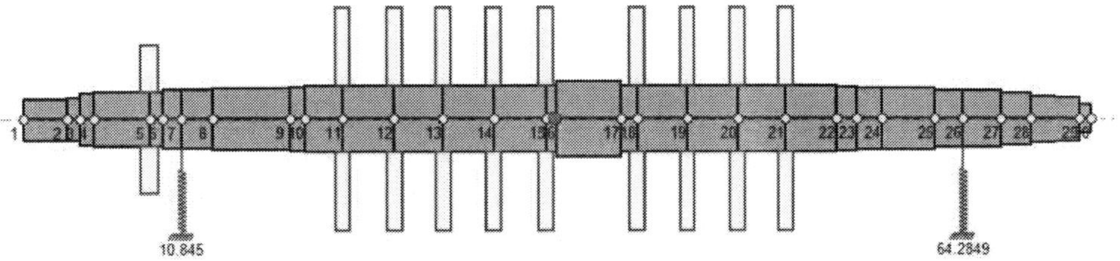

Figure 27 Reduced Bearing Span Shaft Model

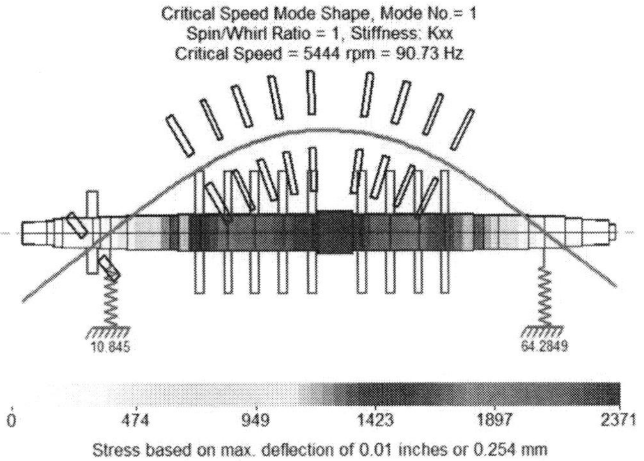

Figure 28 Shaft Modal Stresses of the First Bending Mode

A stability analysis of the original bearing design with the new shaft design (Figure 29) demonstrates how the new rotor accommodates the stiffer bearing design, resulting in a stable first bending mode. Note that the ratio of twice the bearing stiffness to the shaft modal stiffness in this case is 1.7, a significant improvement from the original design.

Since the long squeeze film damper was also considered along with the new rotor design, this case was analyzed for stability as well (Figure 30). Similar to before, however, the addition of the damper without a central groove proves to have little effect on the stability of the system.

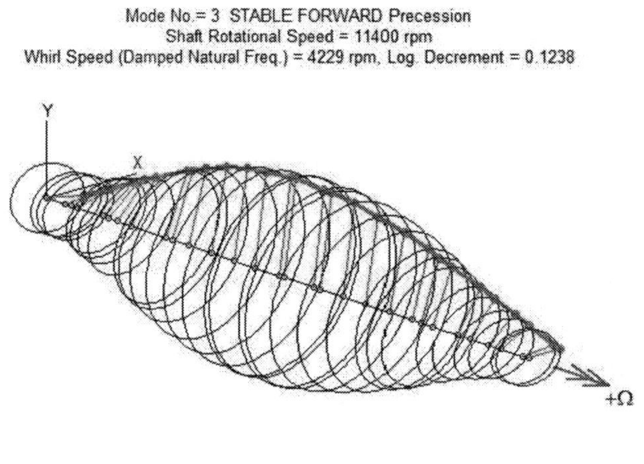

Figure 29 Stability Analysis of the New Rotor Design with the Original Five Pad LOP Bearings

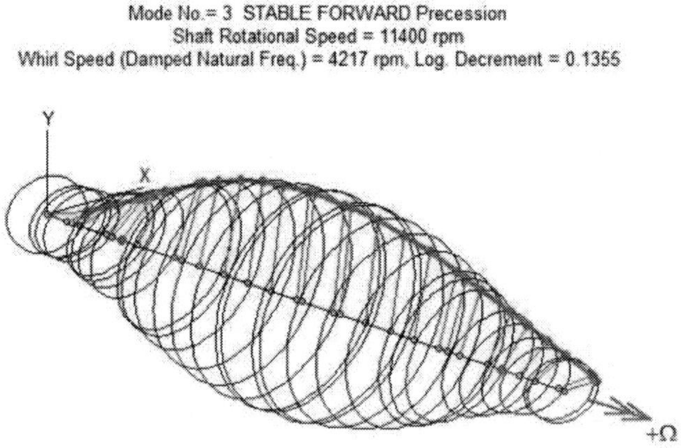

Figure 30 Stability Analysis of the New Rotor Design With the Original Bearing and a Long Squeeze Film Damper

Conclusions

The Kaybob compressor failure of 1971 was an historic case of rotordynamic instability and has taught us many things about the design of these machines. Tilting pad journal bearings have become a new normal in compressor stabilization due to their innate capacity for avoiding self-excited oil whirl, however great care must be taken in the design of these bearings for both ensuring a stable machine and meeting API specifications. Many have been tempted to try and avoid critical speeds and reduce motion at the bearings by increasing bearing stiffness, however this approach can and has often created a stability disaster as demonstrated by the Kaybob case. By designing for stability, however, these modes can also be eliminated from the operating range to the extent that they are not even detected because they are so well-damped.

In particular the Kaybob case has revealed the importance of designing bearings such that the bearing stiffness is tuned in to be in proper proportion to the shaft modal stiffness. By using this simple relationship and designing for proper damping as well, the bearings can properly absorb shaft vibrational energy and produce highly stable modes. While bearing asymmetry has been shown in past cases to produce some increases in stability - particularly in the presence of internal shaft friction or moderate levels of aerodynamic cross-coupling - the bearing stiffness must still be considered as a primary design variable as asymmetry will not benefit a system with a large bearing to shaft modal stiffness ratio. The presented design cases for 5- and 4-pad bearings demonstrate the importance of these concepts.

Squeeze-film dampers have also been a significant source of compressor stabilization in decades past and will continue to do so when properly designed. It was shown in the context of the Kaybob failure that important features such as central grooves can make a large difference in the stiffness and damping characteristics of the damper, which can ultimately have as large an effect on compressor stability as the design of the bearing itself. While the redesign of the compressor rotor also produced a stable machine, this solution was far more costly than the simple use of a properly designed bearing or squeeze-film damper. However, the team at Cooper-Bessemer cannot be faulted for these errors as they did not have the luxury of high-speed analysis tools that are available to designers today. Overall, the failure of the Kaybob compressor and their work to investigate the instability has helped create the knowledge and methods necessary to avoid these kinds of failures in the future.

Acknowledgments

This work was supported by the Rotating Machinery and Controls Laboratory and industrial consortium at the University of Virginia and Rodyn Vibration Analysis, Inc.

References

1. Alford, J., 1965", *"Protecting Turbomachinery From Self-Excited Rotor Whirl,"* Journal of Engineering for Power, pp. 333–334.

2. Allaire, P. E, Nicholas, J. C., and Gunter, E. J., 1977, *"Systems of the Finite Elements for Journal Bearings"*, ASME Journal of Lubrication Technology, pp. 187 – 197

3. Barrett, L. E., Gunter, E. J., and Allaire, P. E. 1978, *"Optimum Bearing and Support Damping for Unbalance Response and Stability in Rotating Machinery,"* Journal of Engineering for Power, Vol. 100, No. 1, pp. 89-94.

4. Chen, W. J. and Gunter, E. J. 2007, *"Introduction to Dynamics of Rotor-Bearing Systems,"* Trafford Publishing, ISBN: 978-1-4120-5190-3, pp. 1-469.

5. Gunter, E. J. 1966, *"Dynamic Stability of Rotor-Bearing Systems,"* NASA SP-113.

6. Gunter, E. J. and Trumpler, P. R. 1969, *"The Influence of Internal Friction on the Stability of High Speed Rotors with Anisotropic Supports,"* Journal of Engineering for Industry, Vol. 91, No. 4, pp. 1105-1113.

7. Gunter, E. J., Barrett, L. E. And P. E. Allaire, 1977, *" Design of Nonlinear Squeeze–Film Dampers for Aircraft Engines"*, Journal of lubrication technology, pp. 57 – 66.

8. Gunter, E. J. 2012, *"The Influence of Bearing Asymmetry on Rotor Stability,"* Proceedings of the Vibration Institute Training Conference, June 19-22, Williamsburg, Virginia, pp. 1-28.

9. Lund, J. W. 1964, *"Spring and Damping Coefficients for the Tilting Pad Journal Bearing,"* ASLE Transactions, Vol. 7, No. 4, pp. 342-352.

10. Lund, J. W. 1974, *"Stability and Damped Critical Speeds of a Flexible Rotor in Fluid Film Bearings,"* Journal of Engineering for Industry, Vol. 96, No. 2, pp. 509-517.

11. Nicholas, J. C. and Kirk, R. G. 1979, *"Selection and Design of Tilting Pad and Fixed Lobe Journal Bearings for Optimum Turborotor Dynamics,"* Proceedings of the Eighth Turbomachinery Symposium, Texas A&M Univ., College Station, Texas, pp. 43-57.

12. Nicholas, J. C. and Kirk, R. G. 1982, *"Four Pad Tilting Pad Bearing Design and Application for Multistage Axial Compressors,"* Journal of Lubrication Technology, Vol. 104, No. 4, pp. 523-539.

13. Smith, K. J. 1975, *"An Operation History of Fractional Frequency Whirl,"* Proceedings of the Fourth Turbomachinery Symposium, Texas A&M Univ., pp. 115-125.

14. Weaver, B. K., Zhang, Y., Clarens, A. F., and Untaroiu, A. 2015, *"Nonlinear Analysis of Rub Impact in a Three-Disk Rotor and Correction via Bearing and Lubricant Adjustment,"* Journal of Engineering for Gas Turbines and Power, Vol. 137, No. 9, pp. 092504-1-092504-8.

ROTORDYNAMICS OF PHASE LAG: ANALOGY TO CHILD'S PLAY

Michael T. Schunk

Niagara Frontier Chapter of the Vibration Institute
Chairman of the Board

Praxair, Inc.
4501 Royal Ave.
Niagara Falls, NY 14303
michael_schunk@praxair.com

Abstract: During introductory courses in balancing, the concept of "phase lag" is typically given a cursory mention without much discussion of what it is or why it happens. In this paper, the concept of phase lag in the context of rotordynamics is explored. In order to make the subject more approachable, the action of the input (rotor's unbalance) and the output (vibration of the rotor) is likened to a simple child's toy. Through this analogy, the procedure of balancing a rotor is better understood. The paper concludes with an appendix of 5 cases with differing operating speeds relative to the system's critical speed. A visual representation of the characteristics of the rotor, including its operation relative to the system's critical speed is given. This is communicated via Bode plots, time domain visualizations of the rotor, and the analogous operation of the said child's toy.

Key Words: Balancing; Damping; Heavy spot; High spot, Lag; Phase; Rotordynamic; Stiffness

During my Vibration Institute course "Balancing of Rotating Machinery" the term *phase* was used to describe a reference point from which to balance a rotor, via weight addition or subtraction. I knew that phase was some quantity measured with a trigger probe (commonly known by the trade name *Keyphasor®*), however I did not have a clear understanding of what *phase* really meant and why it mattered in the effort of balancing. This paper is an attempt to shed some light on that subject, not just for balancing, but for rotordynamics in general.

Before we consider a rotor and examine five basic response cases, we will envision a more intuitive model: a simple child's toy composed of a weight attached to a handle-mounted spring of sufficient stiffness, as shown in **Figure 1** below. In order to facilitate the understanding of this analogy, I recommend watching the following video on YouTube.com either now, or as you read along in the paper: http://youtu.be/V3Xn-_nplbs.

Note: If you are reading a printed copy, it may be easier to go to www.youtube.com and search "Demonstration Video for Rotordynamics of Phase Lag."

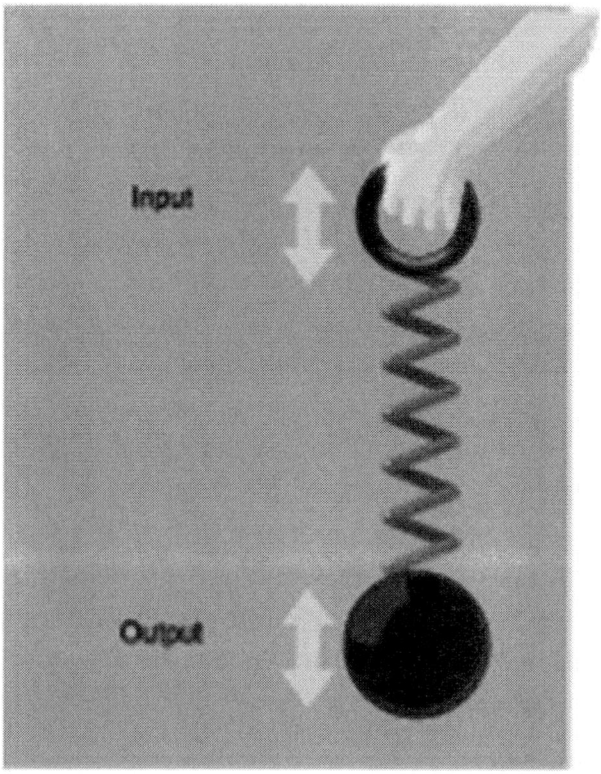

Figure 1: Child's Toy, A Simple Linear Dynamic System [1]

The model representing the said child's toy in a linear oscillatory dynamic system is given in **Figure 2**.

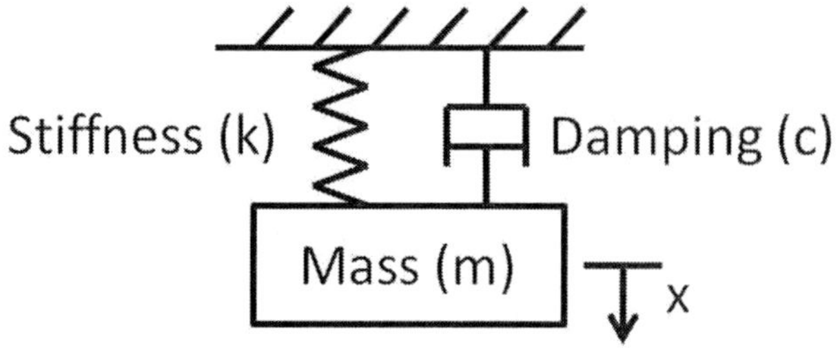

Figure 2: Linear Dynamic System Model, a traditional mass, damper, spring system.

The response of this simple linear dynamic system is related to the weight's *mass (m)*, *damping (c),* and *stiffness (k)*, which are defined in **Table 1** below:

Table I: Linear Dynamic System Characteristics

System Characteristic	Symbol	Definition	Effect on System (all effects are linear	SI Unit	US Unit
Mass	m	A measure of the amount of matter in a body	Resistance to acceleration	kg	lb_m
Damping	c	Absorption of energy	Resistance to velocity (reduces amplitude of oscillations)	Ns/m	lb_{rs}/in
Stiffness	k	Ratio of force required to create a specified deflection	Resistance to displacement (movement)	N/m	lb_r/in

Typically, as emphasized in **Figure 2**, the stiffness and damping elements are schematically shown to exist as separate components of a dynamic system. It should however be noted in this case, that the spring on the child's toy in **Figure 1** is a real spring, exhibiting both stiffness and damping in one component.

Referring to the said child's toy, a child controls the position of the weight by moving the handle up and down. Anyone who has ever played with such a toy knows that if the handle is moved in a more-or-less sinusoidal manner, the weight will start oscillating at the same rate; however, it will generally be out of synch with the handle. Only at relatively low-frequencies where the spring doesn't stretch will the handle and weight move in lock step.[1] As such, the system model shown in **Figure 3** can be constructed.

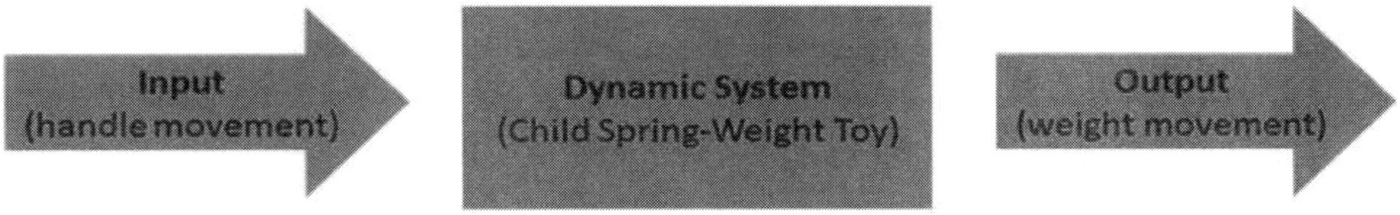

Figure 3: Child's Toy represented as a dynamic system block diagram.

Considering the general case when the motion of the handle and the motion of the weight are out of synch, the dynamic system, namely the toy, transfers the input (handle movement) into an output (movement of the weight) of different *amplitude* (magnitude of movement up and down). The *phase* (timing between the up and down movement of the handle versus that of the weight) has also been modified by the system. As this system can be considered linear (mainly true if the spring is a "simple spring" that follows Hooke's law), there is no change in *frequency* (oscillations per unit time) from the input to the output. This phenomenon of a linear system is illustrated by means of the child's toy example in **Figure 4** below.

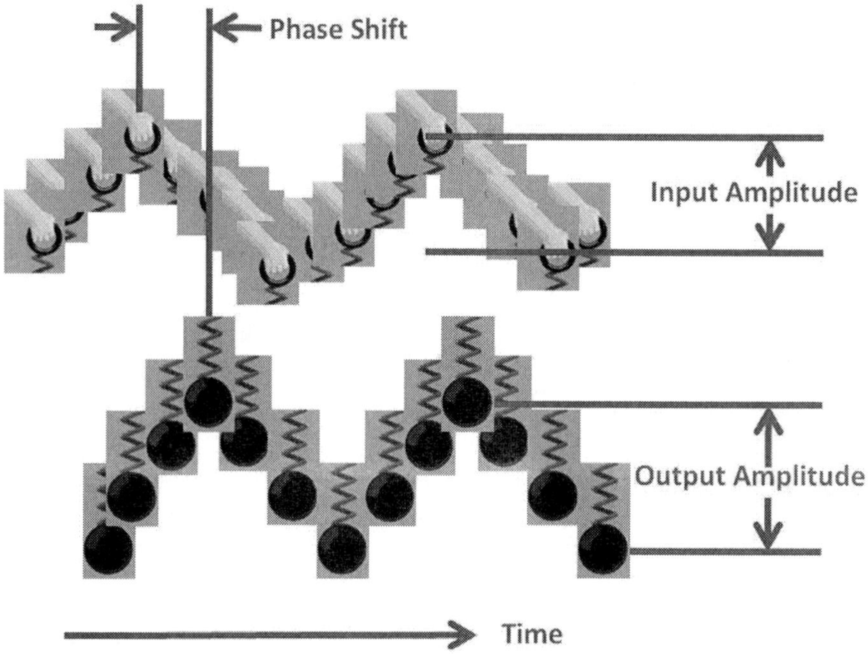

Figure 4: Input Amplitude, Output Amplitude, and Phase Shift as Child's Toy is Excited.

Moving the handle at higher and higher frequencies, the weight will begin to oscillate more than the handle, yet lag further and further behind it. At the *natural frequency* (rate at which an object vibrates when excited then allowed to vibrate freely) of the system, the weight's oscillations will reach their maximum heights. For our purposes we will use the undamped natural frequency, f_n which is a function of the mass of the weight and the stiffness of the spring, per **Figure 5** and **Eq. (1)**.

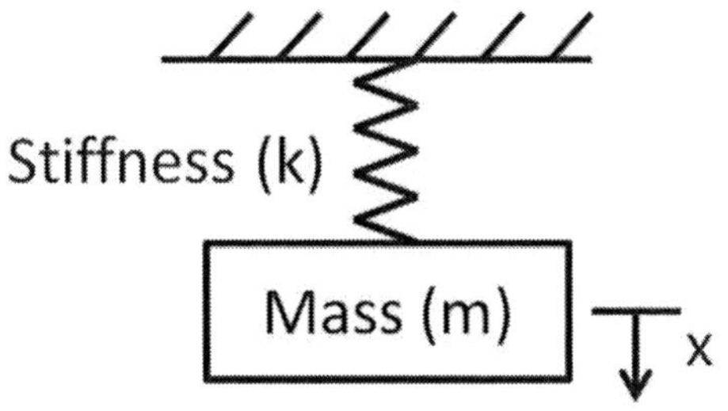

Figure 5 Undamped Spring Mass System

$$f_n = \frac{1}{2\pi}\sqrt{\frac{k}{m}}$$

Equation 1 Undamped Natural Frequency of a System

With increasing frequencies above the natural frequency, the amplitude of the weight's oscillations will decrease and its corresponding phase will grow more negative (that is, oscillations will grow smaller and smaller and lag further behind). At very high frequencies the amplitudes of the weight will be small in relation with the handle, and will move in opposite directions [1]. Such behavior is commonly illustrated by the *Bode plot*, as shown in **Figure 6**.

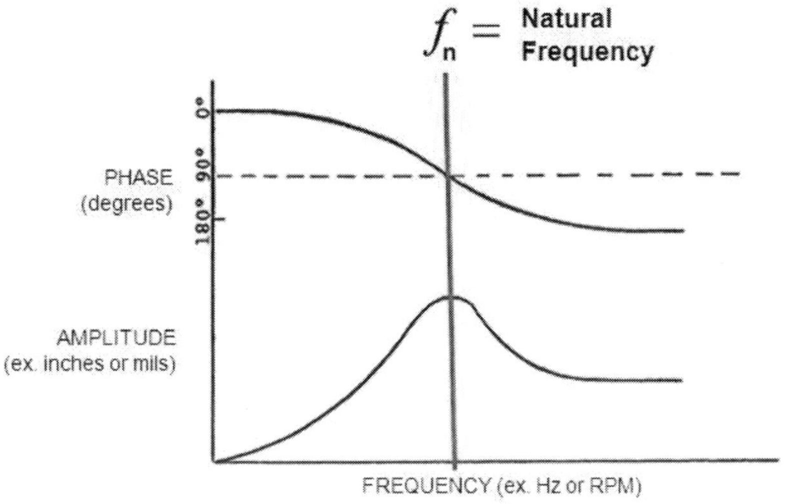

Figure 6: Bode Plot [2]

The Bode plot, sometimes called the "fingerprint" of a system, shows the characteristic of a dynamic system in terms of changes in phase (top plot) and amplitude (bottom plot) response as a function of frequency. The Bode plot in **Figure 6** shows the case of a single natural frequency, as can be seen by the peak in amplitude and 180° of phase shift, with the value of the natural frequency occurring at the peak amplitude coinciding with 90° of phase shift. As mentioned previously regarding linear systems, while the amplitude and phase characteristics do change from input to output, the frequency does not.

Now let's apply this to the rotordynamic system, shown in **Figure 7** below.

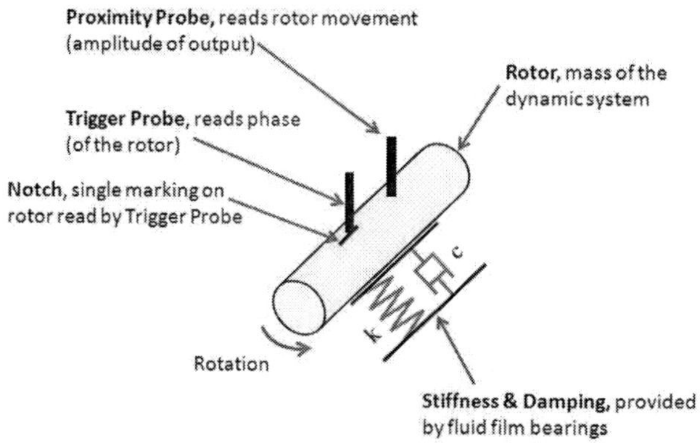

Figure 7: Three-Dimensional view of rotor model, illustrating vibration probe and Trigger Probe positions along rotor.

No rotor is perfectly balanced about its center. As such, the maldistribution of weight can be characterized by an imaginary *heavy spot* on the rotor, per **Figure 8** below. When this heavy spot rotates around the center axis of the rotor, an unbalance force (input into the system) is created. This input force is effectively the handle movement on the child's toy example. Also present are the stiffness and damping characteristics provided by the oil film surrounding the rotor in the areas of its bearings, analogous to the spring on the child's toy. It is worth emphasizing that this rotor model's bearings are fluid film, rather than rolling element (ex. ball bearings).

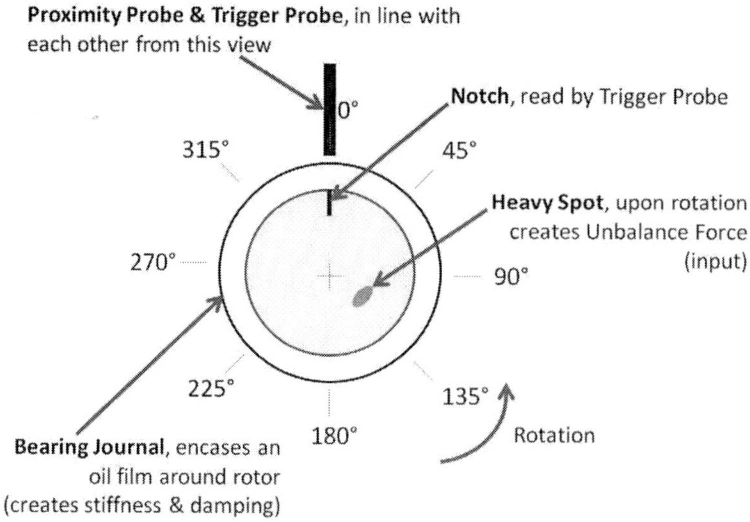

Figure 8: Rotor model viewed "end on"

As such, the diagram for the rotordynamic system is given as **Figure 9**.

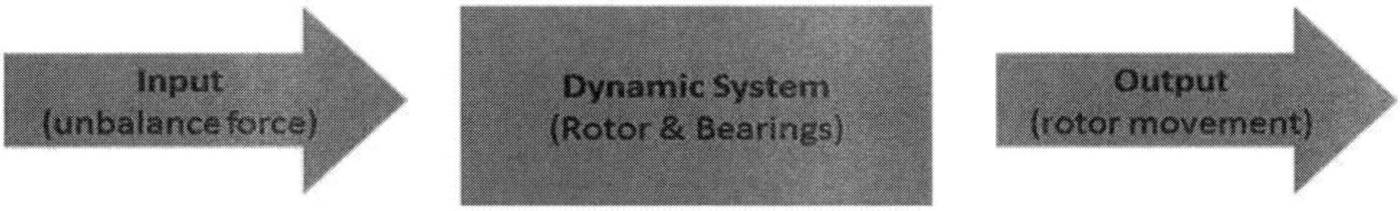

Figure 9: Rotordynamic System represented as a dynamic system block diagram.

The *proximity probe* indicates rotor movement by measuring the distance from the rotor to the tip of the probe. The point in rotation where the rotor is closest to the proximity probe is said to be the *high spot*. The high spot can be thought of as the peak displacement of the weight on the child's toy.

The *trigger probe* reads the position of a notch on the rotor. Demonstrated later, by using the trigger probe, the time lag between input and output can be converted into an angular lag known as *phase lag*, measured in degrees. Phase lag can be thought of as the delay in movement of the weight on the child's toy after the input movement on the handle is given.

In this paper, phase will always be read as a *lagging* amount, meaning that for example, in the counter-clockwise rotation shown, the heavy spot will pass the Trigger Probe 135° after the notch has passed it, per **Figure 8**. Said another way, the phase always reads opposite the direction of rotation. It is worth noting that when the notch is ground into the rotor, the position of the heavy spot is unknown. In this case 135° is used due to the ease of demonstration. The phase angle from the notch to the heavy spot typically will not be such a "nice" or "convenient" number. It may just as well be 17° or 332°.

Please refer to the Appendix of this paper where we will examine the behavior of this rotordynamic system in five cases of increasing speed, relative to its *critical speed*, which is a term used for the rotor speed at which the rotordynamic system is excited at its natural frequency.

While rotordynamics certainly is not child's play, it is hoped that this simple analogy can shed some light on its basic premise of an output lagging an input. Furthermore, it is hoped that this simple introduction will serve as a bridge for the reader to better understand balancing, as well as more advanced rotordynamics topics.

As a wealth of papers and books already exist on the topic of rotor balancing, the subject will not be rehashed here. The reader is encouraged to read the referenced

publications in the paper, do internet searches on the topic, as well as watch the video found at the following URL, where I put the previously outlined principles to work in performing a balance on a rotor kit:
http://www.youtube.com/watch?v=Ql8AKryBSDs&feature=plcp
Note: If you are reading a printed copy, it may be easier to go to www.youtube.com and search "Single Plane Balancing Exercise with Rotor Kit."

REFERENCE

1. VanDoren, V J (2005), "Frequency Domain Analysis Explained," Control Engineering, Vol. 52, No. 10, pp.48-50. Reproduced by permission of the author.
2. Eshleman, R L (2005), "Balancing of Rotating Machinery," Vibration Institute
3. Eshleman, R L (2007), "Rotordynamics and Balancing," Vibration Institute
4. Sabin, S (2000), "Understanding and Using Dynamic Stiffness – A Tutorial," Orbit Magazine, Second Quarter

APPENDIX
(For "Rotordynamics of Phase Lag: Analogy to Child's Play")

CASE 1: ROTOR WELL BELOW CRITICAL SPEED

Child's Toy Analogy: At a relatively low speed movement of the handle, where the spring does not stretch, the child's toy exhibits no dynamic effects. That is to say that the output of the motion of the weight will move in lock step with the input on the handle.

How it Relates to the Rotor: Likewise, when the rotor is running well below its critical speed, the heavy spot (represented by the green oval and vector on the rotor below) and the high spot (shown as an orange vector) will pass a given mark at the same time. Therefore the heavy and the high spots are said to be "in phase" with each other (0 degree phase lag). This is demonstrated in **Figure 10** below, where both the vibration probe reads vibration (the orange sine wave) and the Trigger Probe reads phase (the black line showing a downward spike).

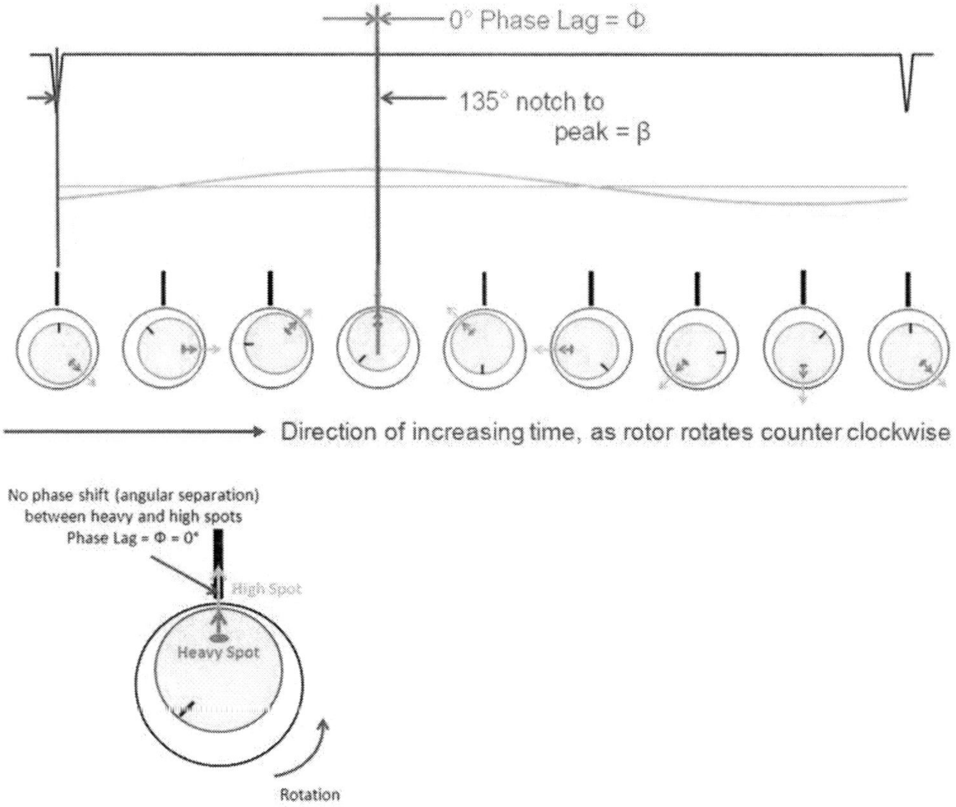

Figure 10: Rotordynamic system with 0° phase lag (no phase lag)

Do not confuse the *notch to peak* reading (β) with measurement of phase lag. The 135° notch to peak represents only the beginning point from which phase lag will be measured. In order to understand how the 0° phase lag is arrived at, the following equation necessary:

$\Phi = \delta - \beta$, where:

Φ = Phase Lag, the angle between the unbalance force vector (heavy spot) and the response movement (high spot). In other words, this is the phase lag between input and output of the rotordynamic system.

δ = Phase reading of the heavy spot. In this example, we know this to be 135° for all of the cases which will be presented. It should be noted that the heavy spot can only be measured indirectly by using the trial weight method, integral to the balancing process. This procedure is not covered here, but there is a wealth of material available on the subject.[2, 3]

β = Phase Measurement, from the notch to the positive peak of displacement

Equation 2: Phase lag between heavy spot and high spot.

Using **Equation 2**, the Phase Lag between the high and heavy spots is calculated as such:

$\Phi = \beta - \delta$

$\Phi = 135° - 135° =$ **0° Phase Lag**

The characteristic of this first case is represented on the Bode Plot by the vertical red line in **Figure 11**. Note that the top plot representing phase, shows no shift between high and heavy spots. The lower plot, showing amplitude gain, shows negligible amplification of movement.

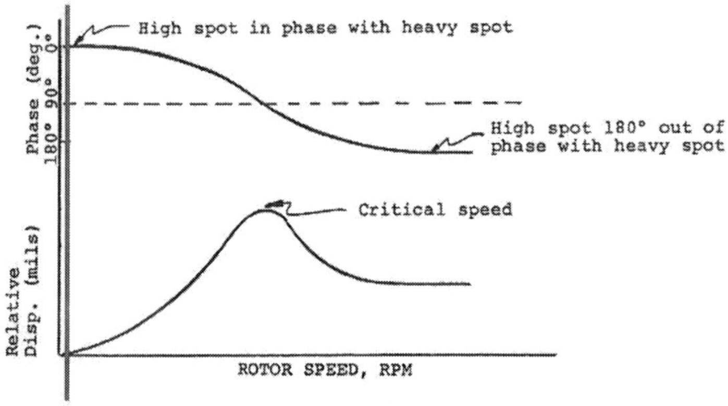

CASE #1

Figure 11: Bode Plot highlighting Case 1, Rotor Well Below Critical Speed.

CASE 2: ROTOR APPROACHING CRITICAL SPEED

Child's Toy Analogy: Once again imagining the child's toy, as the handle is moved up and down with some speed, the spring begins to stretch and the dynamics in the system come into action.

How it Relates to the Rotor: In the same manner as the child's toy, as the speed of the rotor becomes significant relative to its critical speed, there starts to be a phase lag between the heavy spot (input) and the high spot (output). A simple way of thinking of this is that the high spot is having trouble keeping up with the heavy spot. Note also that the amplitude of the displacement sine wave has become greater.

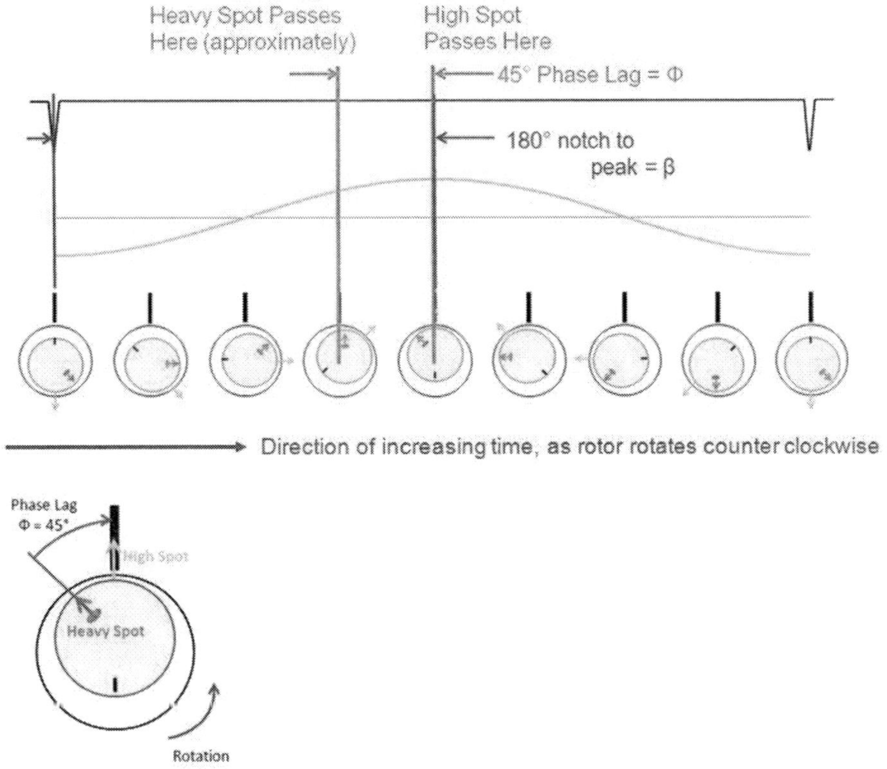

Figure 12: Rotordynamic system with 45° phase lag

$\Phi = \beta - \delta$

$\Phi = 180° - 135° =$ **45° Phase Lag**

The characteristic of this second case is represented on the Bode Plot by the vertical red line in **Figure 13.** Note that the top plot representing phase, shows about a 45° shift between high and heavy spots. The lower plot, showing amplitude, shows significant amplification of movement, hence the growth in amplitude of the displacement sine wave.

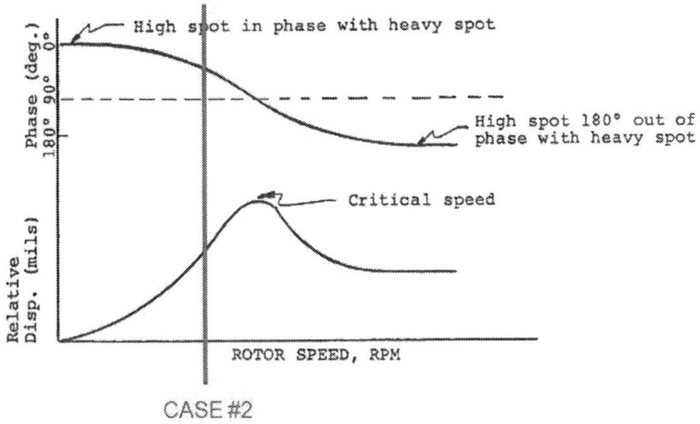

Figure 13: Bode Plot highlighting Case 2, Rotor Approaching Critical Speed.

CASE 3: ROTOR AT CRITICAL SPEED

Child's Toy Analogy: As the motion on the handle of the toy speeds up, a point will be reached where the amplitude of the movement of the weight will be at a maximum and simultaneously the handle will be 90° out of phase. That is to say, during the instant of no movement of the handle when switching from an up to a down input, the speed of the weight will be at a maximum. Conversely in the moment where the weight is still because it is changing directions, the input on the handle will have maximum speed. The toy has reached its natural frequency.

How it Relates to the Rotor: As the rotordynamic system reaches its critical speed, when the amplitude of vibration is at a maximum, the exciting force (input) is acting in a direction perpendicular to the displacement. Without getting deep into rotordynamic theory, it is said that at the critical speed, displacement of the rotor is limited only by damping, hence the maximum vibration, compared to the other cases where in addition to damping, stiffness and mass also play a role.

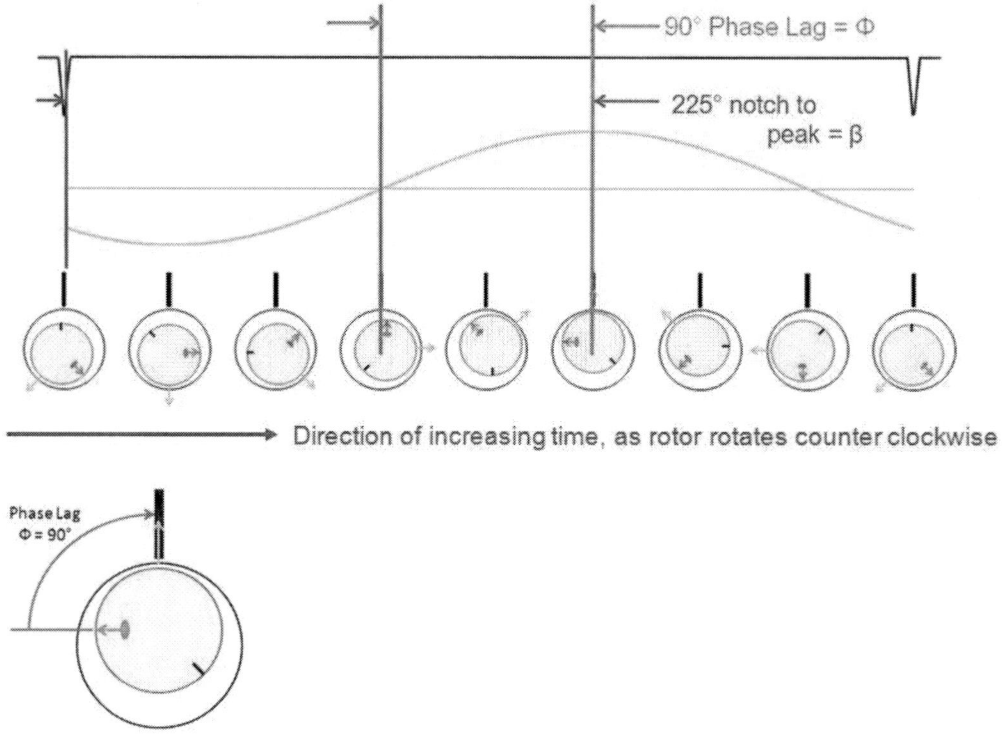

Figure 14: Rotordynamic system with 90° phase lag

$\Phi = \beta - \delta$

$\Phi = 225° - 135°$ = **90° Phase Lag**

The characteristic of this third case is represented on the Bode Plot by the vertical red line in **Figure 15**. Note that the top plot representing phase, shows a 90° shift between high and heavy spots. The lower plot, showing amplitude gain, shows peak amplification of movement, thus the large amplitude of the displacement sine wave. These characteristics of the Bode plot indicate a dynamic system operating at its natural frequency, or in rotordynamics terms, its critical speed.

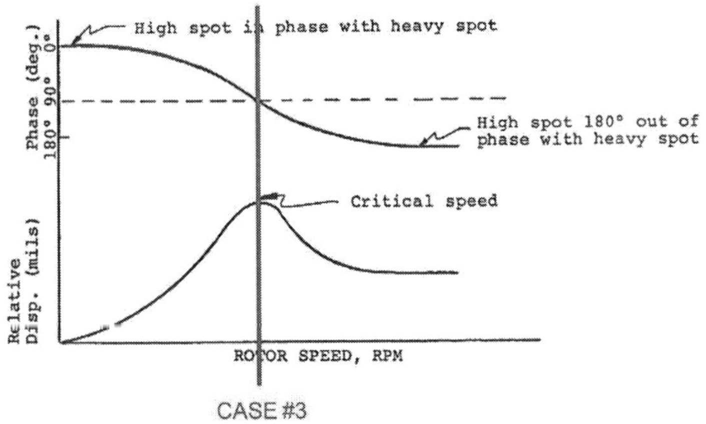

Figure 15: Bode Plot highlighting Case 3, Rotor at Critical Speed.

CASE 4: ROTOR GETTING AWAY FROM CRITICAL SPEED

Child's Toy Analogy: As the handle on the toy is moved more vigorously, the motion of the weight will begin to go in the opposite direction of the handle. The amplitude of the motion of the weight will also decrease.

How it Relates to the Rotor: As the speed of the rotor goes above the critical speed, its vibration will calm down as the displacement of the rotor will start pointing away from the unbalance force. This can be observed as a reduction of amplitude of the displacement sine wave. Meanwhile, the phase lag has once again grown.

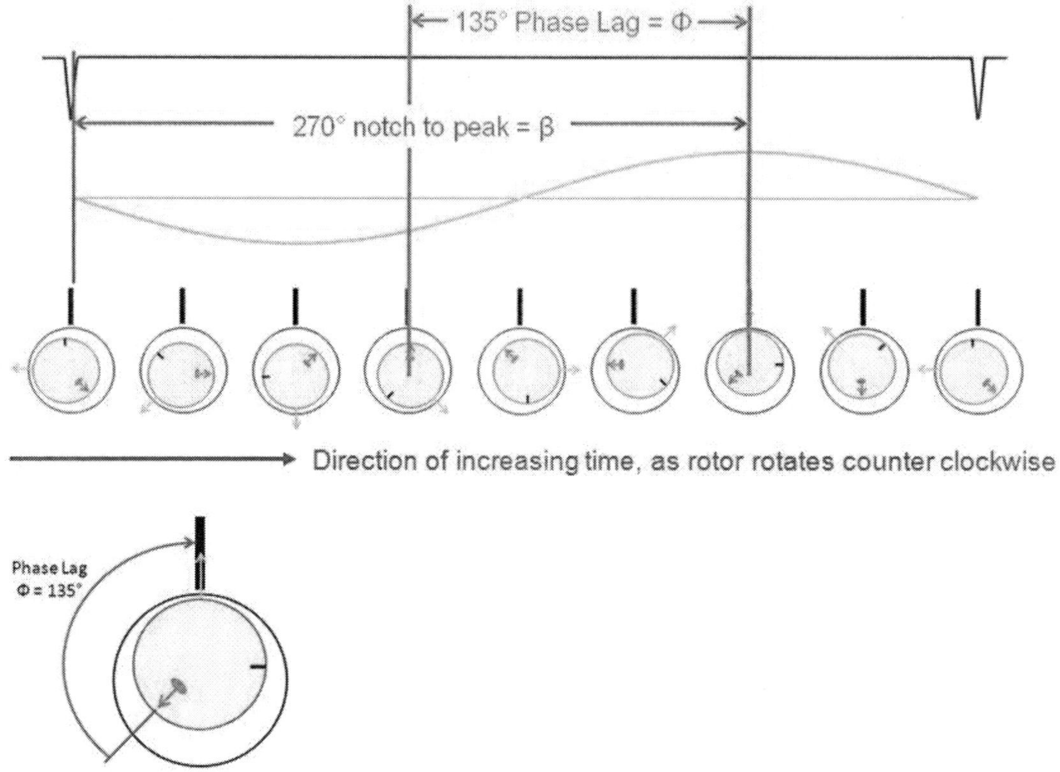

Figure 16: Rotordynamic system with 135° phase lag

$\Phi = \beta - \delta$

$\Phi = 270° - 135°$ = **135° Phase Lag**

The characteristic of this fourth case is represented on the Bode Plot by the vertical red line in **Figure 17**. Note that the top plot representing phase, shows an approximate 135° shift between high and heavy spots. The lower plot, showing amplitude gain, shows the amplification of movement tapering down. This explains the tapering down of the amplitude of the displacement sine wave.

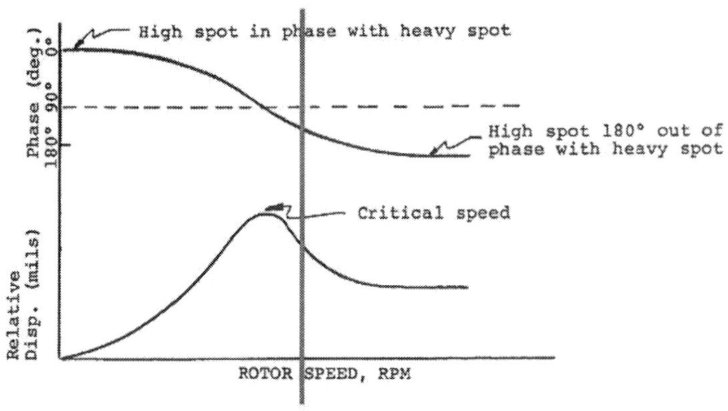

Figure 17: Bode Plot highlighting Case 4, Rotor Getting Away From Critical Speed.

CASE 5: ROTOR WELL PAST CRITICAL SPEED

Child's Toy Analogy: As the handle of the toy is moved as fast as possible, it seems that its motion is barely transferred to the weight. If examined close enough, it is observed that the weight moves in the opposite direction from the input of the handle.

How it Relates to the Rotor: With the rotor well past critical speed, the motion of the rotor is 180° out of phase with the unbalance force. As is the case of the child's toy, the vibration of the rotor has calmed down significantly.

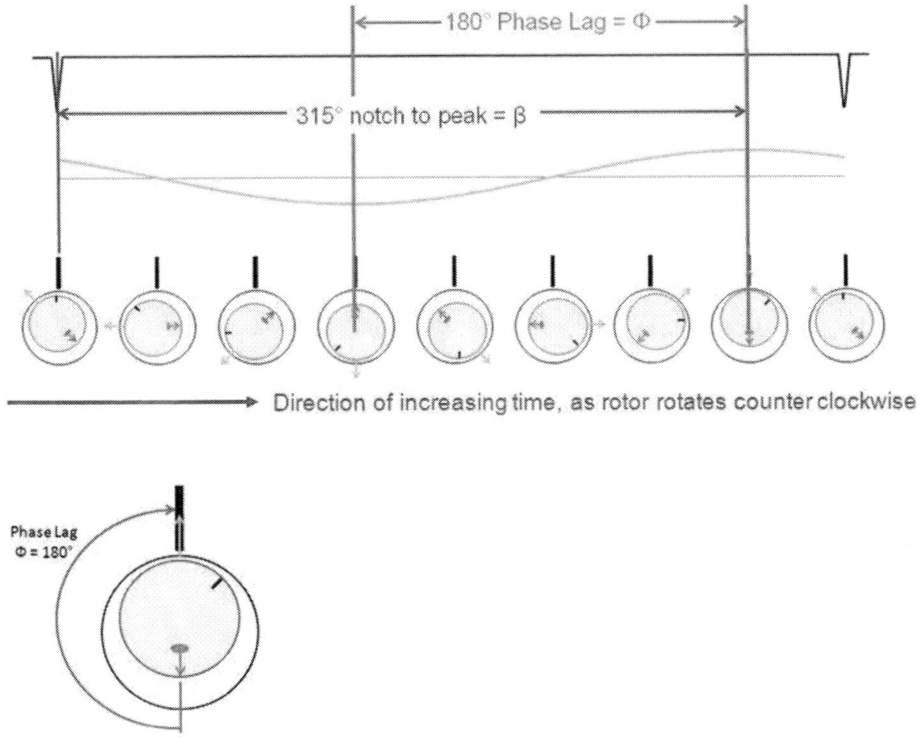

Figure 18: Rotordynamic system with 180° phase lag

$\Phi = \beta - \delta$

$\Phi = 315° - 135° =$ **180° Phase Lag**

The characteristic of this fifth case is represented on the Bode Plot by the vertical red line in **Figure 19**. Note that the top plot representing phase, shows an approximate 180° shift between high and heavy spots. The lower plot, showing amplitude gain, shows the amplification of movement significantly down from its peak at critical speed.

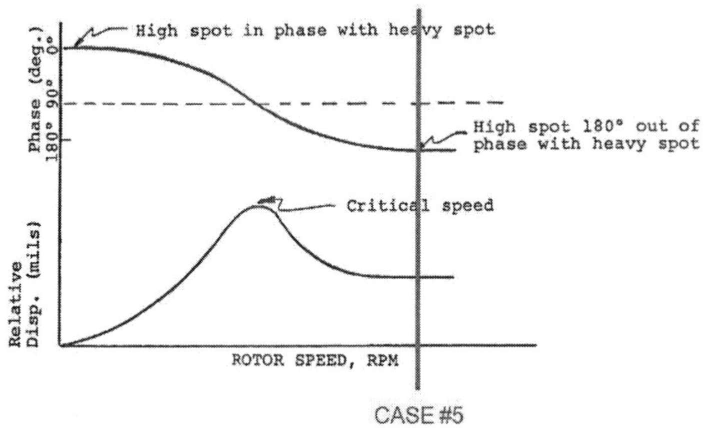

Figure 19: Bode Plot highlighting Case 5, Rotor Well Past Critical Speed.

Biography:

Mike Schunk holds a Bachelor of Science in Mechanical Engineering from the Rochester Institute of Technology and a Masters of Business Administration (MBA) from the State University of New York at Buffalo. He also holds the certifications of Category III Vibration Analyst and Category I Balancing Technician. He has 14 years of experience in working with rotating equipment, performing duties ranging from design, to product testing, to field service troubleshooting with Cameron Compression's Centrifugal Compressor Division (now owned by Ingersoll Rand) and most recently as a Reliability Engineer for Praxair's air separation, liquid hydrogen, and liquid CO_2 market segments.

Mike particularly enjoys developing intuitive visual, and whenever possible "hands on," explanations for machinery phenomena, as opposed to rigorous mathematical ones. He has found such approaches beneficial to himself as well as others during training and problem solving. Through the combination of his work and studies, Mike continuously strives to build "The Business Case for Reliability."

PREVENTING LUBRICATION RELATED BEARING FAILURES WITH ULTRASOUND

Adrian Messer

UE Systems, Inc.
14 Hayes Street
Elmsford, NY 10523
Adrianm@uesystems.com

Abstract

Airborne and Structure-borne ultrasound has evolved into a technology that many would consider to be the cornerstone of an effective condition based maintenance program (Figure 1). From compressed air & gas leak detection, electrical inspection, and various mechanical inspections, ultrasound is truly a versatile tool for anyone's CBM "toolbox." One application in particular that has seen a large increase in usage, is to use ultrasound for condition based lubrication of bearings.

Studies have shown that the most common causes of premature bearing failures are related to lubrication. From over lubrication, under lubrication, using the wrong grease for the wrong application, mixing incompatible greases, to lubricant contamination, they are all related to lubrication. Ultrasound assisted lubrication helps to take the guesswork out of how much lubricant to apply. Additionally, typical lubrication programs are based on timed intervals. With time based lubrication, the tendency is to over lubricate. Lubricant is applied at timed intervals whether the equipment needs to be greased or not. With ultrasound assisted lubrication, the lubrication program moves more towards condition based lubrication rather than time based.

By sensing subtle changes in high frequency sound, ultrasonic instruments can be used to listen to bearings while applying grease. Bearings that are in need of lubrication will have an increase in high frequency sound due to an increase in friction from the lack of lubrication condition. The inspector will notice a decrease in the sound (decibel) while greasing a bearing that is in need of lubrication. The inspector will be able to hear the sound of the bearing while greasing via the headset. A visual readout with a decibel, and in some cases, onboard spectrum analysis can also assist the lubricator while greasing.

By preventing over and under lubrication of bearings, one can potentially reduce a large number of premature bearing failures. The benefits of less rotating equipment failures include less unscheduled downtime, reduced maintenance costs, and reduced motor repair costs, in addition to potentially less lubricant used, and more effective PM's by technicians.

Keywords: Bearings, Bearing lubrication, Condition based lubrication, Ultrasound, Ultrasound assisted lubrication

Keeping a handle on lubrication is easy, right? All one needs to do is make sure the right lubricant is used in the right amount and at the right time. Not so fast; if only it were that simple.

Most bearing failures are lubrication-related. Bearing failures most often lead to unplanned downtime, which can impact production, as well as affect all related components around the bearing. Downtime is costly. While the cost varies by incident and by plant, it can add up. Since the most common cause of bearing failure is lubrication-related, it's clear to see that lubrication is serious business. And for the longest time, that serious business has been conducted in a way that on its face makes perfect sense but in fact borders on haphazardness.

Many technicians, unfortunately, have relied on preventive, time-based lubrication alone, that is, every certain number of months, the grease gun comes out, and the bearings are lubricated. After all, under-lubrication can be lethal, causing equipment failure, costly repairs and replacements, significant unplanned downtime, and lost profits. But by relying solely on time-based lubrication, or even a combination of planned maintenance and temperature readings to serve as a proxy for lubrication status, one runs the risk of something just as bad, if not worse: over-lubrication. In fact over-lubrication can cause premature bearing failure. Relying on time-based, periodic lubrication assumes bearings need to be greased at defined time periods. Often, this evolves into a well-intentioned guessing game at best. Adding more lubrication to a bearing that is already adequately greased is a real risk. By using ultrasound technology, along with standard practices such as removing old grease and replacing it with new, technicians can combine standard, time-based maintenance with condition-based, predictive maintenance, gaining in the process both a clearer picture of what's really going on and better reliability

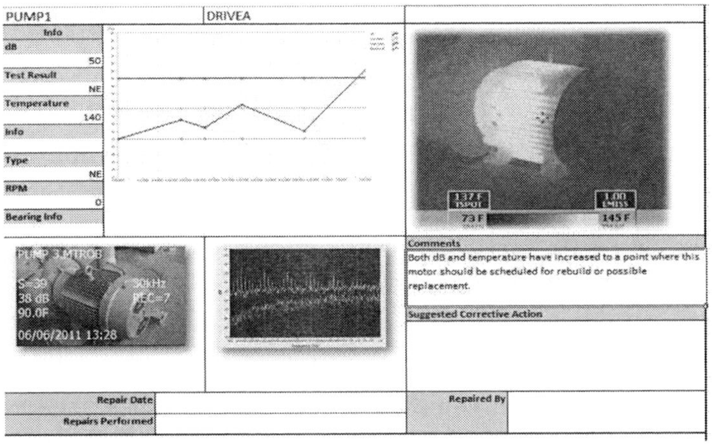

(Figure 1) Advancements in ultrasound instrumentation have led to a greater trending & reporting capability. Reports such as Lube Route Report, Alarm Report, and a point specific 4 Image Report (above) can lead to greater decision making regarding the health of an asset, as well as combining other data such as infrared or vibration analysis.

Ultrasound and Lubrication

Ultrasonic equipment detects airborne and structure-borne ultrasounds normally inaudible to the human ear and electronically transposes them into audible signals that a technician can hear through headphones and view on a display panel as decibel (dB) levels. In some instruments, the received sound can also be viewed on a spectral analysis screen. With this information, a trained technician can interpret the bearing condition in order to determine what, if any, corrective action is needed.

Ultrasonic technology (Figure 2) helps the lubrication technician take a lot of the guesswork out of lubrication needs. Ultrasound is a localized signal, meaning when a sensing probe is applied to a bearing it will not be affected by crosstalk and allows the technician to hear and monitor the condition of each individual bearing. Ultrasound looks at each individually, much the same way medical ultrasound can detect exactly which artery is clogged or which vein is leaking.

As an example of ultrasound's efficacy, consider this. A maintenance manager at a large firm reports that, since adopting ultrasound technology and practicing ultrasound-assisted, condition-based monitoring, rather than running to put out the fire, his plant has gone from close to 30 rotating equipment failures per year to zero in three years. But how does ultrasound work, exactly, vis-à-vis lubrication?

The first step is establishing both a baseline dB level and a sound sample. This is ideally done when moving through a route for the first time by first comparing dB levels and sound qualities of similar bearings. Anomalies will be easily identified. Once established, each bearing can be trended over time for any changes in either amplitude or sound quality. Generally speaking, when the amplitude of a bearing exceeds 8 dB and there's no difference in the sound quality established at baseline, the bearing needs to be lubricated. To prevent potentially disastrous over-lubrication, the technician will then apply lubrication, a little at a time, until the dB level drops. Many departments set up their condition-based lubrication programs by incorporating a two-stage approach. The reliability inspector uses a relatively sophisticated ultrasound instrument to monitor and trend bearings. A report of bearings in need of lubrication is produced. The lube technician then uses a specialized ultrasound instrument that alerts the technician when to stop adding grease. These instruments can be affixed to a regular grease gun or worn in a holster.

(Figure 2) An Ultrasound Assisted Lubrication instrument

To improve efficiencies, it's a good practice for the technician to note when the equipment was last greased and how much grease was used to calculate roughly how much lubrication is used per week. By using ultrasound to lubricate each and every time, the technician produces historical data that can be used as a guide from previous calculations, helping the department to determine whether the lubrication schedule can be modified, perhaps saving man-hours, and whether the manufacturer's suggested lubrication amount is accurate. If less is needed, there's cost-savings potential.

While most of this discussion has focused on the dangers of under and over lubrication (Figure 3 & Figure 4), ultrasound is just as reliable in picking up other potential bearing failure conditions. The technician, using ultrasound, can hear telltale grinding sounds and other anomalies, which are often accompanied by an amplitude increase. With regard to lubrication, the advantage of ultrasound is that it is able to isolate bearings and determine their individual needs, thus reducing the possibility that some bearings are too dry and prevent others from over-lubrication.

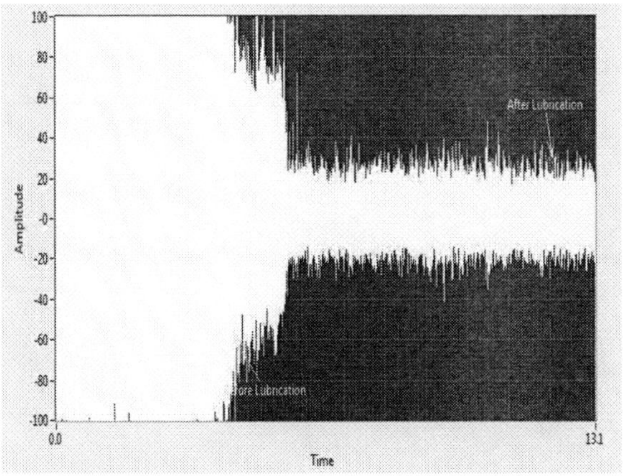

(Figure 3) The above example shows a bearing in the process of being lubricated. There is a very distinct before lube and after lube. Once the dB decreased to the baseline or normal level, lubrication stopped.

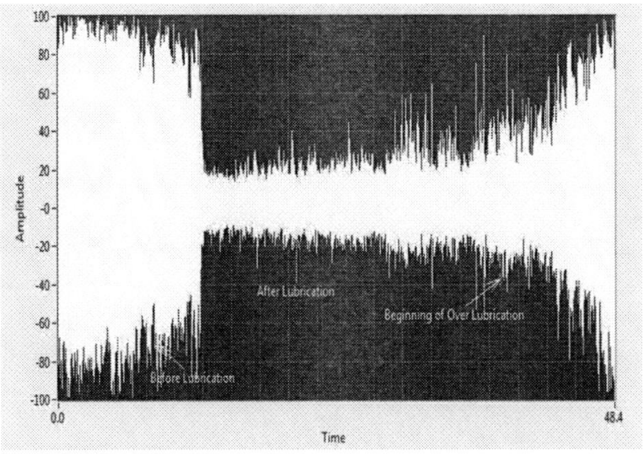

(Figure 4) The above example shows a bearing in the process of being lubricated to the point of over lubrication. Both dB & amplitude gradually increased as more lubricant was added.

Business Case

It is always a daunting proposition to make a new investment in technology. Will it pay off? Will my staff actually have an easy time using it? There are still many plants that are figuratively using crystal balls and outdated methodologies in their maintenance practices. The end result is poor reliability, unnecessary man-hours, downtime, and lost productivity and profit. While technology can't cure all reliability ills, it can offer a valuable and powerful diagnostic tool for technicians' toolkits. When it comes to something as important to reliability as lubrication, the question really becomes, "Can you afford not to use ultrasound technology?"

Biography:

Adrian Messer, CMRP, is the Manager of US Operations for UE Systems, Inc. For more than a decade, Adrian has been helping facilities around the country transform their reliability programs by implementing Ultrasound technology for condition monitoring and energy conservation applications. As a subject matter expert on Ultrasound technology and implementation best practices, Mr. Messer has been a featured speaker at numerous industry events. He is a graduate of Clemson University and maintains close ties to the University assisting current students in a mentorship program through the College of Business & Behavioral Science. Adrian is a Certified Maintenance & Reliability Professional (CMRP) through the Society of Maintenance & Reliability Professionals (SMRP), and he is also a charter member of the Carolina's Chapter of SMRP.

INTRODUCTION TO CONDITION MONITORING FOR RECIPROCATING COMPRESSORS

By: Mary Margaret Chapman
Regional Sales Manager at Windrock, Inc.
1832 Midpark Road, Suite 102
Knoxville, Tennessee 37921

Abstract: Vibration Spectrum Analysis is practiced ubiquitously to analyze rotating equipment in industrial facilities such as paper mills, steel mills, chemical plants and others. In conjunction with oil analysis and other technologies, predictive maintenance personnel can pinpoint machinery failures like unbalance, misalignment, bearing issues, etc. This diagnosis is completed by analyzing the vibration frequency spectrum of an event and its correlation to the machine and conditions. However, spectrum analysis is not the most effective tool for determining defects in reciprocating machinery like engines and compressors. While analyzing this type of machinery, it is better to utilize crank angle-based data, which does not utilize the frequency spectrum, but rather examines events relative to crankshaft positions. By combining pressure, vibration and ultrasonic readings taken based on crank angle, it is possible to determine defects in a reciprocating machine like cross head damage and cylinder leakage (valves, rings and packing). This paper will give a brief explanation of the components of a reciprocating compressor, explain the differences between spectrum analyses and crank angle-based analysis and review basic compressor analysis techniques.

Keywords: Compressor; Condition; Crank angle; Maintenance; Monitoring; Performance; Reciprocating; Vibration

The Overview of Present Day Montoring:

Predictive maintenance practices have been present in industry for more than thirty years. The most widely adopted technology is vibration monitoring using frequency-based techniques. Portable data collectors with pre-defined routes and online continuous monitoring are used to determine the health of machines. Determining the frequency of a vibration can lead the analyst to discovering faults such as bearing failures, unbalance, misalignment, etc. This is done by collecting vibration time waveform samples and converting them through the FFT process to frequency-based data (Figure 1).

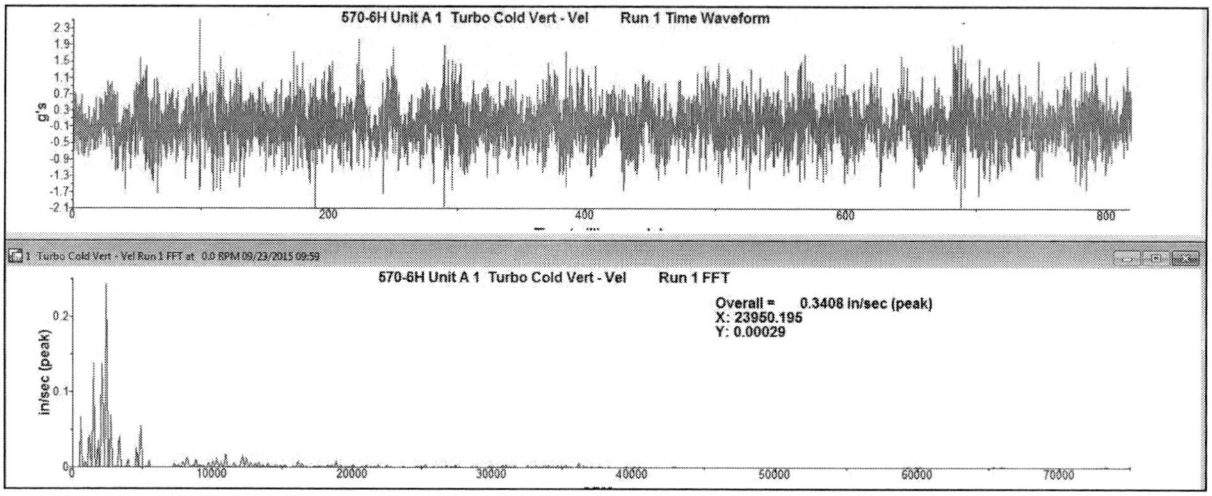

Figure I

An analyst can examine vibration levels and frequencies to determine the root cause of excessive vibration and use that data to predict potential failure modes. For a simple example, consider monitoring vibration on a motor-driven pump with six blades. If high vibration is noted at six times running speed, it leads the analyst to believe there is a vane pass issue. For another example consider a fan running at 900 RPM. Most of the energy should be visible at running speed, which in this example would be 900 CPM (or 15Hz). If the amplitude begins to rise, it could lead one to believe the fan is becoming imbalanced. These are only some of the mechanical faults on rotating machinery that can be detected through frequency-based vibration analysis.

This works well for a rotating piece of equipment but frequency analysis falls short when it comes to determining the condition of a reciprocating machine.

By incorporating reciprocating compressor analysis in a reliability program, the user achieves two benefits. Similar to a rotating equipment monitoring program, a user can determine mechanical issues, thus eliminating failures, allowing for planned maintenance and increasing uptime. The second benefit is measuring the performance of the reciprocating compressor by calculating performance data (e.g., gas through-put, horsepower consumed). First, let's review the components of the compressor before explaining the basics of analysis.

<u>Components of a Reciprocating Compressor:</u>

Compressors are typically engine or electric motor-driven with anywhere from one to ten cylinders, which are also known as throws (Figure 2; Figure 3).

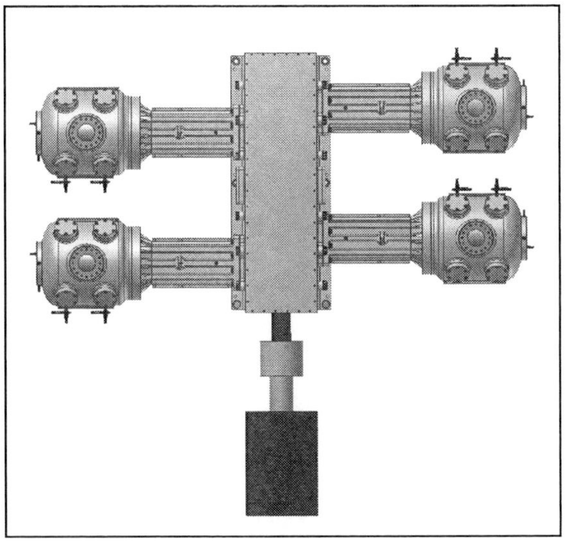

Figure 2

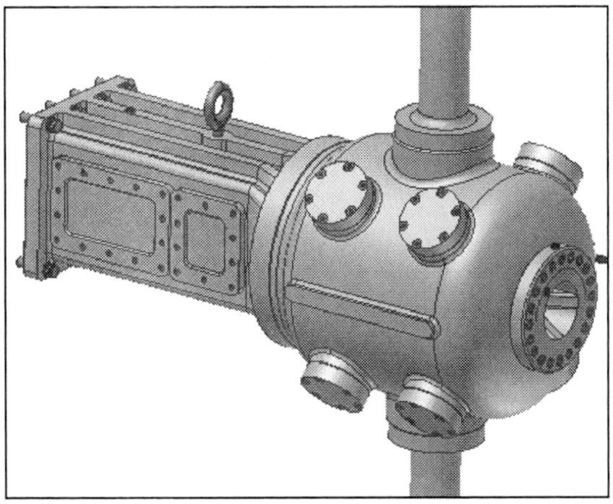

Figure 3

A compressor has a crank shaft to which the cylinders are attached. As the crankshaft turns, it moves the pistons back and forth within the cylinder, compressing and allowing the gas to expand. This is accomplished via a connecting rod attached to a cross head that slides as the crankshaft turns, changing the rotating movement of the shaft into reciprocating movement of the piston (Figure 4).

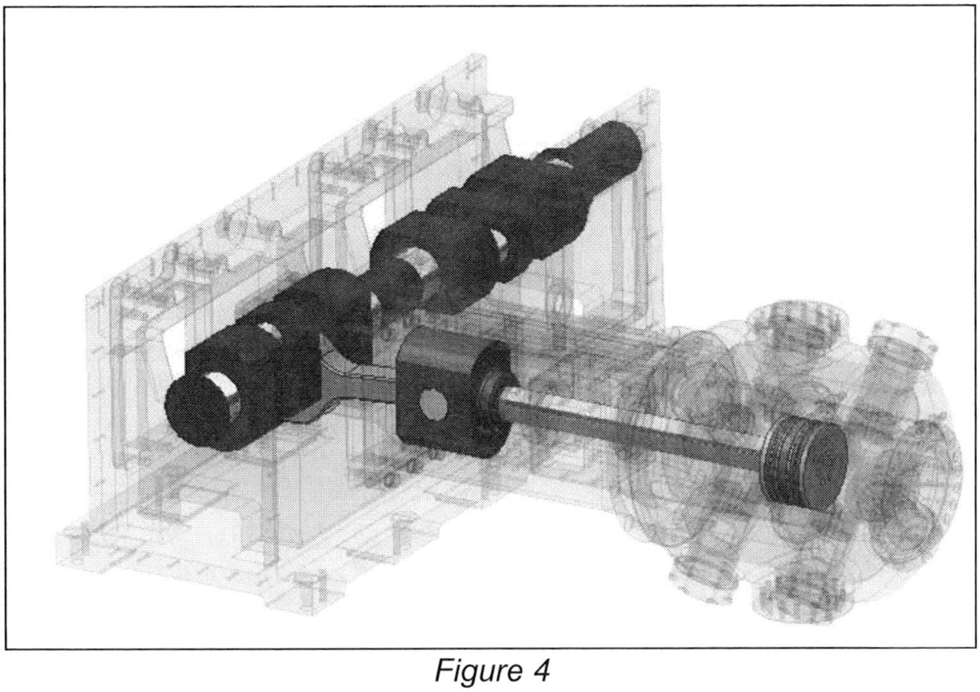

Figure 4

Gas moves through suction and discharge valves based on differential pressure. The piston is considered to be TDC (Top Dead Center) when it is as far away from the crankshaft as possible. When it is as close to the crankshaft as possible, the piston is considered to be BDC (Bottom Dead Center). As the piston is moving from TDC to

BDC, the gas in the head end is expanding. When the pressure in the cylinder gets lower than the inlet line pressure on the suction side, the suction valves will open and fill the cylinder with gas. As the piston is moving towards TDC, it will be compressing the gas in the head end of the cylinder. Once the pressure inside the cylinder becomes greater than the discharge line pressure the discharge valves will open, allowing gas to exit the cylinder (Figure 5). If the cylinder is double-acting, a similar cycle simultaneously occurs in the crank-end of the cylinder, 180 degrees out of phase with the head-end.

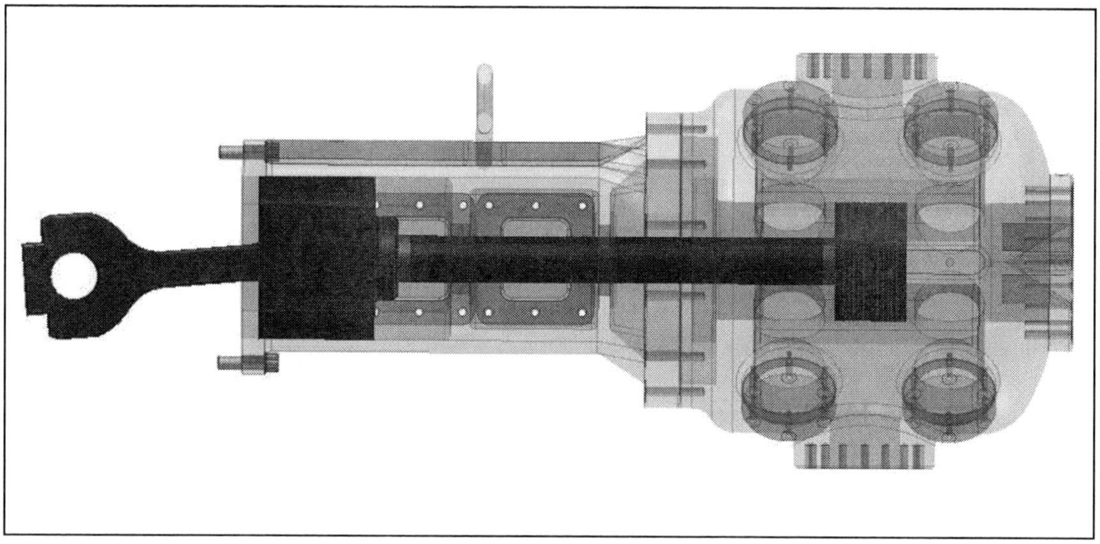

Figure 5

Limitations of Frequency-Based Analysis on Reciprocating Compressors:

Frequency-based analysis does not perform well for mechanical issues like crosshead problems or mechanical faults with the valves. It also does not provide any information about the performance of the compressor such as valve and packing leakage that affects throughput. Mechanical faults are typically more impact-related vs frequency-related which can be seen by taking waveform data. However, without knowledge of the mechanical events happening in the compressor, you cannot determine if the vibration is normal operation or a potential failure mode. Below is time waveform data taken on the crosshead of the compressor showing two definite impacts but, without additional information, the cause cannot be determined (Figure 6). In order to better understand the events in the compressor, vibration data is often examined relative to crankshaft location. Knowing the speed and geometry of the compressor allows the user to know the piston location within the cylinder. By comparing the vibration data to the known mechanical events, an analyst can determine issues and causes in a reciprocating compressor. For this reason, crank angle-based data collection is a better method to show the health of a reciprocating machine.

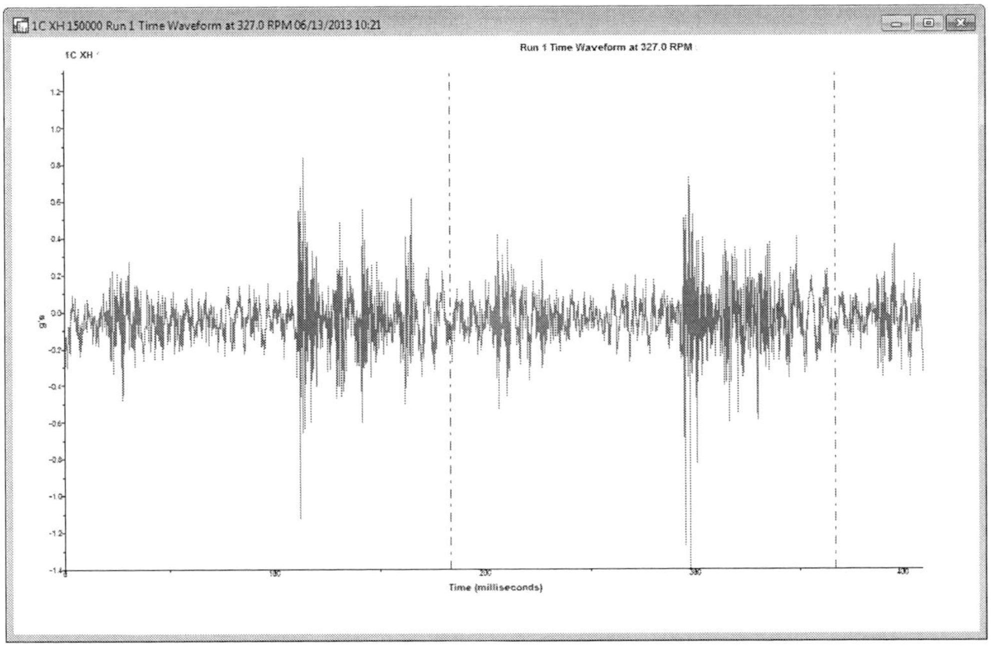

Figure 6

Where to Use Frequency-Based Techniques on Reciprocating Compressors:

There is a role for frequency-based monitoring on a reciprocating compressor. Analysts use FFT to look for movement in the frame or cylinder. Below is the Number 1 compressor throw on a Dresser-Rand compressor. Data taken in January shows higher than normal movement on the cylinder end of 6.76 mils (Figure 7). Personnel inspected the compressor and identified loose mounting bolts. After tightening, the overall vibration was reduced to 2.63 mils. Follow-up data taken in March shows a reduction in movement (Figure 8).

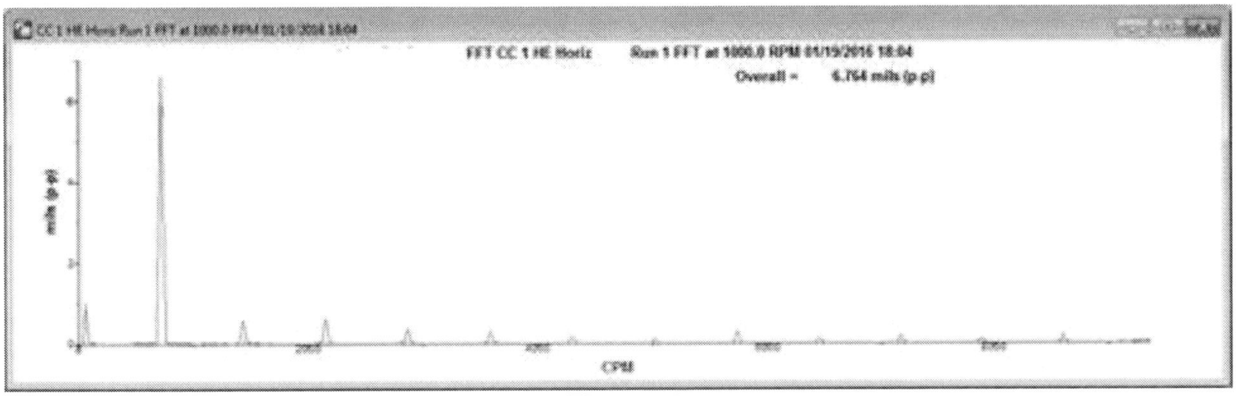

Figure 7

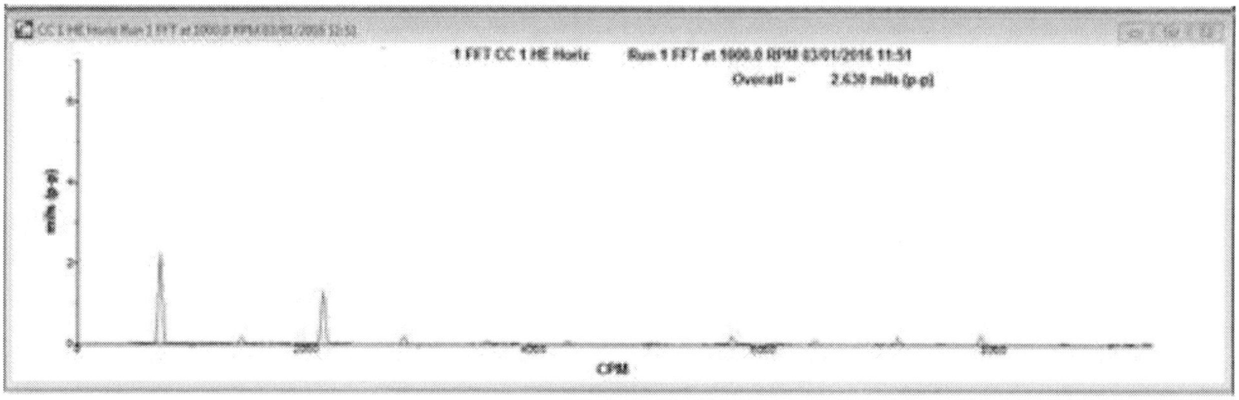

Figure 8

<u>Basics of Reciprocating Compressor Monitoring:</u>
The core variables of monitoring a reciprocating compressor are pressure in the head and crank ends, vibration, and ultrasonic data.

Once TDC is determined and marked on the compressor, the analyst wirelessly connects an encoder or magnetic pickup to the portable analyzer. With this crank-angle information, all data collected can be related to the position of the crankshaft and piston.

The analyst can then determine mechanical and performance issues with the compressors. The most common mechanical faults detected are bearing issues, crosshead problems, and valve operating faults. By knowing the geometry (bore size, piston size, stroke length, etc.), data collected can be compared to a theoretical model to determine compressor performance. Typical faults that affect performance are leaking suction and discharge valves, cylinder rings and rod packing.

<u>In-Cylinder Pressure:</u>
To take pressure readings, the compressor needs to be outfitted with pressure indicator valves in both the head and crank ends as pictured below (Figure 9). This allows for a portable pressure sensor to be mounted, the valve opened, and pressure readings taken.

Figure 9

The two most common methods to view pressure data are a pressure trace which is pressure versus crank angle (Figure 10) and pressure volume which is pressure versus swept volume (Figure 11).

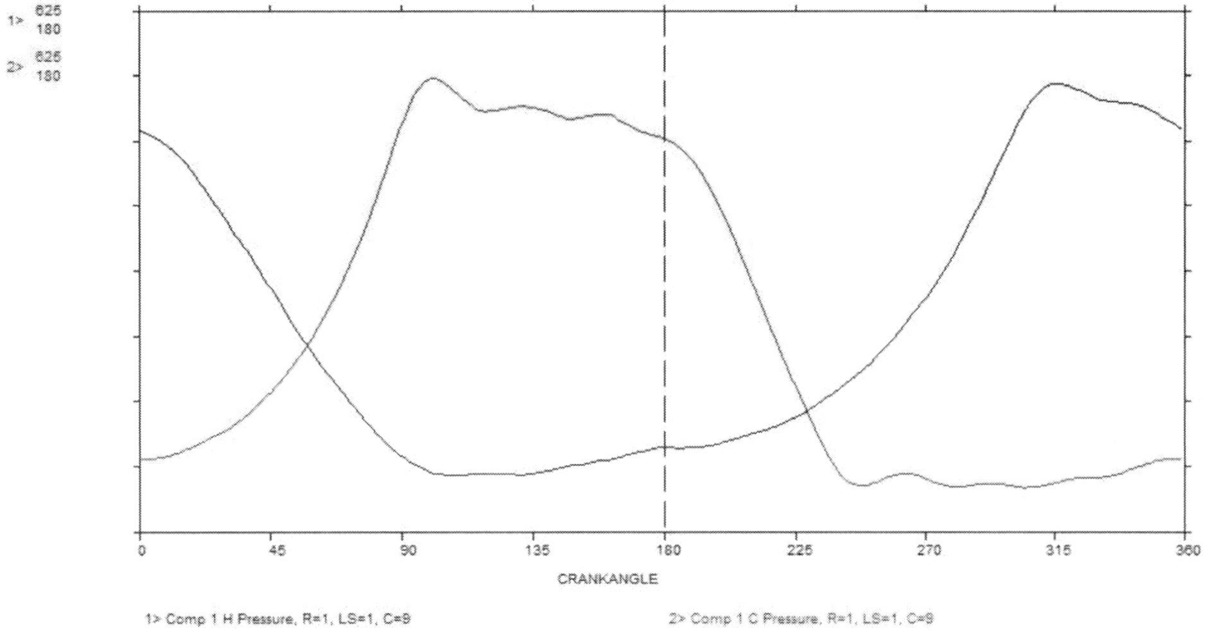

Figure 10

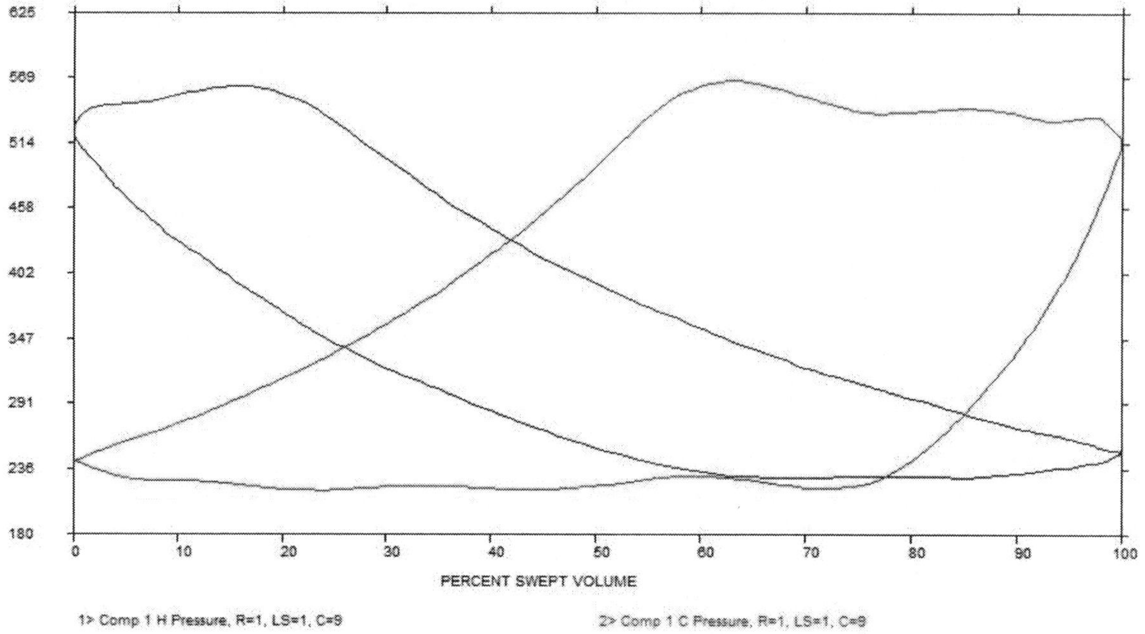

Figure 11

Knowing the geometry of the compressor throw allows a "theoretical" curve to be generated and displayed. The theoretical curve (represented by the dotted line) represents how the cylinder operates with no faults (Figure 12). It indicates the ideal compression, expansion, suction and discharge of gas in the cylinder. By examining the Pressure Volume curve with the theoretical overlaid, one can determine the health and performance of that throw. If the machine has a leaking suction valve, it will compress the gas slower and expand faster (Figure 12).

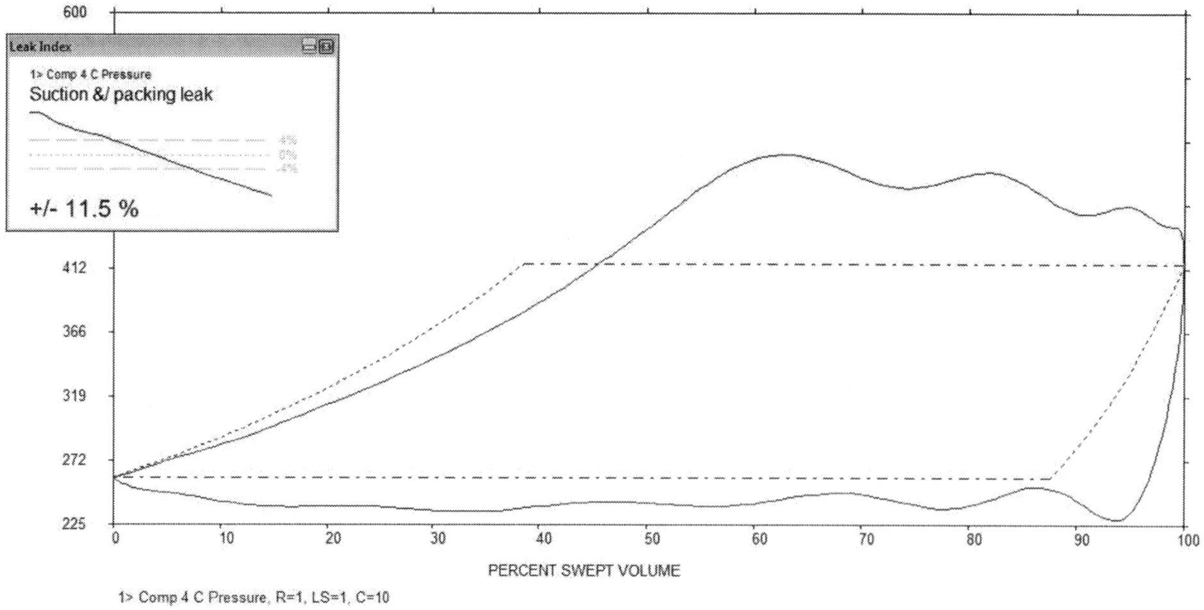

Figure 12

A leaking discharge valve will compress the gas faster and expand slower because of the recirculation of gas. In this example, the pressure is compressing faster than it should and expanding slower than it should (Figure 13).

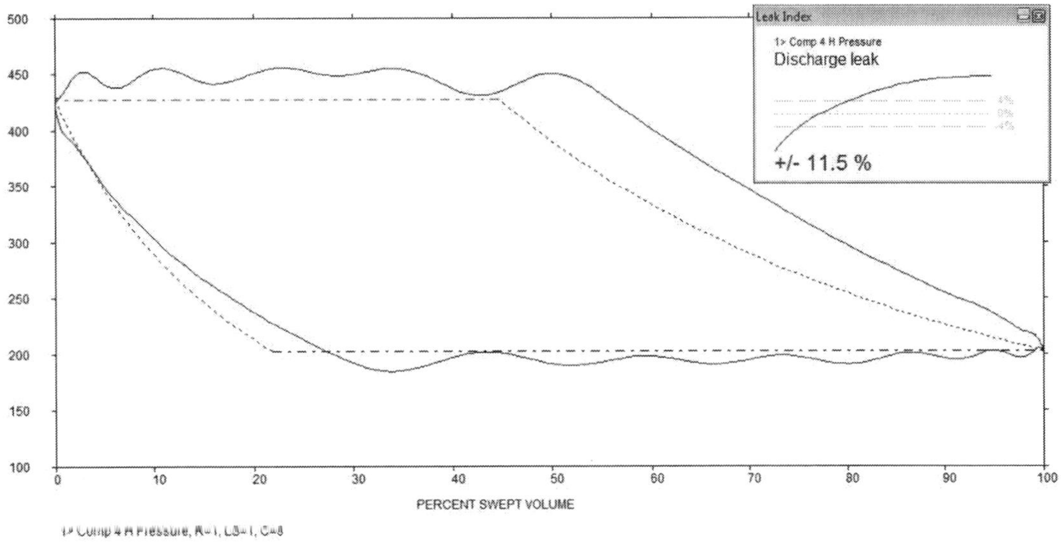

Figure 13

Ring issues can also be present if gas is leaking from the head end to the crank end and vice versa. Notice that the gas is compressing faster than it should at the beginning of the compression cycle. Then, as differential pressure changes between the head and crank end, the compression curve changes and compresses slower than it should (Figure 14).

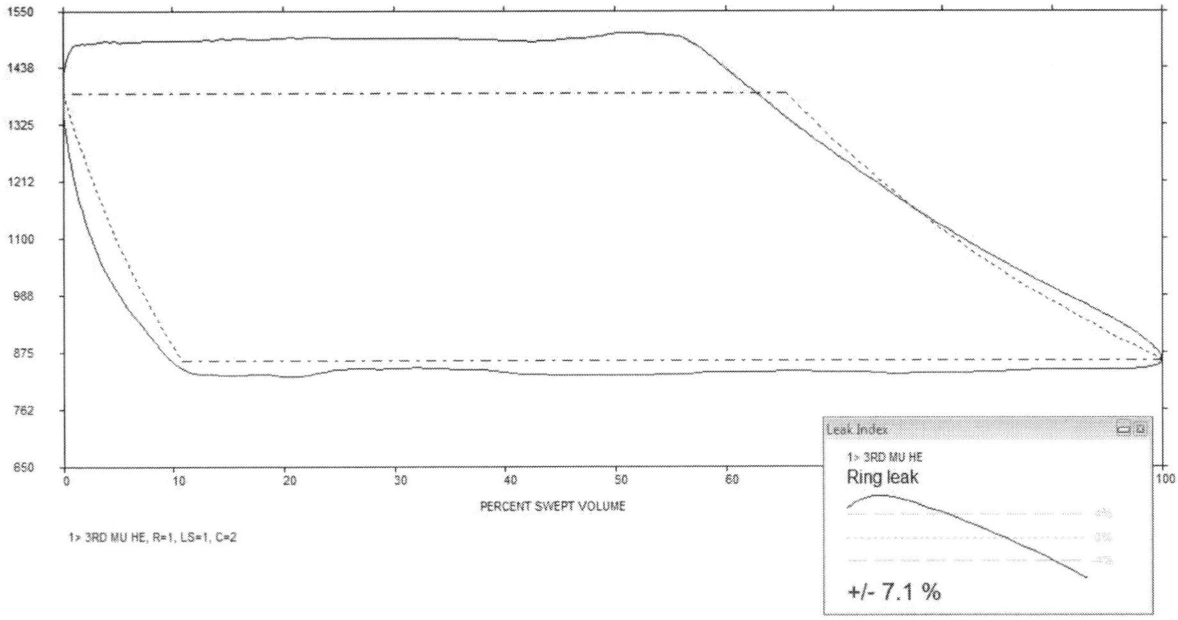

Figure 14

All of these issues affect the performance of the machine. Since the system is not operating efficiently, throughput is lost and/or the machine is using more horsepower to get the throughput needed. Consumed power can be calculated by the area inside the Pressure Volume curve. This shows the energy required by the compressor to obtain the gas throughput (Figure 15).

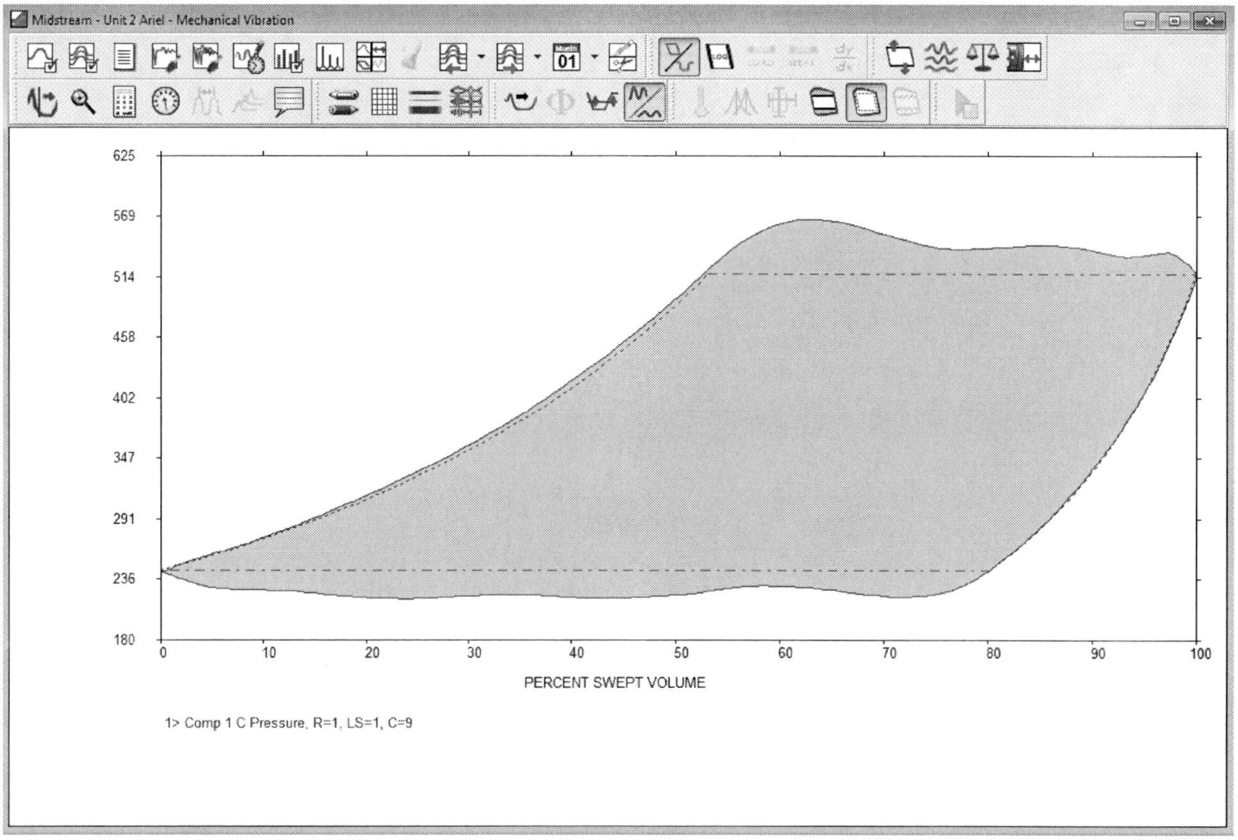

Figure 15

By comparing horsepower to throughput, the analyst can determine cost associated with moving gas. Figure 16 shows losses due to recirculating gas.

```
Compressor Costs:                                        Est. Compression Ratio: 1.27

No Power/Valve loss cost considered.
Gas Recirculation cost:   163.84 $/day


Total Compressor loss:     163.84 $/day
                          4423.73 $/month
                         53858.93 $/year
```

Figure 16

Vibration:

Accelerometers (Figure 17) are typically used to take data on crossheads, cylinder heads, valve caps and frame.

Figure 17

Filtered vibration levels are acquired at every degree of crank angle. This is represented in the smaller plot in Figure 18. In order to provide easily recognizable shapes and patterns, the vibration is mirrored and plotted as shown in the bigger plot of Figure 18.

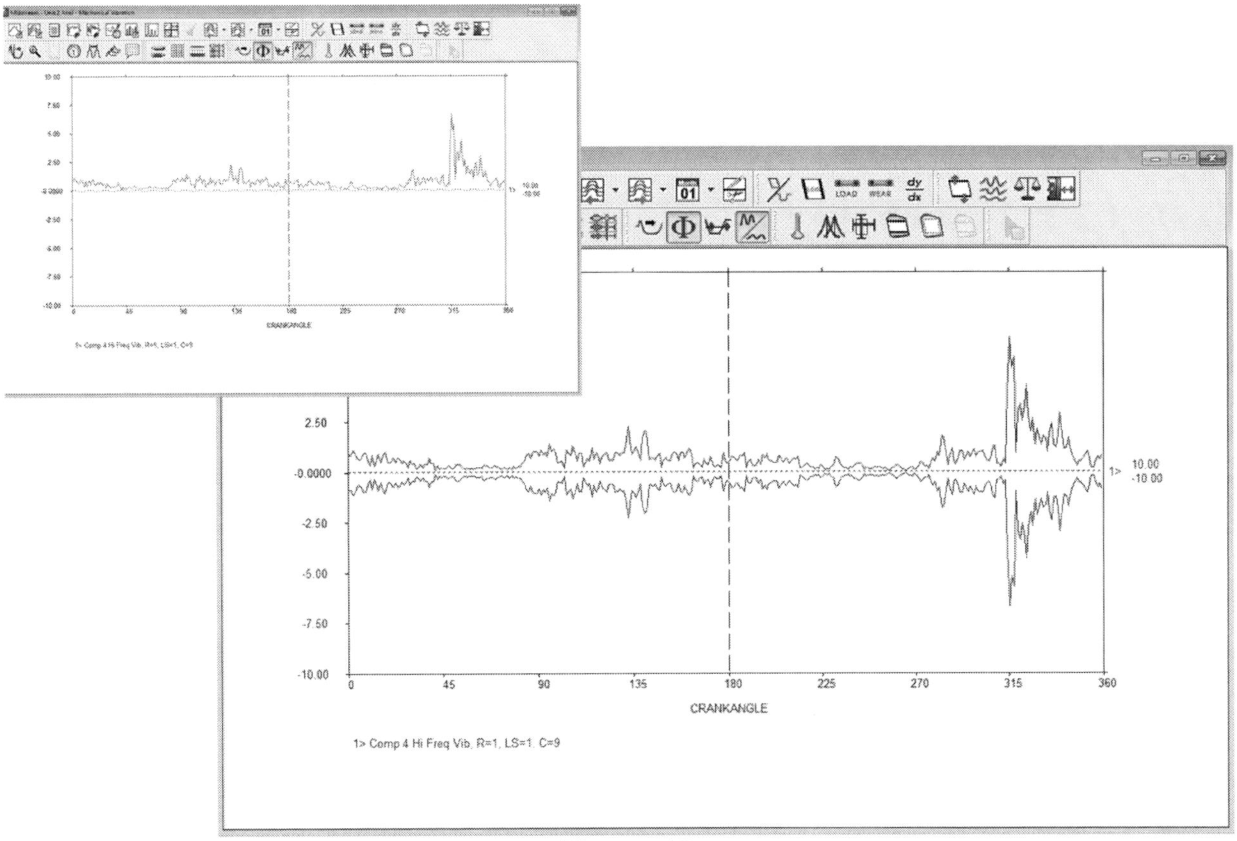

Figure 18

By looking at the data related to crank angle, the analyst can pinpoint the event causing the vibration. In Figure 19, there is a knock or vibration when the head end discharge valve is opening. This could be due to a sticky valve, improper springs, etc. This high vibration can lead to premature failure of the valve.

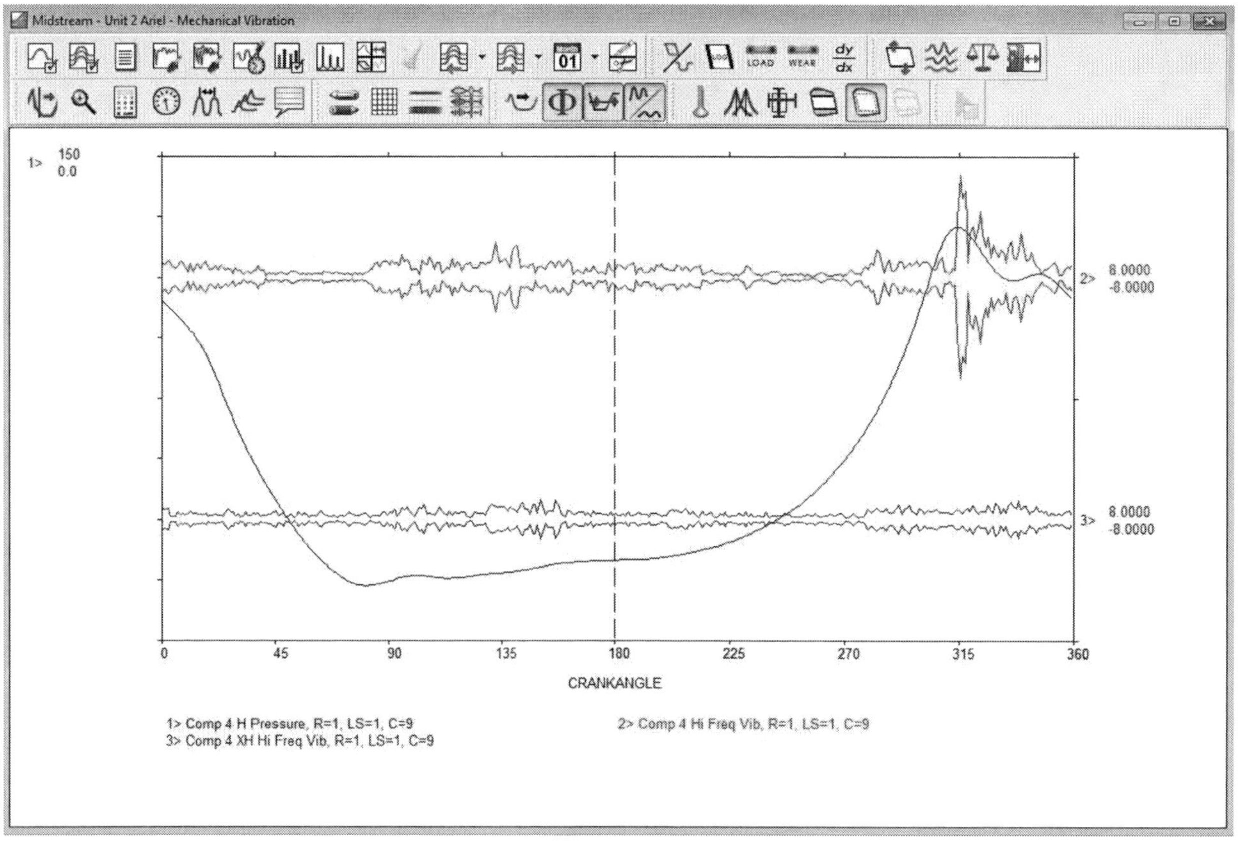

Figure 19

Ultrasonics:

Ultrasonic readings are typically taken on the cylinder wall and valve caps. By analyzing ultrasonic data, the analyst can find or verify leaks or unusual noises. Just the same as the vibration readings, the data is taken vs crank angle and mirrored (Figure 20; Figure 21).

Figure 20

Figure 21

Figure 22 is an example of a leaking discharge valve. The left side are properly working valves. When the crank end discharges, the discharge valves opens and a rush of gas leaves the cylinder, which can be seen in the ultrasonic data. The right side shows a cylinder with leaking discharge valves. A valve should be quiet while closed. Notice the noise being generated while the valves are closed, indicating a leaking discharge valve.

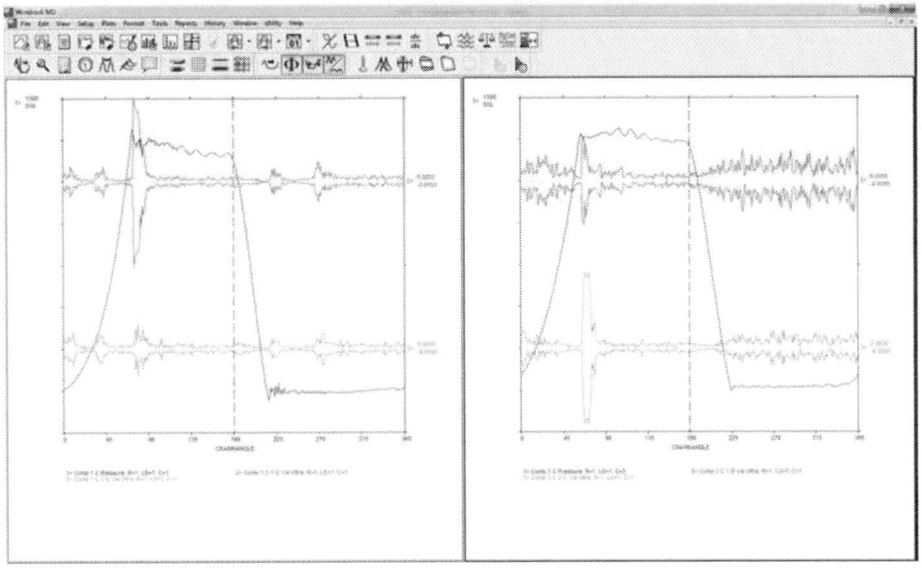

Figure 22

Case History #1:

A refinery was losing capacity on a critical hydrogen compressor. When the compressor is offline, the facility loses 10% of their production, a significant economic loss. Rather than bringing the machine down for component inspection to determine the cause of lost capacity, the company utilized reciprocating analysis to pinpoint the cause. Crank angle-based pressure, vibration, and ultrasonic data was collected and analyzed.

Previous data taken had showed slight ring leakage (Figure 23), but when the new data was collected, it was evident that the ring leak was much worse (Figure 24).

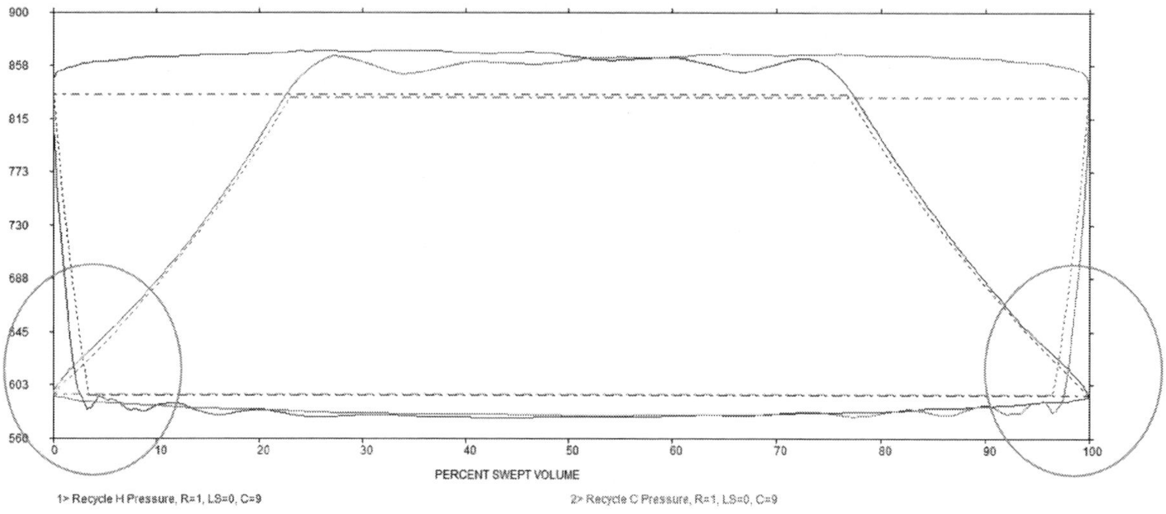

Figure 23

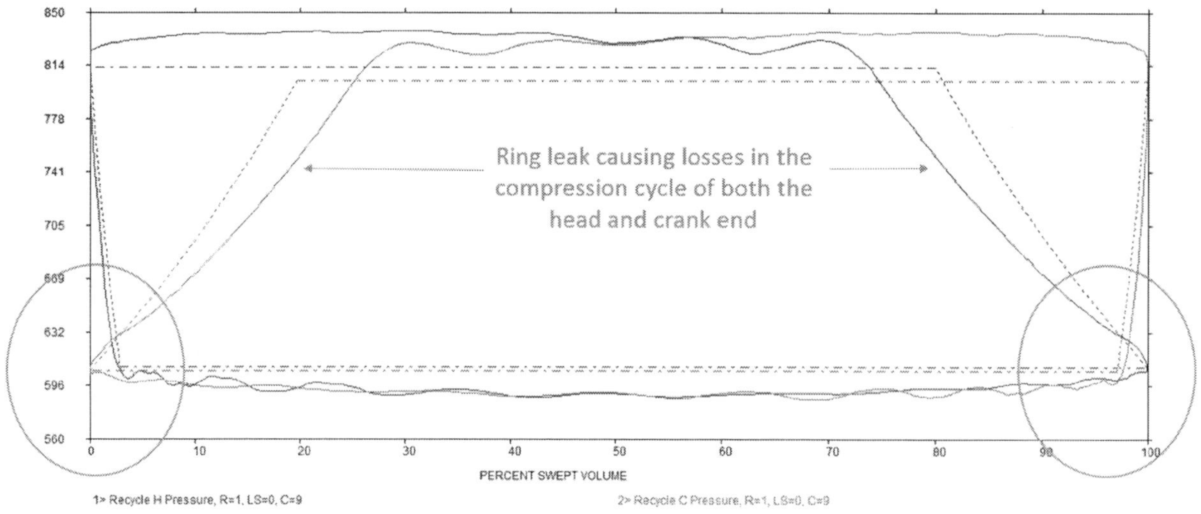

Figure 24

With this information, the outage was planned knowing that the cylinder would be removed and the rings replaced. This knowledge significantly reduced the downtime hours, therefore reducing the production loses.

<u>Case History #2:</u>
Here is an example of the limitation of only looking at time waveform and spectrum data on a compressor. Data was taken in May on a crosshead. As seen on the two plots, there are impacts happening, translating to a mound of energy in the spectrum data (Figure 25). Impacts are expected in reciprocating machinery and, since the crank angle position is not known, one cannot determine if the impacts are normal or abnormal. Data taken a month later shows a different waveform signature and more energy in the spectrum data. Looking at the amplitude, it is evident that the compressor has higher vibration but the cause and severity cannot be determined.

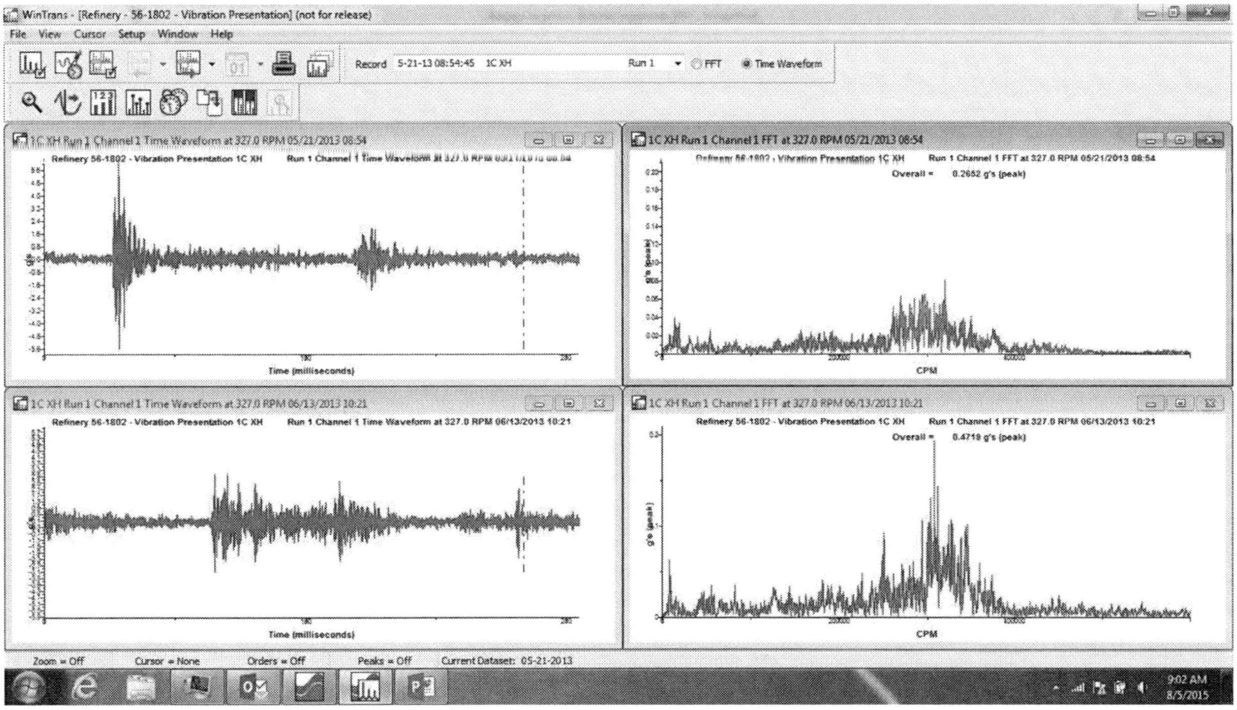

Figure 25

A quick look at the crank angle pressure data immediately shows the issue (Figure 26). The left side of the plot is the data taken in May. A month later, a restriction is apparent on the crank end of this cylinder. The discharge valve opens but the pressure is still increasing as gas is flowing.

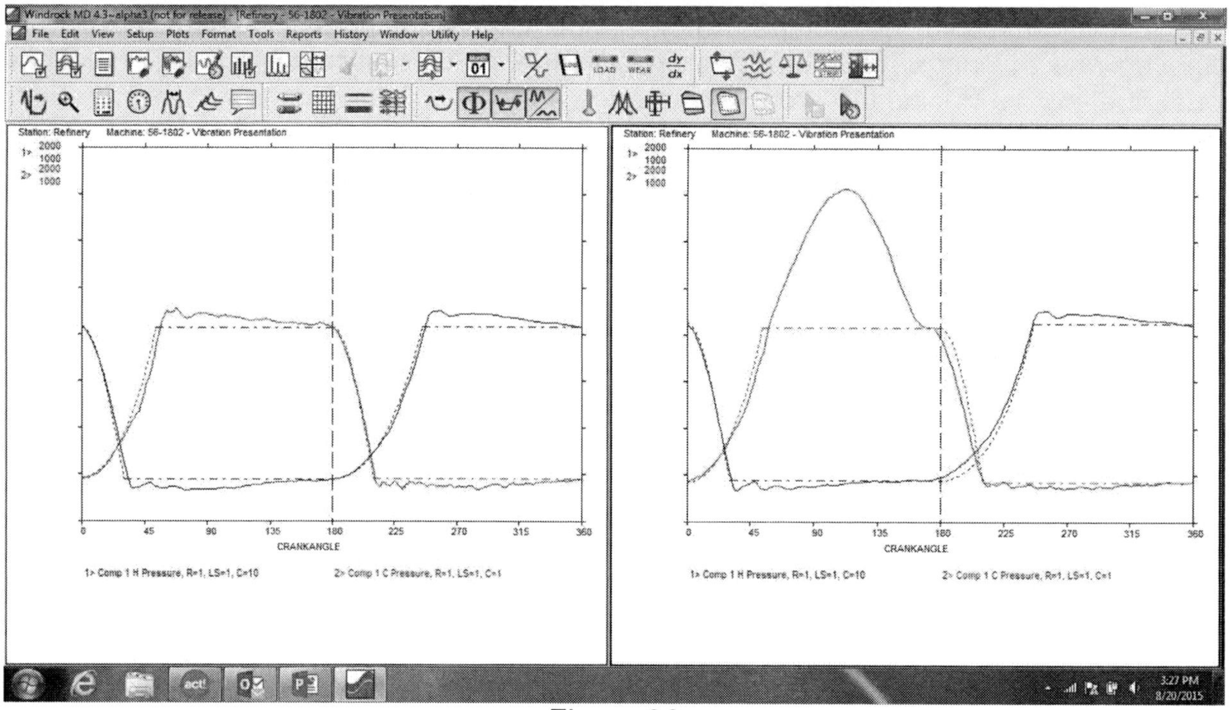

Figure 26

By looking at the vibration and ultrasonic data related to the crank angle, it is simple to identify the increase in energy being created when the crank end discharge valve is open (Figure 27). This could lead to mechanical issues in the future.

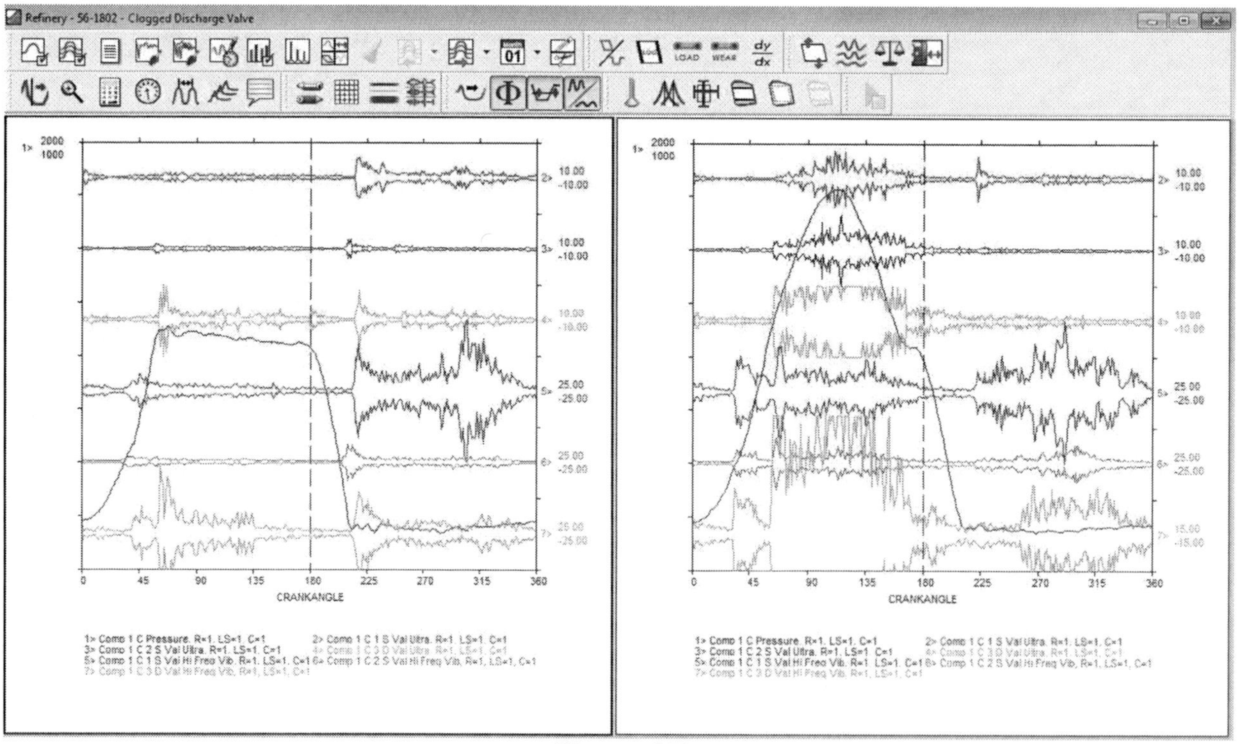

Figure 27

The compressor was dismantled and it was discovered that the rider bands had deteriorated and partially clogged the discharge valve openings.

The horsepower calculated to move the gas in May during normal operation was 729HP to move 38.2037 MMSCFD (Million Standard Cubic Feet / Day). A month later, with the partially clogged valves, horsepower increased to 1030HP and throughput dropped to 36.0429 MMSCFD. That is an increase of 301HP (29%) to move 2 MMSCFD (6%) less gas than the previous month (Figure 28). At an electricity cost of $0.06/kWh, the deficiency was costing $317.86/day (or nearly $10,000 per month.)

Compressor Cylinder ID	IHP @ RPM	IHP/ MMSCFD	Capacity MMSCFD	Date	Time
1> Comp 1 H Pressure	813.3 @ 327.8	19.31	42.1168	5-21-13	09:42:10
2> Comp 1 C Pressure	729.5 @ 327.8	19.09	38.2037	5-21-13	09:43:20

Compressor Cylinder ID	IHP @ RPM	IHP/ MMSCFD	Capacity MMSCFD	Date	Time
1> Comp 1 H Pressure	826.8 @ 327.8	19.85	41.6644	6-13-13	09:55:58
2> Comp 1 C Pressure	1030.4 @ 327.9	28.59	36.0429	6-13-13	09:54:52

Figure 28

As shown in this example, although the frequency based data did show something was wrong, it is unlikely that one could determine the actual cause based on this alone. By looking at the pressure, vibration, and ultrasonic data, it was very easy to determine the issues with the machine and the economic impact. By identifying, quantifying, and fixing this problem, the customer is able to document a savings of $10,000 per month of operation and ensure adequate capacity for the refinery unit.

Conclusion:

For the last 25 years, reciprocating monitoring programs have proven effective in determining reciprocating machinery health and performance. Event-based monitoring using data collected in conjunction with the crank shaft position, allows the analyst to determine not only mechanical issues like knocks and bearing looseness, but also determine the performance of the machine by detecting faults like leaking valves. By monitoring reciprocating compressors, a facility can realize the same benefits as typical frequency-based monitoring programs on rotating equipment. Some of these benefits include planned maintenance, eliminating failures, utilizing manpower, and increasing uptime.

Biography:

Mary Margaret Chapman has been involved in the predictive maintenance and reliability industry for 27 years. She spent 24 years focused on selling equipment and services related to rotating machinery such as electric motors, fans, pumps, etc. At the beginning of 2013, she began her career at Windrock Inc., a company that focuses on condition monitoring of reciprocating machinery. She is a Regional Sales Manager covering Arkansas, Iowa, Kansas, Minnesota, Missouri, Nebraska, North Dakota, Oklahoma, and South Dakota and helps companies with their reciprocating monitoring programs. Mary has a degree in Computer Science with a minor in Business from the University of Tennessee and currently resides in Knoxville, Tennessee.

DIFFICULT AND UNUSUAL BALANCE CASE HISTORIES

Nelson L. Baxter
ABM Technical Services, Inc
2800 E. Shady Oak Lane
Mooresville, In. 46158
NELSONBAXTER@ATT.NET

Abstract: Many balancing jobs are straight forward. The steps involve: 1-Taking an initial set of amplitude and phase readings. 2- The installation of a trial weight. 3- Retaking amplitude and phase readings to determine the effect of the trial weight. 4- Determination of a solution through either a graphical approach or via numerical calculations. This paper presents a number of case histories that illustrate how other complicating factors can make balancing a much more complex endeavor. Thermal vectors in motors and generators can significantly complicate the balancing process. Overhung masses and large differences in support stiffnesses can create situations where the cross effects are many times larger than the direct effects. Loose interfaces can result in non-linearities making both graphical and numerical solutions ineffective. Very high levels of unbalance can also create non-linear systems. Finally, rotor bows and rubs can result in machines that are very difficult or impossible to balance.

Keywords: **Balancing; Cross effect; Fans; Motors; Non-linear response; Over-hung mass; Rotor bows; Turbines**

Awhile back some balancing training was given to an individual. He was supplied with equipment and a very nice balancing program on a notepad computer that had a complete tool box of applications that made the process as painless as possible. After balancing a few fans this individual came back to his boss and told him that he had "completely mastered" the subject of balancing and felt that he was ready to take a leadership position. When his boss told me this, it occurred to me that maybe something should be written that would illustrate the fact that not all balance jobs are easy and some of them can actually be quite humbling. This series of case histories gives several examples where there were one or more complicating factors that made the balance solution difficult to obtain.

CASE HISTORIES:

CASE 1- 5000 HP 890 RPM SCRUBBER BOOSTER FANS: Following the installation of two new scrubber booster fans, difficulties were encountered while field balancing both fans. The manufacturer had added several balance shots over a two week period with no success. The time was fast approaching to when these fans would be needed in order to allow the 450 Megawatt unit to operate without being in violation of EPA standards. Contractual penalties were being reviewed and the lawyers on both sides were positioning themselves for battle. The fans had been exposed to and covered by a sticky liquid that had backed up into the duct work from the scrubber unit. It was believed by the manufacturer that they had not been thoroughly cleaned; so, high pressure washers were utilized to clean both sides of the blades. More attempts were made following the clean to balance the fans, but in each case they were unsuccessful.

In order to study the problem more thoroughly, instrumentation that is commonly utilized when balancing large turbines was brought on site. The fans were outfitted with both casing pickups and proximity probes. The plots below clearly showed what was happening. After the fans were started, the amplitude and phase would change very slowly. Over a 5 to 10 minute period there did not seem to be much of a change; so, the person doing the balancing would take a set of readings and then use those readings to calculate a correction shot and repeat the process over and over without success. The major issue was that the vibration would continue to slowly change over a period of from 2 to 4 hours (See Figure 1).

Figure 1-Casing & Prox Probe Polar Plots

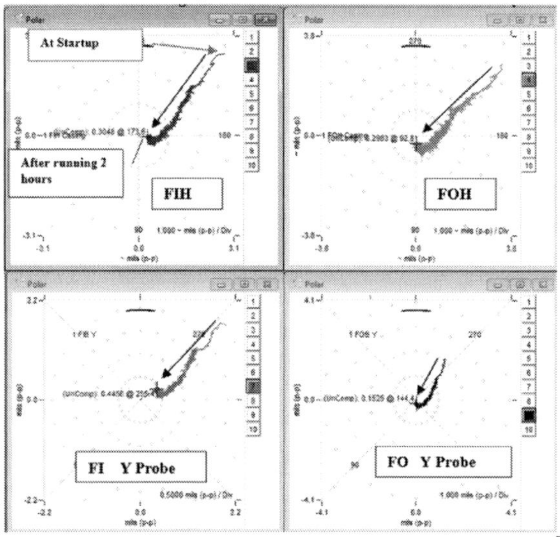

Figure 2 -Startup & Shutdown Slow Roll levels

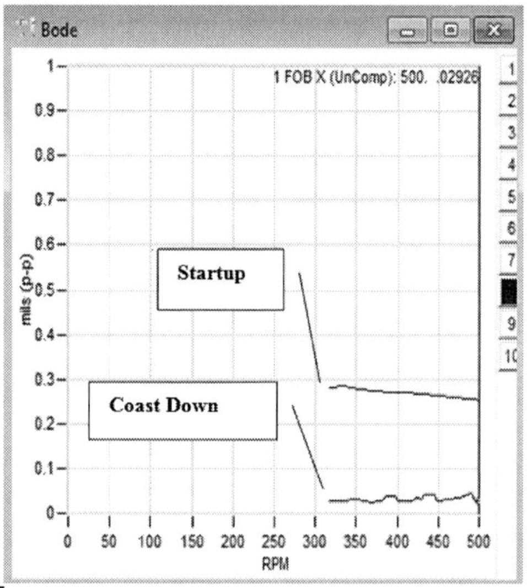

Figure 2 is a plot of the startup versus the shutdown slow roll values taken from the fan's proximity probes. The root cause of the problem was that each time the fans started up, they were found to have acquired a bow. The fans operated near a critical speed so the bow caused a large static in phase response. It took at least 2 hours for the bow to clear so accurate amplitude and phase readings could be obtained. The following data shows the gravity of this problem.

<u>As Found Stable Condition a few months later that required another balance shot</u>

Inboard Horizontal	5.8 mils @ 42 degrees
Outboard Horizontal	7.2 mils @ 28 degrees

Weight Installed- 50 oz @ 115 deg

Amplitude and Phase upon initial startup
Inboard Horizontal 9.5 mils @ 90 degrees
Outboard Horizontal 13.6 mils @ 84 degrees Very high upon startup
After 30 minutes
Inboard Horizontal 4.1 mils @ 87 degrees
Outboard Horizontal 6.5 mils @ 78 degrees
After 60 Minutes
Inboard Horizontal 1.7 mils @ 57 degrees
Outboard Horizontal 3.1 mils @ 52 degrees
After 90 Minutes
Inboard Horizontal 1.0 mils @ 337 degrees
Outboard Horizontal 1.5 mils @ 354 degrees
After 120 Minutes
Inboard Horizontal .9 mils @ 325 degrees
Outboard Horizontal 1.4 mils @ 343 degrees Very low after bow cleared

The balance shot was ultimately determined to be in the correct location, but the fan started up with over 13 mils of vibration. Had the behavior of this fan not been understood, the individual balancing the unit may have panicked and shut the fan down and tried another balance solution, which would not have worked.

Discussion: The complicating factor to balancing these fans is that when sitting still, they developed a slight bow which excited the first mode of unbalance. It took hours for this bow to clear before accurate repeatable amplitude and phase data could be obtained. The solution to this problem was patience in obtaining stable amplitude and phase readings along with the use of some good instrumentation. It was recommended to the customer to install turning gear on these fans so that they would not bow while the scrubber unit was down.

CASE 2- 8000 HORSEPOWER 2 POLE INDUCTION MOTOR: A large feed pump motor was sent out for a normal inspection and cleaning. After returning from the motor shop, it was put into operation and after 45 minutes, high vibration destroyed its bearings. It was returned to the motor shop where it was rebalanced. When it was returned to the plant, it again destroyed the bearings. It was then sent back to the manufacturer and put in a high speed balance pit and balanced at speed. When it was reinstalled back at the plant, it destroyed the bearings for a third time. Due to the nature of the problem, proximity probes were installed on the motor. When it was first started, the vibration was normal. However, as the motor was loaded, the vibration level increased to the point that the shaft motion was 7.8 mils which was nearly equal to the 8 mil bearing clearance. A good rule of thumb is that the shaft motion should use up less than 25% of the bearing clearances. When the shaft motion exceeds 50% of the bearing clearance the dynamic loading gets so high that the babbitt will start to fail. In this case, since almost the entire bearing clearance was being used up by the dynamic motion, it is understandable why the bearing failures occurred so quickly.

It was determined that the motor had a thermal vector. The solution was to balance the motor in the loaded condition. It operated with this thermal compromise shot installed for several years. It was discovered later that the motor shop had dropped the rotor on its first visit. This had damaged the laminations causing a hot spot to develop. This hot spot on one side of the rotor caused the rotor to bow as it heated up thereby producing the sensitivity to load.

Figure 3- Polar plot showing thermal unbalance vector change

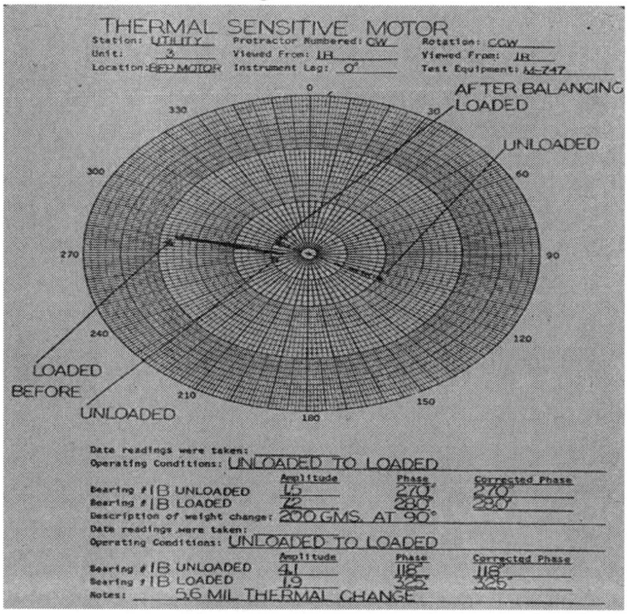

CASE 3 – ANOTHER CASE OF A LARGE MOTOR WITH LOAD RELATED THERMAL VECTOR: During a telephone conversation with an engineer at a refinery in Venezuela, a problem was described that sounded a bit unusual. The discussion went something like, “ever since we balanced our large motor, it doesn’t run well any more”. Further conversation helped clarify the situation. When the plant was brought down for an outage, the motor was run unloaded. During the unloaded operation, it was noted that the vibration on the proximity probes was over 3 mils. Based upon this amplitude, the plant personnel had elected to balance the motor. After balancing, the amplitude was approximately 1.5 mils. Everything seemed to be going well until the plant was restarted and load was applied to the motor. As the motor was loaded, the amplitude went up to nearly 4 mils.

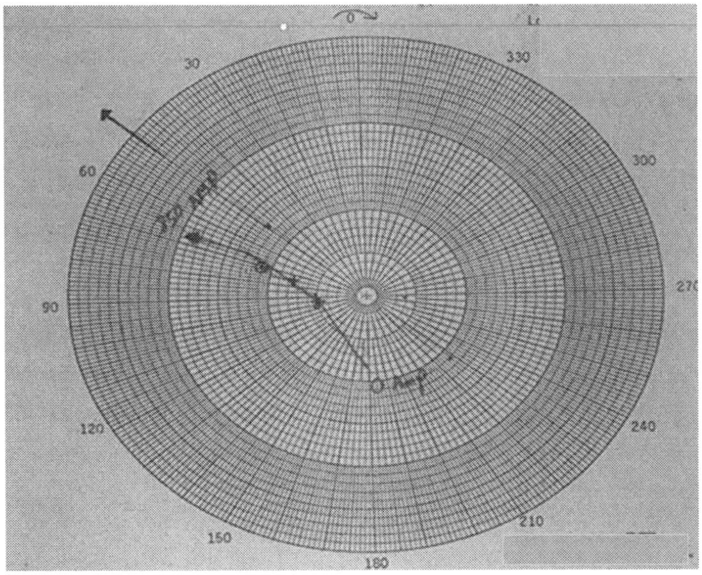

Figure 4- Change In amplitude and phase with increase in load

In order to analyze the vibration, the plant engineer tape recorded the proximity probe signals along with a tach pulse signal. That data was sent back to the U.S. where it was analyzed on a Bently Nevada ADRE system. The analysis of the data showed that the motor had a large thermal vector. What had occurred was that in the past the motor had been compromise balanced. The compromise shot had been designed to let the motor have high levels of vibration in the low load condition, where it seldom operated. This compromise shot was designed so that when the thermal vector, which was 180 degrees out phase from the low load vibration, took effect that the motor would operate with low levels of vibration at full load. Since the motor spent the vast majority of time at full load this had been a logical approach. What had occurred was that when plant personnel had balanced the motor at low load during the plant outage, they had undone the previous compromise balance shot.

This story would have been interesting enough if it had ended at that point, however, the saga continued. Due to the presence of the large thermal vector, plant management elected to purchase another motor. Nothing was specified regarding thermal vectors in the new motor and only a maximum proximity probe vibration limit of 2 mils was mentioned in the purchase specifications. When the new motor was tested, it was found that it also had a thermal vector. In fact, the manufacturer stated that at the factory where the motors were manufactured they routinely had to put in compromise balance shots on that particular motor design. To date they had performed approximately 128 such operations.

The moral of the story is that when dealing with 2 pole motors above 1000 HP be aware of the fact that the 1X running speed vibration may vary significantly from low to high load. It is imperative that phase angles be recorded along with the amplitude to identify this type of situation. For instance, a motor might have 2 mils at 90 degrees phase unloaded and 2 mils at 270 degrees in the loaded condition. If only amplitude is recorded, there would appear to be no problem. However, if phase is taken into account, then it becomes apparent in the above case that there has been a 4 mil change from low load to high load. This is therefore a situation where you do not want under any circumstances to balance out the 2 mils at low load, because if such an action is taken, then the vibration will increase at full load.

Cases 2 and 3 are similar in that they both involve large two pole motors. Due to their size and speed, large 2 pole motors often run above a critical speed. This seems to make them more susceptible to thermal vectors. In Case 2, the thermal vector was the result of damage done during the shipping or disassembly process. Case 3 is an example of the problem being present ever since the motor was manufactured. In any case, thermal load related vectors are common in 2 pole motors that in size are above 1000 HP. Motor shops and field balance personnel need to be aware of this issue and must know that no load balancing of this class of motors is often not effective in insuring that the machine will operate within tolerances in the loaded condition. During my fairly long career as a vibration analyst, this same issue comes up time and time again. Motor shops all over the country will bring large 2 pole motors in and balance the rotors in a shop balance machine only to have them run poorly when loaded.

CASE 4 – 650 MW GENERATOR WITH 11 MIL THERMAL VECTOR: The vibration level on a 650 Megawatt generator was found to be very sensitive to load. Further testing found that the sensitivity was actually associated with the amount of electrical current being supplied to the generator rotor. This is determined by both the Megawatt output and the MVAR loading.

Figure 5- 650- Megawatt Generator Rotor

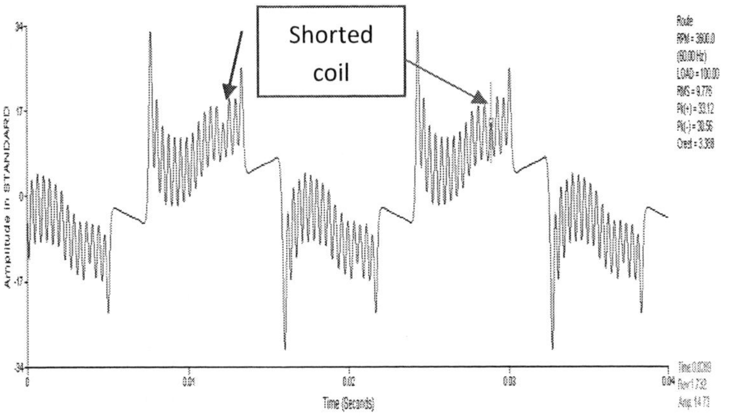

Figure 6- Flux Probe output showing shorted coil

Figure 6 shows example of a shorted coil that makes the unbalance vector sensitive to the amount of amps of excitation current being supplied to the generator field.

In order to solve this problem, amplitude and phase readings were taken over the entire load and field current range. Compromise balance solutions were then calculated and installed in the generator rotor. These shots, which were installed over time, allowed the turbine generator to run until the generator could be repaired. By the time the rotor was removed, the thermal vector was 11 mils. At low load, the amplitude was 6 mils. At high load, the amplitude was 5 mils 180 degrees out from the low load phase.

Discussion: Shorted coils occasionally occur in generator rotors. They are most easily detected by either monitoring the amplitude and phase while changing the MVAR loading or by the installation of a flux probe that measures the field strength. The flux probe plot shown in this case is typical and in some cases there may be more than one shorted coil. The shorts usually occur under the end rings. There has been a correlation found between the time a generator is on turning gear and the tendency to have shorted coils. It is thought that friction and movement between the end rings and the coils as the rotor slowly turns is a contributor to the insulation failures.

CASE 5- 5000 HP 720 RPM FAN THAT COULD NOT BE BALANCED: This Fan had above normal levels of unbalance. Several attempts were made to balance the fan, all of which were unsuccessful. After plant personnel attempted several balance shots in the I.D. fan, the casing vibration levels were reduced to around 2-3 mils, but would not go any lower. Since there was difficulty in balancing the fan, shaft stick measurements were taken to determine the absolute motion of the shaft. The shaft movement was discovered to be over 17 mils. Since the bearing clearance was only 8 mils, there was a strong indication that the bearing was moving in the housing. A large plunger bolt at the top of the bearing was tightened. After tightening, the casing vibration increased to 21.5 mils. Following tightening of the plunger bolt, the fan was easily balanced to below the 1 mil level desired by the plant personnel.

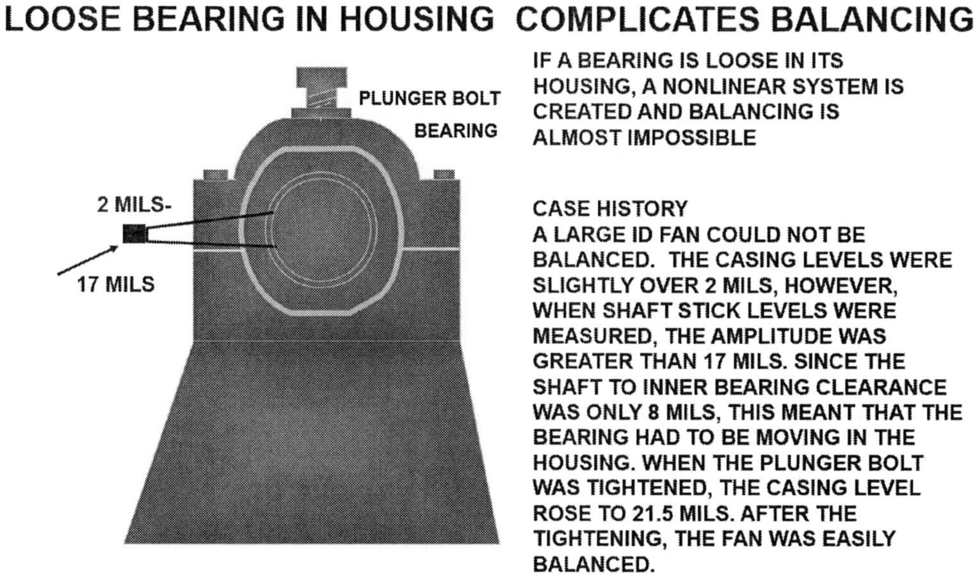

Figure 7- Diagram showing loose bearing in housing

Discussion: What had been occurring was that with the bearing moving in the housing a non-linear system was present. This made balancing all but impossible. In the field of balancing, looseness is one of the most common factors that will result in complicating the balancing process.

CASE 6 - MILL MOTOR THAT COULD NOT BE BALANCED IN MOTOR SHOP: When analyzing motors, it is easy to read more into a problem than actually exists. This case history illustrates the need to always look for the simple solution first. While at a power plant working on a large fan, the vibration analyst requested that a large mill motor be examined. The motor had been sent out to motor shops twice. In each instance, the vibration remained high when the motor was put back into operation. Attempts were made to field balance the motor, however, they were unsuccessful. The motor continued to have an amplitude of 8 mils of vibration, all of which was at its operating speed.

The first thing that was checked was the tightness of the motor to its base plate. All the bolts were tight. The next thing that was verified was the tightness of the base plate to the concrete pedestal. The bolts felt tight, however, motion between a bolt head and the base plate was apparent. What had happened was that the bolt had bottomed out. This meant that even though the bolt felt tight to the mechanic with a wrench, no force was applied between the bolt head and the plate. To solve the problem, a washer was installed. After the installation of the washer, the vibration dropped from 8 mils down to less than 1 mil. Over $30,000 had been spent on this large motor at the repair shops and the problem had been nothing more than the need to add a washer to a bolt.

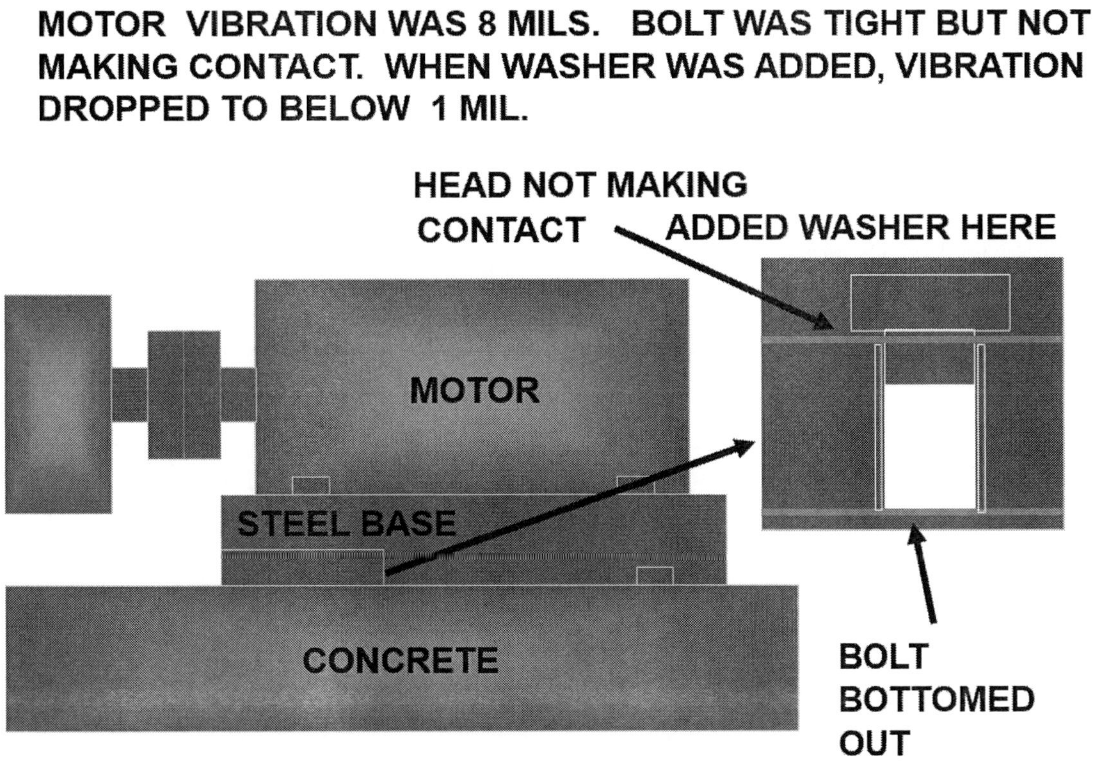

Figure 8- Diagram of bottomed out hold down bolt

Discussion: Cases 5 & 6 are both situations where looseness existed. In the first case, the looseness was between the bearing and the bearing housing. This can occur because of several reasons. The first is wear and fretting between the bearing outer shell and the case. This often happens on older machines. The second is what was shown in the fan case history. Large fans have a plunger bolt that lock the bearing to the case. If these plunger bolts are not properly torqued, then the bearing will move within the housing. Some machines even require a certain amount of crush fit between the bearings and the housing. Proper tightness and clearances are vital to prevent a rotor bearing system from having a non-linear response. A non-linear response makes balancing whether by graph or numerical methods difficult at best.

CASE 7 – A STUBBORN CENTRIFUGE WITH LIGHT SWITCH EFFECT: Centrifuges are used throughout the several industries primarily to separate liquid from suspended solids. They occasionally need to be balanced and it is normally a straight forward process. This particular case did not fall into the "normal" balance category. The balancing was being performed by a strobe light instrument which in this case turned out to be an advantage. Several balance attempts had been made with poor results. Strobe light balancers use a strobe light that is triggered by a filtered vibration signal. The first step is to tune the strobe light to the operating frequency. This visually freezes the shaft and a then band pass filter allows only the vibration energy that is centered on the operating speed to pass. This filtered vibration signal is then used to trigger the strobe thereby visually freezing the shaft at a fixed point. The shaft having been painted with phase angles then freezes at the high spot of the motion. The amplitude and this phase angle are then used in the balance calculations just like it is done with today's instruments that digitally display the phase.

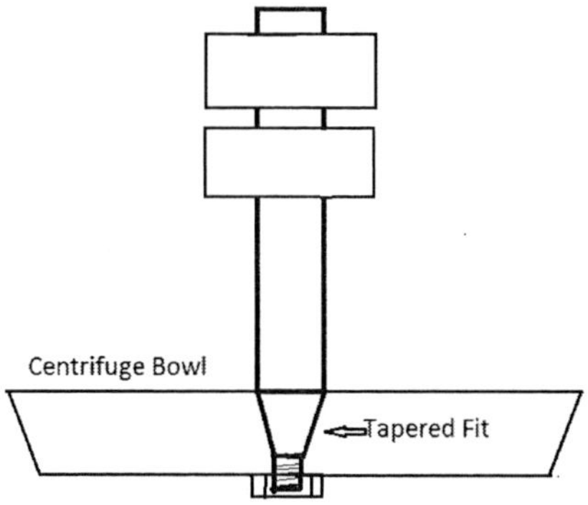

Figure 9- Vertical Centrifuge with loose tapered shaft to bowl fit

While taking the readings it was noted that the frozen phase angle on the shaft would suddenly shift 180 degrees. The problem was as follows. Suppose the vibration level was 5 mils at 270 degrees. What would happen is that with the addition of a trial weight, the vibration would drop to say 2 mils at 270 degrees. Then when a trim shot would be added to take out the remaining 2 mils of vibration, the vibration when the machine got back to full speed would suddenly jump back up to say 3.5 mils at 90 degrees. If attempts were made to remove the 3.5 mils at 90 degrees, the phase would jump back to 270 degrees. The problem was traced to a loose fit between the centrifuge bowl and the vertical shaft.

Discussion: The above situation is what could be referred to as the light switch effect. If you push a light switch towards the center, it will suddenly snap to the other position. If you push it back the other direction it will snap to the other side. You can never get it to stay in the center. When there is looseness between a rotating element and its supporting shaft the same thing can happen. The author has seen this same effect in centrifuges, motors, and fans. When it is present, balancing to low amplitudes will generally not be successful. This is non-linearity in its worse form.

CASE 8 – NO PHASE READINGS WERE AVAILABLE: A large utility called needing a compartment cooling fan balanced on one of their gas turbine generators. Unfortunately the entire unit was covered and there was no exposed shaft, nor any method of measuring the phase. The only measurement that could be obtained was the amplitude. In order to solve this problem, the four run no phase method was utilized to balance the fan.

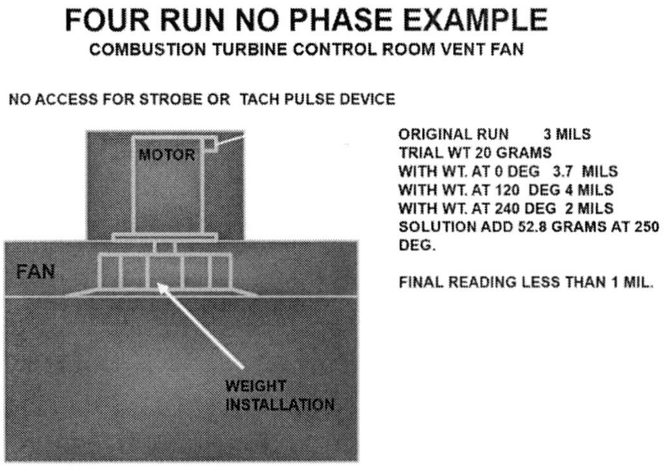

Figure 10- Fan with no access to read phase

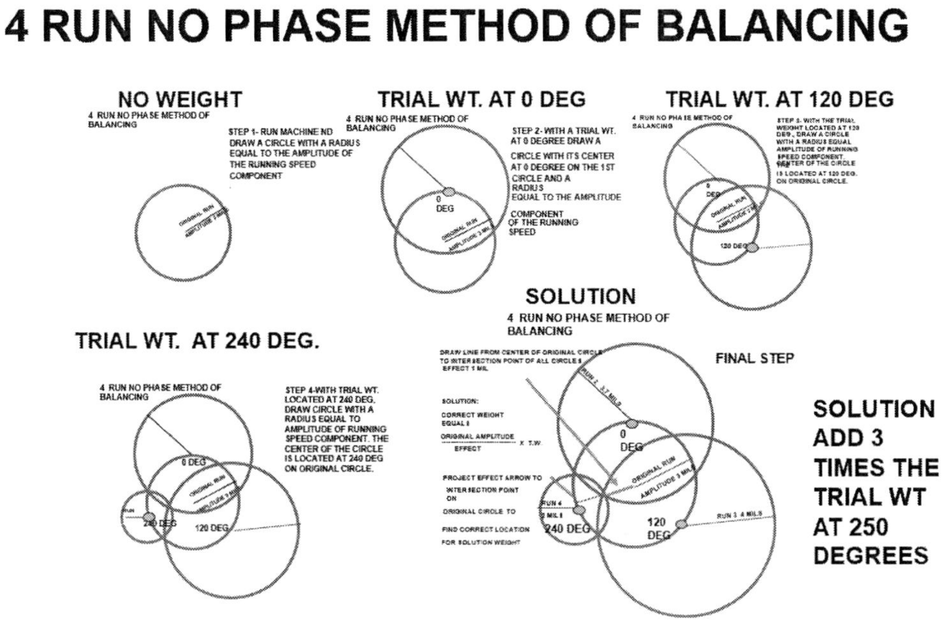

Figure 11- 4 run no phase graphical solution

DISCUSSION: Use the amplitude from the spectrum, not the overall amplitude, so only the unbalance level is recorded. At some point in this process the trial weight will be at or close to the unbalance location, so the level will get high and that might drive the system non-linear, thereby distorting the solution. If care is taken in placing the trial weight at the proper angles and the circles are carefully drawn, then this technique can be effective in single plane balancing.

CASE NO. 9- HIGH CROSS EFFECT FROM FAN TO MOTOR: A motor driving a fan had high levels of vibration. The amplitudes were higher on the motor than on the fan itself. The representative from the motor shop that had worked on the motor tried several times to balance the motor, with no success. Past experience on other large fans had shown that this might have been fan unbalance instead of motor unbalance. When the phase angles were measured, the fan angles were leading the motor angles. Based upon this data, balance shots were installed in the fan even though it had lower vibration amplitudes than the motor. Balancing the fan brought the motor levels down to very acceptable levels. The thing to remember is that the fan rotor will generally have a much greater amount of polar rotating inertia than a motor. It is quite common for a large fan to shake a motor, but much less common for a motor to shake a fan.

	MOB	MIB	FIB	FOB
BEFORE	2.3 78 deg	2.1 56 deg	1.5 22 deg	.76 11 deg
AFTER	.36 72deg	.4 90 deg	.6 114 deg	.66 132 deg

Above are the before and after levels on the motor and the fan. Note that before balancing that the fan phase angles led the motor phase angles.

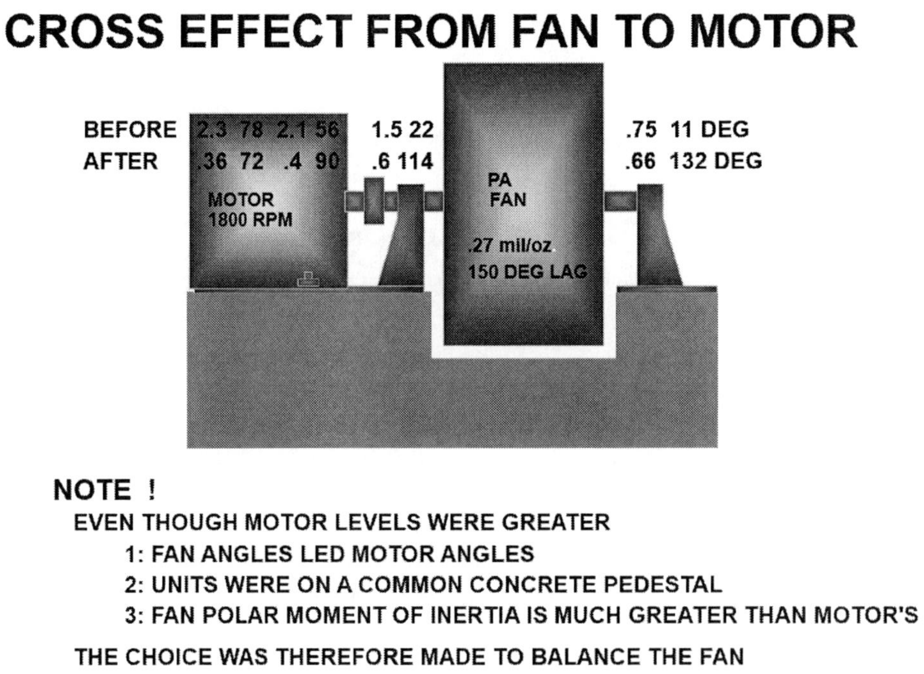

Figure 12- Diagram showing cross effect from fan shaking motor

CASE 10 – CROSS EFFECT BETWEEN TWO ROLLS RESULTING IN NEED FOR SYNCHRONUS TIME AVERAGE BALANCING:

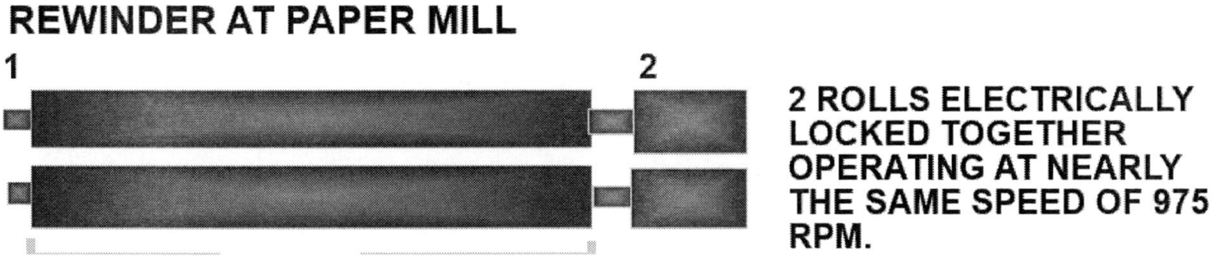

Figure 13- Two rotors running near the same speed causing beat

A re-winder at a paper mill consisted of two 20 ft long rolls. The roll drive control system controlled the speed of the rolls so that one roll operated at a slightly higher speed. This was done in order to keep the proper amount of tension on the paper during the winding process. The rolls were located on the second story of the building. While standing on the floor, it was noticed that the vibration would gradually build up to a high level; then, the amplitude would go almost to zero. This increase and decrease would repeat over a fixed time interval. This type of pattern is indicative of a beat. A beat exists when two signals that are slightly different in frequency add to and subtract from one another. One way to think of it is to have a fixed vector that represents one machine's unbalance. The cross effect from the other machine then acts like a rotating vector that is attached to the end of the first machine's vector. As the second vector rotates because of a slight speed difference, the vector that results from the addition of the two signals increases and decreases and the resulting phase angle oscillates.

Figure 14- Formula for the addition of two signals

BEATS OCCUR WHEN TWO OR MORE SIGNALS ARE ADDED TOGETHER

FORMULA FOR THE ADDITION OF TWO SIGNALS

$$Y = A * \sin(2\pi f_1 t) + B * \sin(2\pi f_2 t)$$

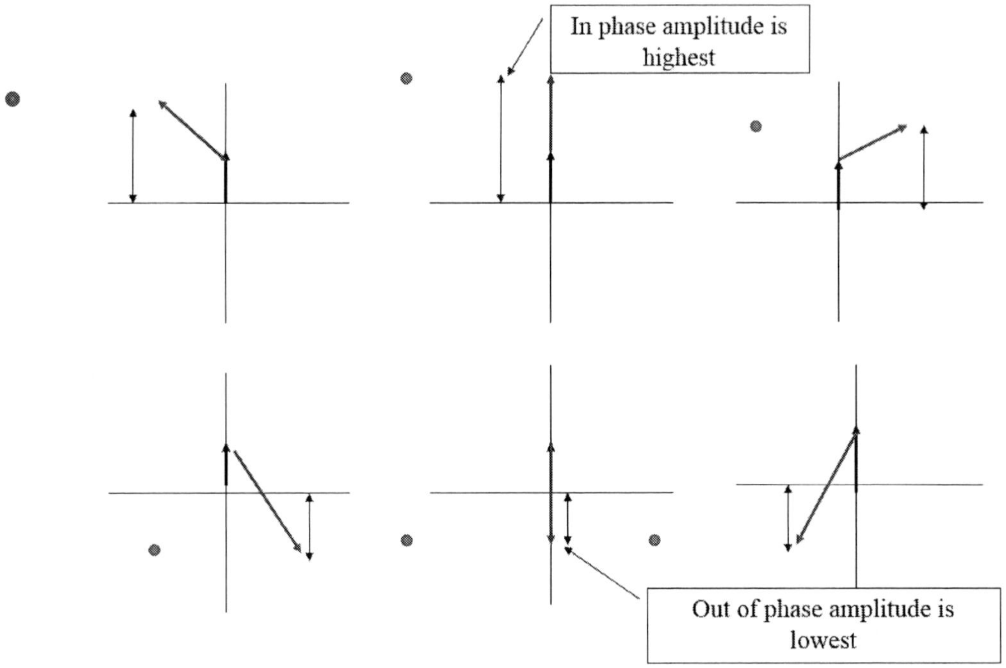

Figure 15- Illustration of two signals beating causing variations in amplitude and phase

As can be imagined, such a situation that ends up having continuously changing amplitude and oscillating phase angle greatly complicates the balancing process. When asked how long the problem had existed, plant personnel stated that it had been present for eight years, but that recently there had been concern that it was affecting the quality of the product. They indicated that balancing technicians had tried to balance out the problem with no success. The solution to the problem was to use synchronous time averaging to obtain the balance data. By averaging in the time domain, a signal is just as likely to be positive as negative so if the combined signal is synchronized to a reference signal from one component, then the signals from the non-synchronized machine will with time average to zero thereby leaving only the amplitude and phase from the synchronized machine. Using this technique, the rolls were both balanced to a level of .2 mils solving the floor vibration problem. When calculations were made, it was found that the amount of force going into the floor was 4400 lb when the two rolls' unbalance vectors were in phase with each other.

Discussion: When beats are present, synchronous time averaging is a powerful tool that can separate the signals so each machine can be attacked on a one on one basis. Cooling tower fan balancing is an excellent application of this technique. Low acceleration levels combined with noise and cross effect from adjacent cooling tower cells make the use of synchronous time averaging a much appreciated tool for this application.

CASE 11 – COUPLING UNBALANCE IN TWO HIGH SPEED COMPRESSORS:

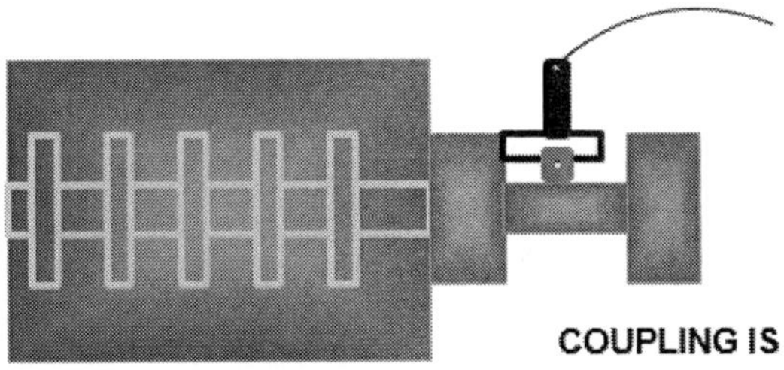

Figure 16- Overhung coupling

A 6,000 rpm turbine driving a chiller compressor was in alarm following an overhaul. The level was only high on the proximity probes on the inboard bearing next to the coupling. Since the vibration on the other end was low and the critical speed vibration was also low, it was elected to add weight to the coupling. It took only 3.8 grams/mil to balance out the problem. The level on the probes ended up below 0.5 mil. A few months later, the unit was down for another inspection and the coupling was disassembled. It was discovered at that time that on the previous overhaul that the coupling had not been assembled with the match marks lined up. The match marks were put in line and the unit was restarted. The vibration was high like it had been after the previous overhaul. It was suspected that the reason that the balance weights had been installed in the first place was the match mark problem, so the balance weights were removed. Following the removal of all the balance weights, the unit was restarted and it was again under 0.5 mil. This is a case that illustrates the importance of taking care to match mark couplings on high speed units. Couplings represent overhung weight. This combined with the fact that the proximity probes are often located near the coupling can result in high speed machines being very sensitive to any changes in coupling unbalance.

In another case involving a 10,000 rpm ammonia compressor, the owner had spent over $100,000 in trying to reduce the vibration to below the alarm setting. The unit had been disassembled three times and finally was sent back to the factory for a stack balance. When it came back, the problem was the same as in the past. The data showed that when the rotor went through its critical at 6,000 RPM, there was very low response. However, from about 8,000 to 10,000 RPM, the vibration on the coupling end probes increased to well above the alarm level. The interesting thing is that there was no phase shift during the 8,000 to 1,000 rpm range. It took one eleven gram washer on one of the coupling bolts to solve the problem. Several years later the customer called with a similar problem on the same unit. The compressor was balanced in the coupling over the phone using the same sensitivity and lag angle that had been calculated for the previous coupling shot.

CASE 12 – A VERY SICK TURBINE SAVED BY CROSS EFFECT: This case history involved a turbine with a troubled history. It was a 100 Megawatt five bearing GE D-2 Turbine. It was located in an older power plant along with two smaller turbines. The situation was fairly desperate because if this machine could not run, then it endangered the viability of keeping the entire station in operation. If it could be made to run for another year, then there would be money available at that time to replace its rotor and the station could remain in operation thus saving the jobs of several people. To complicate things, this turbine was a five bearing machine which means that the two turbine rotors had three bearings and the generator had two bearings. Anyone who has ever balanced machines where there are only three bearings supporting two rotors knows that cross effects can be an issue. In this instance, the cross effect was what saved the day.

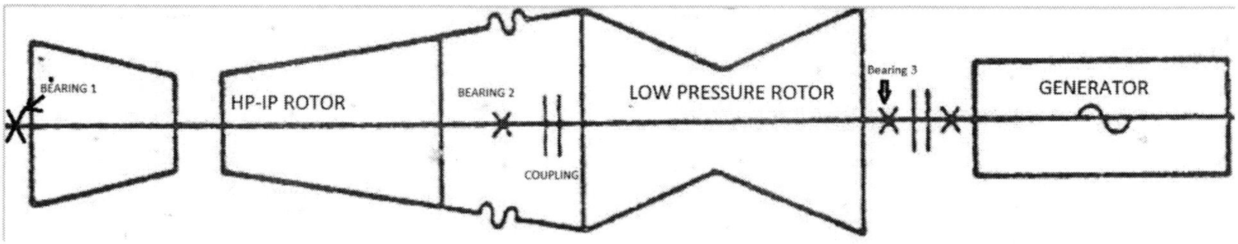

Figure 17- D-2 Turbine layout two turbine rotors supported by three bearings

Over the span of several years, the low pressure turbine had several balance shots installed. The most recent shot left the vibration on the No. 1 end at 5.2 mils @ 238 degrees. The number 1 bearing is the bearing that supports the HP rotor. Based upon data from other D-2 units, it was a known fact that addition of weight in the No. 1 balance plane had very little effect on the vibration levels on the No. 1 bearing. This was thought to be because the radius at which the balance weights were installed was very small on the HP rotor. That issue combined with the fact that the heavier LP rotor was the dog and the lighter HP rotor was the tail made adding the small HP rotor balance weights at the short radius where they were located ineffective in lower vibration on the No. 1 bearing. The only solution was to see if a shot could be installed in the LP rotor that would not disturb it and at the same time the cross effect could be used to lower the vibration on the HP rotor. Complicating this process was the fact that a large portion of the weight grooves in the areas of interest on the LP rotor were already filled with weights. In order to determine if a shot in the LP rotor could reduce the vibration on the HP, past data from another D-2 machine at a different utility was studied. It was noted that the balance shots in the LP rotor on the other D-2 machine had very little effect on the vibration on the number 2 bearing so the only readings that needed to be considered were the direct effects on the number 3 bearing and the cross effects of an LP balance shot on the No. 1 bearing. The balancing program that was available at the time had no provision with dealing with two balance planes and three responses.

The approach that was used was to plot the amplitude and phase angels on the No. 1 and No. 3 bearings on polar paper then using the cross effects to see what would happen when various balance shots were tried. In a few minutes, a solution was worked out that would allow weight added in the LP rotor balance planes to lower the vibration on Bearing 1 without having an adverse effect on bearings 2 or 3. The only problem left was that there was already a long arc of weight in the location where the weight was needed. By adding large amounts of weights at the end of the arcs of weight already installed on the LP rotor it was, however, theoretically possible to solve the problem. On the turbine end of the LP rotor 26 oz was added on the ends of the arc of weights at 80 degrees and 21 oz at 230 degrees. Because of the spread of the weights, the 47 oz of weight only had an effect of 13 oz at 133 degrees. On the generator end of the LP rotor, 20 oz of weight was added at 260 degrees and 15 oz at 120 degrees. The combined effect of that 35 oz of weight was 12.9 oz at 211 degrees. The entire process was a bit of an act of desperation, but when the above actions were taken all the bearings were operating below 3.5 mils. The turbine survived until the rotor could be replaced and the plant remained in operation for 15 more years. By the time it retired, many of the plant personnel were able to retire and the remaining employees were allowed to take jobs at a new combined cycle gas turbine facility that was being built on the same site.

Discussion: This case history shows that all possible solutions need to be examined. This problem was solved by a few sheets of polar graph paper and an understanding of how to split balance weights to get the desired effect.

CASE 13 – CROSS EFFECT IN A DRAG LINE MG SET TRAIN:

Only someone who has ever worked on a drag line can appreciate how difficult these machines are to work on. Large drag lines are used to remove the soil that is above a strip of coal. Sometimes the soil layer is 100 ft. or more thick. The drag lines may weigh 5,000 to 7,000 tons. Their buckets can remove 100 or more cubic yards of dirt at a time. Typically there are several motor generator sets that generate D.C. current for the drag and hoist motors. This particular drag line had 5 of these very large Motor Generator sets. A motor generator set consisted of a large synchronous motor and two generators on one side of the motor and three generators on the other side. All the motors and generators were solidly coupled together and there was only one bearing per generator (the motor had two bearings). Since the motors are synchronous, that means that all the units operate at exactly the same speed, which in this case was 1,200 RPM. To make things even worse, all the MG sets were located on the same metal deck. In this case this meant that there were 30 rotating elements (5 synchronous motors and 25 generators) operating on one metal deck all rotating at the exact same speed.

Figure 18- Photo of Drag Line

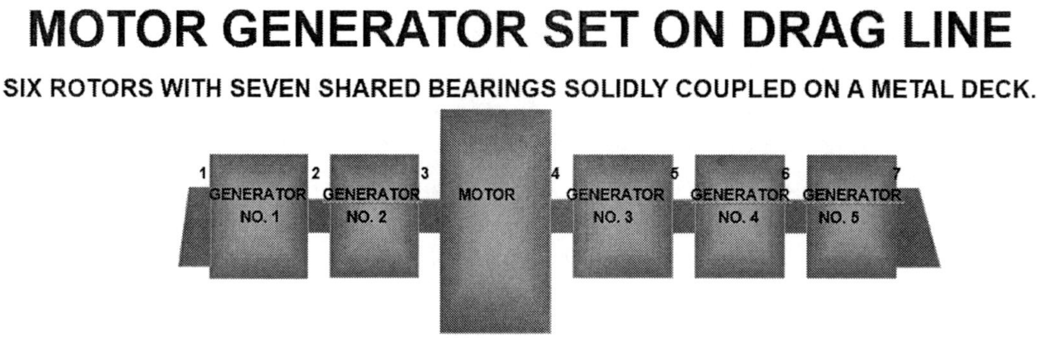

Figure 19- Drag line MG Set Six rotors supported by seven bearings

Six rotors were solidly coupled together, sharing bearings and five of these units on a metal deck, were all operating at exactly the same frequency. As would be expected, the vibration problem occurred at 1,200 cycles per minute. This was a balance person's nightmare. The whole structure, even the boom, was shaking at 1,200 cpm. However, with a little investigation, it was found that the vibration was higher on one of the MG sets. Efforts were then made to balance that unit. The highest reading was 11 mils in the vertical direction on the outboard generator of that unit. The first balance shot was installed in that rotor. The levels went down in the horizontal direction and up in the vertical direction. Therefore the influence coefficients indicated that it would not be possible to balance the unit in that rotor. In an attempt to simplify the problem, two of the generators were uncoupled from the motor. Even uncoupled, because of the common metal deck, the generator vibration on the outboard generator was still 7.9 mils. Since the largest component in the MG train was the synchronous motor, it was elected to try a balance shot in that rotor. When a trial shot was added, all the response vectors indicated that a solution could be obtained by adding the proper amount of weight at the correct angular location on the synchronous motor's rotor. Fifty ounces of weight had to be added at approximately a 30-inch radius on the motor. When this was done, the vibration level on the whole unit went back to normal. When in doubt, add the trial weight to the rotor with the largest polar moment of inertia.

CASE 14 – HP ROTOR WITH BOW REQUIRES MID PLANE BALANCE SHOT TO GET THROUGH CRITICAL AND COUPLE SHOT TO LOWER LEVEL AT FULL SPEED OPERATION: This case study involved a 500 Megawatt HP-IP Rotor with a high critical speed static unbalance and high running speed couple unbalance. During the startup of a 500 Megawatt turbine, the vibration at 1,420 RPM when the rotor went through its critical speed reached a level of over 15 mils on the HP-IP rotor. The vibration was almost 7 mils when the machine reached 3,600 RPM. In order to gain information as to the correct balance approach, a Bently Nevada ADRE system was connected to the supervisory instrumentation. The location of the Keyphaser and the zero mark on the shaft were verified. It was determined that the true phase angle was being detected by the left probe and that 90 degrees would have to be subtracted from the right side proximity probe to get an accurate phase angle. The figures below show the resulting runout compensated Nyquist plots from the left probes at the bearing 1 & 2 locations. The plots show that when the HP-IP rotor reached the critical speed of 1,420 RPM that the phase angles were nearly identical. This is what would be expected for a first mode response. The amplitude was 15.4 mils on bearing 1 and 13.1 mils on bearing 2.

BEARING 1 RUNOUT COMPENSATE NYQUIST BEARING 2 RUNOUT COMPENSATED NYQUIST

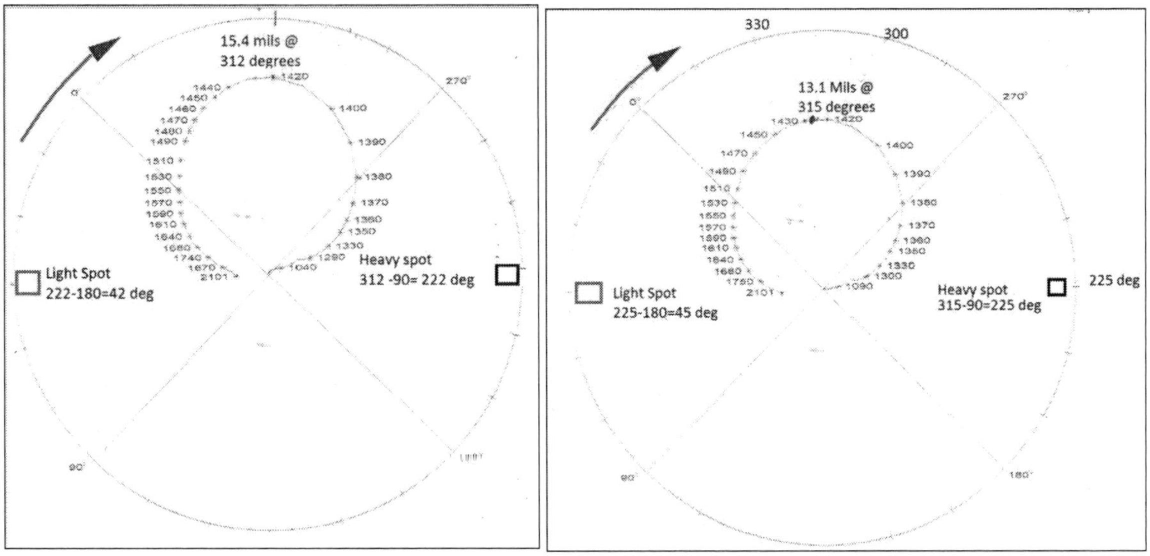

Figure 20 Nyquist plots showing 1st critical speed response

Based upon the large response, it was elected to try and solve the critical speed vibration issue, then attack the at-full-speed problem later. Based upon past experience it was known that putting weight in the end planes would not reduce the critical speed response. The decision was therefore made to install the weights in the mid plane of the rotor. This required that both an inner cylinder and outer cylinder plug had to be removed. The temperature at the mid plane point was 1,000 degrees F and the pressure was 2,400 psi, so this decision was not one that could be made without critical considerations. In order to make sure the balance plugs did not get stuck half way in the threaded mid plane balance hole, they were set on the casing and allowed to heat up. Three balance plugs weighing 9 oz each were installed. Since the lag angle at the critical speed was 90 degrees, that number of degrees was subtracted from the critical speed phase to locate the source of the unbalance and then the correction weight was installed 180 degrees from that location. Following the installation of this mid plane balance shot, the critical speed response was below 5 mils. The lag angle was calculated to be 104 degrees and the sensitivity was 2.2 oz/mil.

Figure 21-500 Megawatt HP-IP Rotor Mid Span Temperature 1000 degrees F

The next step in the process was to bring the turbine to full speed. The readings on the HP-IP rotor were:

Bearing 1 Left Probe 6.3 Mils at 260 degrees
Bearing 2 Left Probe 6.9 mils at 80 degrees

Past experience showed that the sensitivity of couple unbalance(phase is at or close to 180 difference from end to end) was 1.5 oz/mil at full speed and that the couple lag angle was 68 degrees. 9.8 oz was added in plane 1 at 28 degrees and 9.8 mils was added in plane 2 at 200 degrees.

Final results:

Bearing 1 left probe 1.4 mils at 165 degrees
Bearing 2 left probe 2.4 mils at 0 degrees

This case history is a good illustration of what can happen when there are two types of unbalance in a rotor that goes through a critical and then approaches a second critical speed.

Discussion: In this case, the static (in phase) unbalance that was seen when the rotor reached its first critical speed had to be removed by a mid-plane balance shot. The author had attempted on a previous job to add a static shot in the end planes of an identical rotor design. Thirty-six oz was added in the end planes and it had no effect on the critical speed vibration. That piece of knowledge meant that a mid-plane shot was the only solution available to lower the critical speed response. Once the turbine was at the 3,600 RPM operating speed, the second mode was excited as evidenced by the 180 degree phase difference between the vibration on the two bearings. That problem was ideally suited to a couple shot (equal weights located 180 degrees from each other).

CASE 15- BALANCING A GE FRAME 7 GT GENERATOR WITH 2nd MODE EXCITATION USING NYQUIST PLOT INFORMATION:

The figures below show Nyquist plots taken from a 100 MW generator rotor. These plots show how important it is to obtain this type of data. Two important things were learned from these plots:

1: The second mode is being excited. Note that the phase angles from one end of the rotor to the other end are almost 180 degrees apart. (The vectors point in the opposite direction).

2: The rotor goes through approximately 85 degrees of second mode phase shift.

These two pieces of information tell the analyst to install a couple shot (equal amounts of weight 180 degrees apart) and that the second mode lag angle is 85 degrees.

Using these two pieces of information, it was possible to place the weights in the correct location on the first balance attempt. This was in spite of having no previous history on this unit.

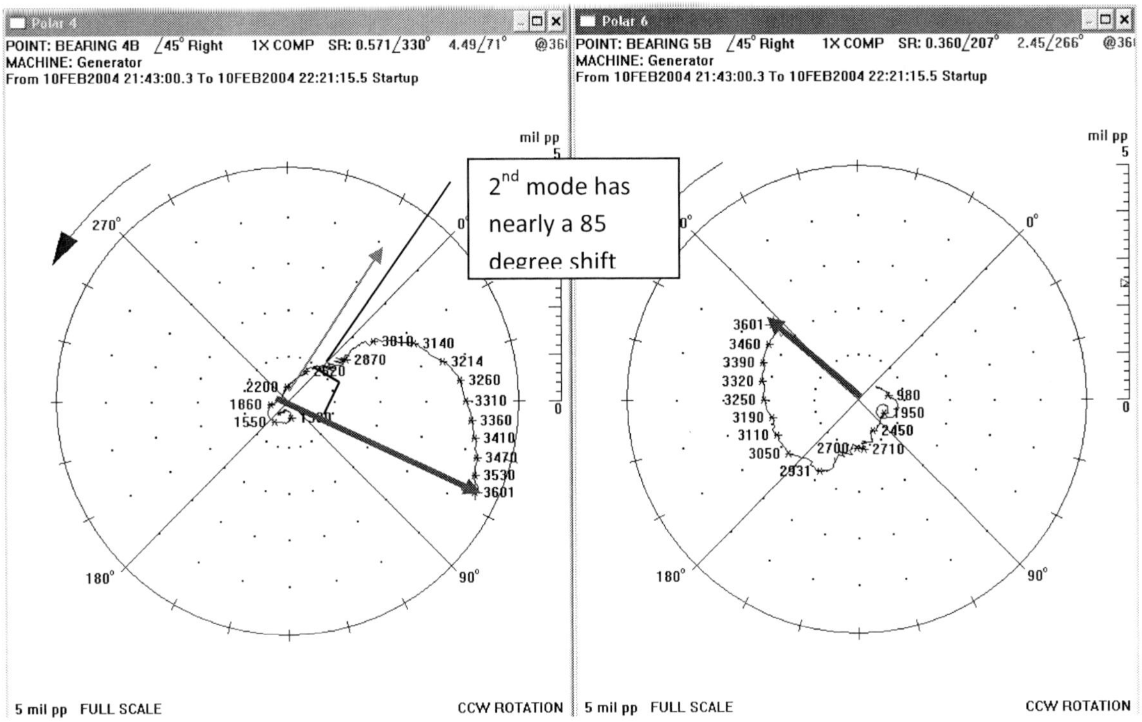

Figure 22- Nyquist plot showing 2nd critical speed response in a generator

Discussion: By looking at the phase relationship between the two ends of the shaft the kind of shot (in this case a couple pair) could be determined. In addition, the location of the couple pair could also be calculated. If a standard approach of adding a trial weight to each end would have been used, then the amplitudes on the opposite ends could have been driven high enough to trip the unit. In addition, even if that approach would have worked, it would have taken a minimum of three runs. Generators are filled with hydrogen gas. This means that to do a balance shot that the hydrogen has to be purged on each shot with carbon dioxide. The carbon dioxide has to then be purged to allow personnel to get into the generator to install the weights. After the weights are installed, the opposite purge cycle has to be performed. Each balance shot can therefore take 24 hours to complete. At several hundred thousand dollars a day down time on a large base load generator, the advantages of the approach described in this case history are obvious.

CASE 16- CROSS EFFECT ON A 501 GAS TURBINE:

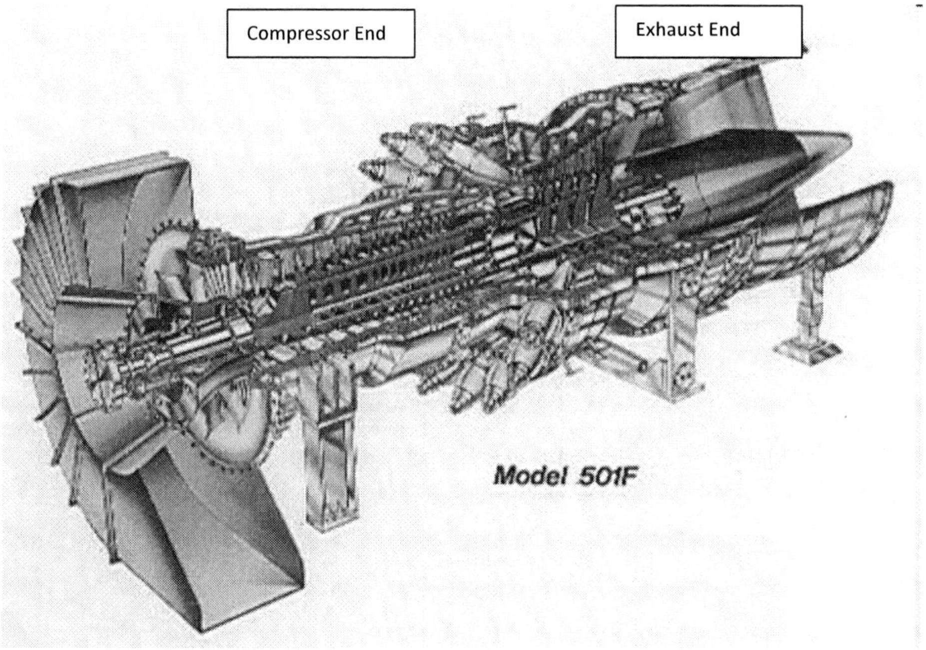

Figure 23- Drawing of 501 Gas turbine

When balancing gas turbines, if you have not balanced one before, the best thing to do might be to disregard your common sense. The reason for this is because the cross effects from one end to the other are often much greater than the direct effects. To state this more clearly, if the amplitude of the unbalance is high on one end of the turbine, then it is quite likely that the solution weight will need to be installed on the opposite end. The following are the lag angles and sensitivities from a Westinghouse 501 Gas Turbine.

Exhaust end balance shot on exhaust end	18.0 oz/mil	25 degree lag
Exhaust end balance shot on compressor end	4.7 oz/mil	46 degree lag
Compressor end shot on compressor end	18.0 oz/mil	353 degree lag
Compressor end shot on exhaust end	8.6 oz/mil	30 degree lag

In this example, if weight was added to the exhaust end, then it would have nearly 4 times as much cross effect on the compressor end as it had direct effect on the exhaust end.

Discussion: Large steam turbines and generators which have symmetrical rotors and relatively symmetrical bearing stiffnesses respond well to static-couple balance techniques. Gas turbines are a totally different animal. Their rotors are not symmetrical nor are their bearing stiffnesses. The best approach to balancing large gas turbines is to get prepared to use a true dual plane approach. On the first try, add weight in the end opposite to where the amplitude is the maximum. After reading the response, determine if it is possible to reduce the levels on both ends to acceptable values by making adjustments to the trial weight. If not, then add weight to the opposite end and perform the dual plane calculation.

CASE 17 - 7000 HORSEPOWER 885 RPM ID FAN ON A VARIABLE SPEED DRIVE:
This case history illustrates how a balance job that was already somewhat difficult was complicated by human error.

This large fan had been operating with 16-18 mils of vibration when it was brought down to have the blades re-tipped. The work was completed with the organization that re tipped the blades being hired to balance the fan. After several tries over a period of three days they requested help.
This fan had two things working against it, the first was that it was extremely out of balance following the blade re-tipping and the second was at full speed this fan ran on a resonance. The one positive thing was that it was on a variable frequency drive. In order to not destroy the fan, it was balanced at 600 RPM instead of trying to do the work at the full 885 RPM speed. That was a good thing because when back calculations were made, if the fan would have initially been brought to full speed, the amplitude would have been 92 mils. When the data from the fan rework organization was analyzed, it was apparent that they were not adding enough trial weight. The entire crew was exhausted and had been on the job site working long days getting the blades re-tipped.

After reviewing the past balance attempts, a trial weight was installed and the readings taken at 600 RPM. A solution was calculated and the weights were attached. Based upon the effect, a solution shot was calculated and the weight was installed. It was, however noticed that the lag angle was not in the expected range for the 600 RPM balance speed of around 40 degrees. The solution weight was installed and the fan restarted, but the results were not as expected. This was puzzling, but a new solution was calculated and the individuals went in to install the new weight. This was a very large fan, so it had to be rotated with a wench. The weight had to be installed in a direction clockwise 45 degrees from where the fan stopped but it was noticed that the workers were rotating the fan counterclockwise. That brought up a red flag so the worker was questioned. He had been thoroughly instructed as to the numbering system and the direction of fan rotation and swore he was doing it right. The safety people would not let the author into the fan to verify the installations because entry required confined space training along with SCBA training. This was strange because the fan is a wide open space and the plant was off line, but in any case that was their position. After explaining things to the person installing the weight three or four times, a light bulb came on and he realized that he had been using the wrong orientation. He was very apologetic and the poor guy was physically and mentally exhausted. Once everything was straightened out the fan was easy to balance. It ended up taking 23 pounds of weight. Safety has become a real issue in that the balance people are often no longer allowed to go into these large fans to verify weight installation. It is now our standard practice to number the shaft on the outside and then rotate to the position where the weight needs to be installed to the bottom 6:00 o'clock position rather than number the blades inside the fan. (See Figure 24). That way the location of where the weight is located can be verified from the outside.

Figure 24- Exterior marking of fan shaft used for weight orientation

Discussion: If you cannot get into a fan or turbine because of safety or other considerations, then clearly mark the shaft so you can see the angular locations while standing outside. Once this is done, tell the person installing the weight to put it for instance "directly on the bottom". This approach avoids dealing in terms like "left and right or clockwise and counter-clockwise". Those terms can cause confusion because they change depending upon which end you are looking from.

CASE 18 – HUMAN AND INSTRUMENTATION ERRORS: This paper has been about difficult balance problems. The primary reasons for balance jobs being unusually difficult are usually mechanically related. There are other factors that can contribute to things not working out well in the balancing arena. Instrumentation and human error can complicate what might have been a relatively simple problem. The following are some of the problems that the author has encountered over the years.

1. Tach pulse hardware that moves with time. When an unbalanced machine is ran for the first time or after an unsuccessful trial weight, the vibration can be quite high. Most tach pulse photocells are mounted on a magnetic pickup with a flexible connection. When high vibration is present, the magnet can slide or the connection move. If not noticed, this can result in a change of phase that was not due to the trial weight. The solution that is calculated is therefore not valid.
2. Tach pulse failures that can be the result of RF interference from variable frequency drives, static discharge, or excessive water vapor in the air.

3. Inadequate pulse width to allow for calculation of phase. This can happen on high speed machines that have too narrow a piece of reflective tape or too narrow of a key phaser notch.
4. Sloppiness in installing a protractor on a shaft. Accurate installation of both the trial and solution weights is vital.
5. Personnel getting mixed up in regards to the direction of rotation of the shaft. This is a particular issue when installing weight in opposite ends of a rotor.
6. Fatigue. This can be the single most important issue.
7. Being rushed. Many times balancing can get on the critical path to getting a facility back into production.

SUMMARY:

Balancing can be a fairly simple task, there are however many instances where things can get quite complicated. Thermal vectors in motors and generators, rotor bows, rubs, looseness, mode shapes, asymmetric rotors and support stiffnesses, internal and external cross effects and beats along with instrumentation and human errors can significantly complicate the balancing process.

The following is the author's attempt to establish different classes of balancing from simple to nearly impossible:

Description	Difficulty	Comments
1: Simple single plane balancing	1	Always check rotation and base bolts
2: Simple dual plane balancing	2	Record responses carefully
3: Cross effect from another machine	2	Use synchronous time averaging
3: Thermal vector present	3	Balance at load
5: Shaft bow	3	Wait until readings stabilize
6: Nonlinear response	3	Two steps first lower level, then restart process
6: Operates near resonance	4	If possible balance off resonance
7: Multiple modes	4	If possible use static couple technique
8: Asymmetric properties	4	Most likely multi-plane balance program
9: Flimsy low damped support	4	Make small gradual moves- accept higher levels
10: Loose rotating component	5	Forget it
11: Fatigue	5	Get rest or consult with someone else

There are of course nightmare scenarios. For instance, you could have a large electric motor that runs near a critical, that has a thermal vector, cracked rotor bars that cause a beat, and also has bearings loose in the housing and a rotor loose on its shaft combined with sitting on a bad base. I think I will ask the new guy that inspired this paper "who had mastered balancing" to take that one on.

Biography:

NELSON L. BAXTER is President of ABM Technical Services, Inc. and co-founder of Machinery Health Monitoring and Electro-mechanical Diagnostic Services. His 34 years of experience in the utilities industry include trouble-shooting rotating equipment, structural testing, and balancing. He is a contributing editor of Sound & Vibration Magazine, a Registered Professional Engineer, a Level IV Certified Vibration Analyst and Instructor for Level IV Class. Baxter holds an M.S. in Nuclear Engineering from Purdue University. He is a member of the Institute's Board of Directors and ASME.

FINITE ELEMENT ANALYSIS: AN OVERVIEW OF APPLICATIONS TO VIBRATION

Robert J. Sayer, PE
Applied Structural Dynamics
Medina, Ohio
sayerinc@bright.net

Abstract: Finite Element Analysis (FEA) is a numerical technique that can be used to approximate the structural dynamic characteristics of vibrating mechanical systems, the understanding of which is paramount to any root-cause failure study involving excessive vibrations. The finite element technique can be used for structural dynamic studies of existing equipment or to evaluate the dynamic characteristics of machines and structures prior to fabrication. FEA models can be used to approximate "stress stiffening" effects in rotating mechanical components (i.e. fan wheels, pump rotors, motor shafts). This paper introduces the basic principles of the method and the factors that affect the accuracy of the numerical solution. Applications of FEA to vibration analysis are presented.

Key Words: Finite element analysis, matrix methods, numerical analysis, resonance, shape functions, stress stiffening, structural dynamics.

Introduction:

Part of any root-cause vibration study should include a solution that addresses the problems identified by the study. Solutions for vibration problems involving the excitation of a natural frequency (resonance) are frequently difficult to obtain based solely upon experimental data. Although an intuitive interpretation of the experimental data may suggest stiffening a component of a machine, the details and the effectiveness of a modification could be difficult to access. Furthermore, in many cases it is difficult to determine the effect of a modification on other natural frequencies that may not currently be a problem, but could become one after implementation of the modification.

The finite element method is a numerical technique that is suited to perform analysis of complex structural-mechanical systems. A finite element model is typically much more detailed than an experimental model (operating deflection shape or experimental modal analysis). Finite element models contain thousands of degrees-of-freedom versus an experimental model that usually include hundreds of degrees-of-freedom or less. The finite element model of a centrifugal fan rotor included in Figure 1 contains over 100,000 degrees-of-freedom. Because of this level of refinement, a finite element model provides the vibration analyst with an unparalleled approach at evaluating the effect of even the finest details of any mechanical design on the structural dynamic characteristics thereof.

The finite element method focuses on calculating the behavior and response of a continuum that consists of an infinite number of points. In a continuum problem, a field variable such as displacement or velocity contains an infinite number of possible values, since it is a function of each point in the continuum. The task of solving the continuum problem is simplified using a finite element representation that divides the continuum into a finite number of subdivisions called *elements*. The elements are connected at *nodal points* into a mesh or finite element model. The process of dividing the continuum into a finite number of elements makes the solution provided by the finite element model an approximation to the theoretical solution.

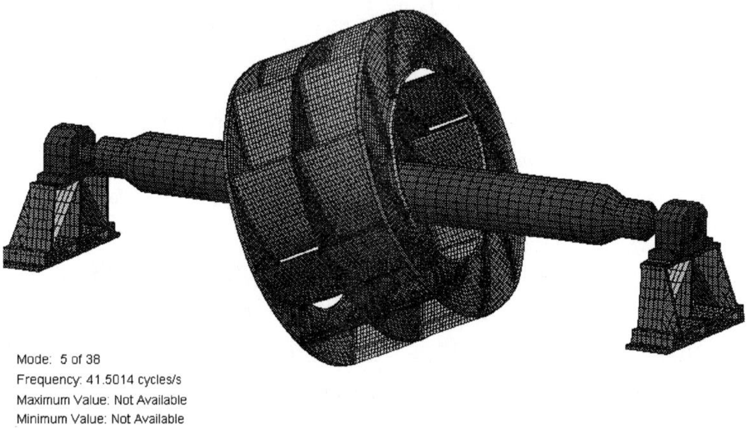

Figure 1: FEA Model of Centrifugal Fan

The stiffness and inertial properties of each individual element is defined by a mathematical displacement or interpolation function (linear, quadratic, etc.), the elastic material properties (modulus of elasticity, Poisson ratio, and material density), the size of the element and its connectivity and relationship to all other elements in the model. The assemblage of all elements produces an estimation of the stiffness and inertial properties of any complex structural-mechanical system.

<u>Principles of FEA:</u>

Finite element analysis involves the solution of the differential equations of motion using concepts of linear algebra. It consists of discretizing a continuum into a finite number of points, from which stiffness and mass matrices are formed. The solutions require matrix manipulations that, although computationally intensive, easily lend themselves to computer solutions.

For static analysis, the FEA solution is obtained from the knowledge that force equals stiffness multiplied by displacement. In matrix form, this is:

$$[K]\{x\} = [F] \qquad \text{or} \qquad \{x\} = [K]^{-1}[F]$$

where [K] is the stiffness matrix, $[K]^{-1}$ is the inverse of the stiffness matrix, {x} is the displacement vector, and [F] is the force matrix. A global stiffness matrix is defined by the assemblage of finite elements, each having its own individual stiffness matrix. The force matrix is defined by the magnitude of all forces and their placement within the finite element model. The only unknowns are the displacements.

The matrix formulation for dynamic analysis is more complex than for static analysis. The solution is obtained from the following linear algebra formulation:

$$[M]\{d^2x/dt^2\} + [C]\{dx/dt\} + [K]\{x\} = [F(t)]$$

where [C] is the damping matrix, [M] is the mass matrix, {x} is the displacement vector, {dx/dt} is the velocity vector, $\{d^2x/dt^2\}$is the acceleration vector, and [F(t)] is the dynamic force matrix.

For modal analysis, the matrix formulation is:

$[K] + \{\omega^2\}[M] = 0$ where $\{\omega^2\}$ is a vector of natural frequencies (eigenvalues).

<u>Accuracy of Approximate Solution:</u>

It is important to realize that the finite element method is an approximate numerical technique. The accuracy of solution obtained by a finite element analysis is dependent upon several factors, including but not limited to the following:

1) Degree of refinement of the finite element mesh/Number of elements
2) Appropriateness of the finite element types used to model a machine/structure
3) Shape function of the finite element
4) Distribution of Mass (Lumped versus Consistent)
5) Boundary conditions used at the limits of the finite element model.
6) Resolution between modeled conditions and actual condition.
7) Material Property variation from normally accepted values.
8) Aspect Ratio & Skewness of the Elements

NUMBER OF ELEMENTS:

Generally, the accuracy of the finite element solution increases as the number of elements increases. Figure 2 illustrates the effect that mesh refinement has on the accuracy of a finite element solution. Three separate finite element models were developed to estimate a complex double curvature mode shape of a beam. Each model contained more elements than the previous. The 1st model was constructed from only two linear elements (3 nodes), the 2nd model used four elements (5 nodes) and the 3rd model used eleven elements (12 nodes).

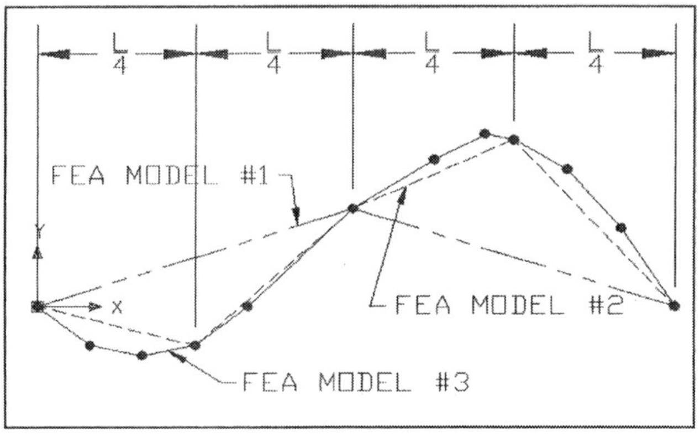

Figure 2: FEA Approximations

The first model that used only two elements provided a grossly inaccurate estimation for the complex mode shape. It does not contain a sufficient number of degrees-of-freedom to describe the desired mode. Since this model contains only one internal nodal point, located at mid-span of the beam, it is not capable of providing the correct mode shape, providing a mode that resembles the simple flexural mode of the beam instead of the complex double curvature mode. The inadequate number of elements used in the first model to represent the flexural characteristics of the beam resulted in a visual aliasing.

The 2nd finite element model provided a better approximation of the complex mode shape. This model had nodes located at the quarter points of the beam and was able to approximate the double curvature characteristics of the mode shape, albeit not as smoothly as the 3rd finite element model. In addition to providing a better evaluation of the mode shape, the 3rd finite element model will also provide a more accurate estimation for the natural frequency associated with the mode. The number of elements required to provide good results is dependent upon many factors, including the complexity of the mechanical system, the complexity of the mode shape being studied, element types, material properties, non-linearities, etc.

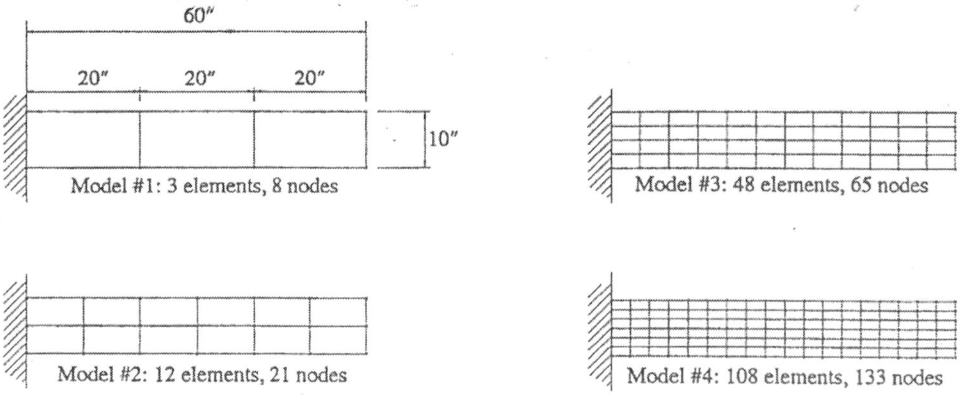

Figure 3: FEA Models of Cantilever Beam

Consider the FEA model for a cantilevered slab. Figure 3 shows four different models, each having more elements than the previous; Model #1 has 8 nodes and 3 elements, Model #2 has

21 nodes and 12 elements, Model #3 has 65 nodes and 48 elements, and Model #4 has 133 nodes and 108 elements. The estimated natural frequency for the initial three flexural modes is summarized in Table 1:

TABLE 1

MODEL	# ELEMENTS	MODE #1	MODE #2	MODE #3
1	3	140.9 Hz	880.8 Hz	2499 Hz
2	12	104.3 Hz	611.1 Hz	1656 Hz
3	48	92.2 Hz	525.5 Hz	1354 Hz
4	108	88.7 Hz	508.6 Hz	1295 Hz

The theoretical solutions for the three natural frequencies are 87.6 Hz, 49.0 Hz and 1260 Hz. This example shows that the accuracy of the FEA solution increases as the quantity of elements that are used to model the beam increase. Also, the number of elements required to provide an accurate estimate for the natural frequency increases with the complexity of the mode shape. For the simple principal flexural mode, the natural frequency estimate provided by Model #1 (140.9 Hz) is around 1.59 times larger than the estimate of Model #4 (88.7 Hz). For the more complex third mode, the estimate provided by Model #1 (2499 Hz) is around 1.93 times larger than the estimate of Model #4.

CONSISTENT versus LUMPED MASS:
The manner in which the mass of an individual finite element is considered is dependent upon the formulation of that element. One formulation distributes the mass amongst the node points within the element and simply lumps the distributed mass at each nodal point. The resultant mass matrix obtained from this formulation is called a lumped mass matrix.

The other formulation distributes the mass throughout the element. The resultant mass matrix for this type of element is called a consistent mass matrix. The complexity of the element formulation that uses a consistent mass matrix is more rigorous than an element using a lumped mass matrix. However, the consistent mass matrix has advantages over the lumped mass matrix approach.

The results contained in Table 1 were obtained from FEA models with consistent mass matrices instead of a lumped mass distribution. The estimation for natural frequency converges from above the theoretical solution. In all cases, the FEA solution is larger than the actual theoretical solution, getting closer as the mesh is refined by increasing the number of elements. Convergence from above is always guaranteed when using a consistent mass distribution. This is important when evaluating the results of a finite element analysis. If a consistent distribution of mass is used instead of a lumped mass distribution, then the actual natural frequency will always be less than the calculated natural frequency (subject to boundary condition effects).

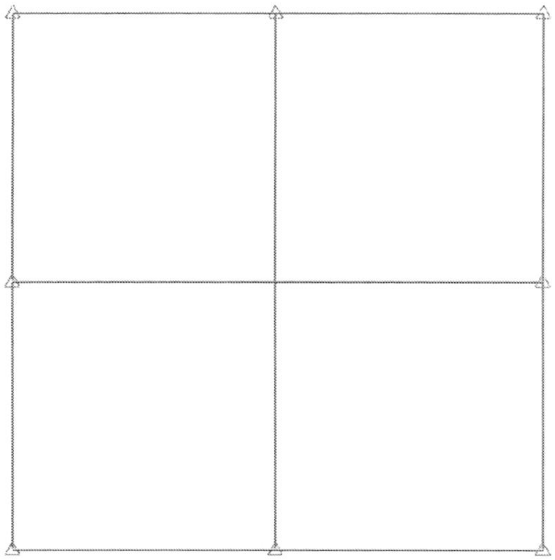

Figure 4: FEA Model of Rectangular Plate

To illustrate the difference between the lumped mass versus distributed mass approaches consider a simple FEA model for a 100"x100" square plate, rigidly constrained along all of its boundaries. This model, shown in Figure 4, employs a very coarse 2x2 element array (4 linear elements with a lumped mass formulation), clearly not enough to provide an accurate solution for the natural frequency of even the principal flexural mode. The theoretical solution is:

$$f_n = 5.728\,[D/\rho t a^4]^{1/2} \quad \text{where } D = Et^3/12(1 - v^2)$$

The dimensions of the plate are 100" x 100". The properties are t = 1", E = 30E06 psi, v = 0.30, and ρ = 0.0007 lb-sec^2/in^3/in. Substitution of the dimensions and material properties yields the theoretical solution of 35.34 Hz for the natural frequency associated with the principal flexural mode. The FEA model provided an estimation of 30.63 Hz for the same natural frequency. In this case, the FEA does not overestimate the natural frequency and does not converge on the actual theoretical solution from above. It underestimates the natural frequency because the lumped mass approach results in a significant amount of mass being placed at mid-span of the plate. There is only one node that is internal to the boundary of the model.

The inaccuracy introduced by the lumped mass approach is minimized in models that use a large number of elements and divide the continuum into many small individual pieces. In these cases, the use of lumped masses does not provide a solution that deviates significantly from that provided by elements using a consistent mass matrix. For example, consider a model of the same square plate using an 8x8 array of elements (64 total linear elements) as shown in Figure 5. This more refined FEA model estimates the natural frequency at 35.96 Hz. The more refined model overestimates the natural frequency. Even though each element uses a lumped mass formulation, the overall distribution of mass approaches a consistent mass formulation, since it is spread over a large number of elements.

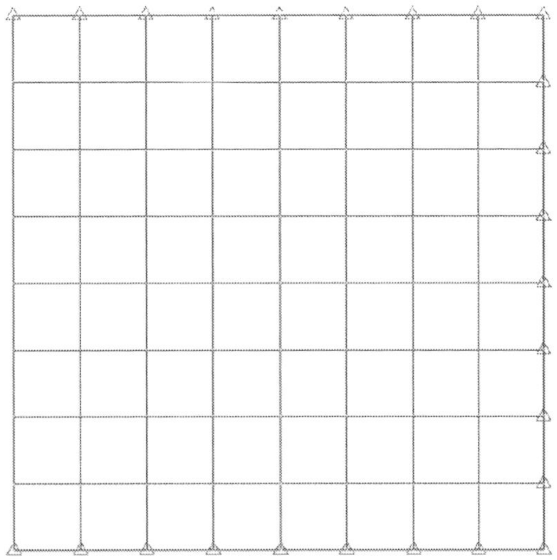

Figure 5: Another FEA Model of Rectangular Plate

Next, the model was refined even more by using a 10x10 array of elements (100 total linear elements). This model provided an estimate of 35.80 Hz for the natural frequency, which is closer to, but still overestimates the actual natural frequency. The solution is now converging from above.

SHAPE FUNCTION:

The accuracy of the finite element solution is affected by the shape function used in the development of the element. An *isoparametric* element uses the shape function to conformally map the element to fit the finite element model. The shape function is also used to calculate the displacement at the nodal points based upon the solution determined for the integration points within the element.

The shape functions used for elements in commercially available FEA codes are developed by the authors of the codes and are typically transparent to the user/analyst. Elements with mid-side nodes, in addition to corner nodes, indicate a higher order shape function. Figure 6 shows both a 4-noded element that uses a linear shape function and an 8-noded element that uses a quadratic shape function.

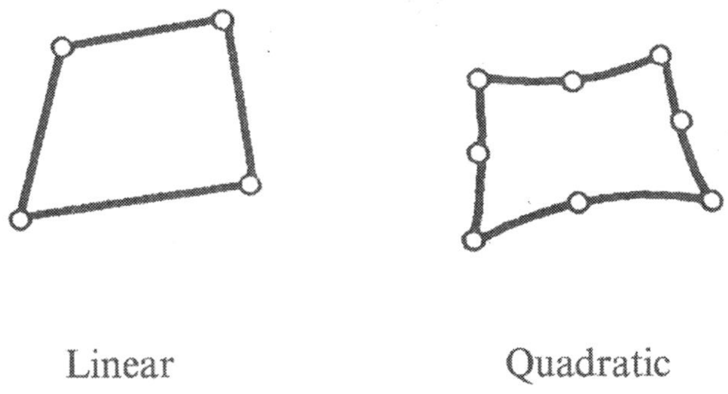

Figure 6: Elements with Different Shape Functions

The accuracy of the approximate solution can be improved by increasing the complexity of the shape function used in the formulation of the individual finite element. To illustrate this, consider the first two FEA models in Figure 3 for the cantilevered slab. Table 2 shows the results obtained by using elements with a linear shape function and a quadratic shape function.

TABLE 2

MODEL	ELEMENTS	SHAPE FUNCTION	MODE #1	MODE #2	MODE #3
1	3	Linear	140.9 Hz	880 Hz	2499 Hz
1	3	Quadratic	89.2 Hz	513 Hz	1380 Hz
2	12	Linear	104.3 Hz	611 Hz	1656 Hz
2	12	Quadratic	87.8 Hz	497 Hz	1261 Hz

BOUNDARY CONDITIONS:
Another factor that affects the accuracy of the finite element solution is boundary conditions. Boundary conditions are mathematical constraints applied to the finite element model to account for items that are not included. For example, the modal characteristics of a rotor are usually estimated by a finite element analysis of the shaft and impeller. The bearing conditions are entered into the model as either rigid constraints or elastic springs. The accuracy of the finite element solution is dependent upon the accuracy of the bearing stiffness used in the analysis.

MODELED CONDITIONS vs. ACTUAL CONDITIONS:
How to mathematically formulate boundary conditions is not always a clear-cut decision. To illustrate the difficulty of defining boundary conditions, consider the structural mechanical system for the pump skid shown in Figure 7. It consisted of a single-stage centrifugal pump, directly driven by an induction motor, both directly supported on a base supplied by the manufacturer of the pump. The pump base was welded to the top plate of an equipment skid. Furthermore, several anchor bolts were installed that passed through the equipment skid, into the concrete floor slab. The top plate of the equipment skid was welded to structural channels around its perimeter. Furthermore, there was an interior grid of structural channels framing between the perimeter channels. The top plate was set on and was structurally supported by these interior channels, but was not welded to them. Two of the interior framing channels were located directly underneath the pump base. The question was "what should be included in the FEA model, and what are the boundary conditions?"

Figure 7: Pump Skid on Equipment Base

An impact test was performed on the actual installation which identified a natural frequency at 52.3 Hz. An experimental modal test determined that this natural frequency was associated with a rocking mode of the motor in the horizontal plane. This information was used to correlate the FEA model and evaluate whether boundary conditions were appropriate. Figure 8 contains the results of a finite element analysis of an FEA model that treated all of the underlying framing channels as though they were solidly attached to the top plate of equipment base. Boundary conditions were defined that assumed only the perimeter channels of the equipment base and the anchor bolts were rigidly fixed to the concrete floor. This FEA model provided an estimate of 63.7 Hz for the natural frequency that field tests determined to be 52.3 Hz. Clearly, the boundary conditions used in this FEA model, although seemingly appropriate, overestimated the stiffness of the actual boundary conditions.

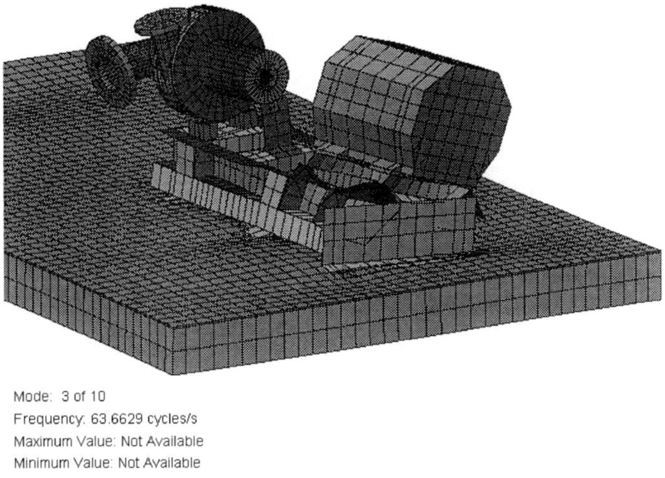

Figure 8: Predicted Mode Shape (FEA Model 1)

Next, the top plate of the equipment base was completely released from the interior framing channels, and the channels were completely removed from the model. This finite element model completely neglected the potential support that could be provided by the underlying structural channels, even though several were located directly under the pump base. It provided the result shown in Figure 9. The calculated natural frequency of 51.8 Hz was much closer to the actual conditions (52.3 Hz). This result was surprising, but not completely unexpected. Since the top plate of the equipment base was not rigidly attached to the structural framing members below it, resistance to the upward movement of the pump and motor base was not provided by the underlying channels. The resistance to downward movement could only be achieved if there wasn't any gap between the top plate and the framing channels. For this case, the finite element model would only correlate to actual data if the structural support of the equipment base was completely ignored in the analysis. This would not be intuitively obvious.

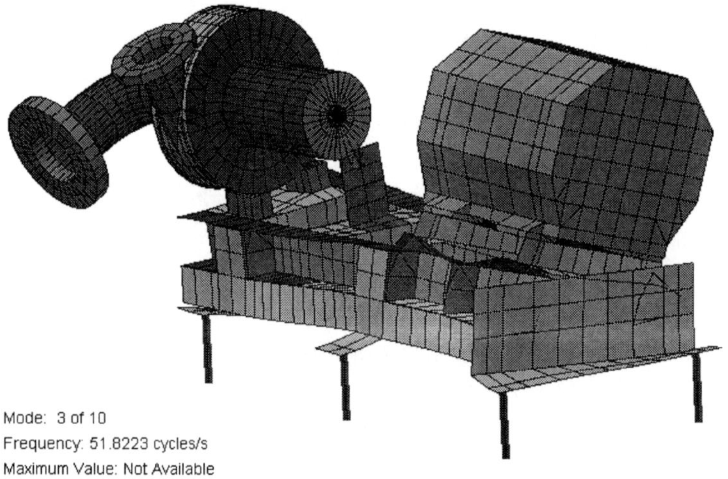

Figure 9: Predicted Mode Shape (FEA Model 2)

MATERIAL PROPERTY VARIATION FROM PUBLISHED DATA:

Standard equations of motion indicate that the natural frequency of a structural-mechanical system is dependent upon its mass and stiffness. The static stiffness is dependent upon the modulus of elasticity of the materials which the system is constructed from. The modulus of elasticity is determined from a stress-strain test that is typically performed with a rather slow application of force. Many materials respond differently when subjected to dynamic forces at high strain rates.

The effective modulus of elasticity of a material, when subjected to high strain rates, can increase significantly from the published value. If not accounted for, this can result in a finite element analysis underestimating natural frequencies. Two examples are provided below to illustrate the tendency of finite element analysis to underestimate the natural frequency associated with mode shapes of centrifugal fans. These modes have complex deformations patterns and occur at higher frequencies.

Figure 10: DWDI Centrifugal Fan

To illustrate this effect, consider a large centrifugal fan similar to that shown in Figure 10. This type of fan wheel will have several modes of natural frequency, referred to as the n-nodal diameter modes, where the shape consists of a complex radial distortion. The degree (or number associated with the mode) depends on the number of inflection points it has. Figure 11 shows the deformation pattern of two of the n-nodal diameter modes as viewed from the end of the fan wheel. A 2-nodal diameter mode will have two peak-positive and two peak-negative radial displacements. A 4-nodal diameter mode will have four peak-positive and four peak-negative radial displacements. The natural frequencies of these modes are high, for the mode shapes shown in the figures below they are 128 Hz and 247 Hz, respectively. The high frequency and complex deformation pattern of these shapes during resonance results in application of stress at a high rate.

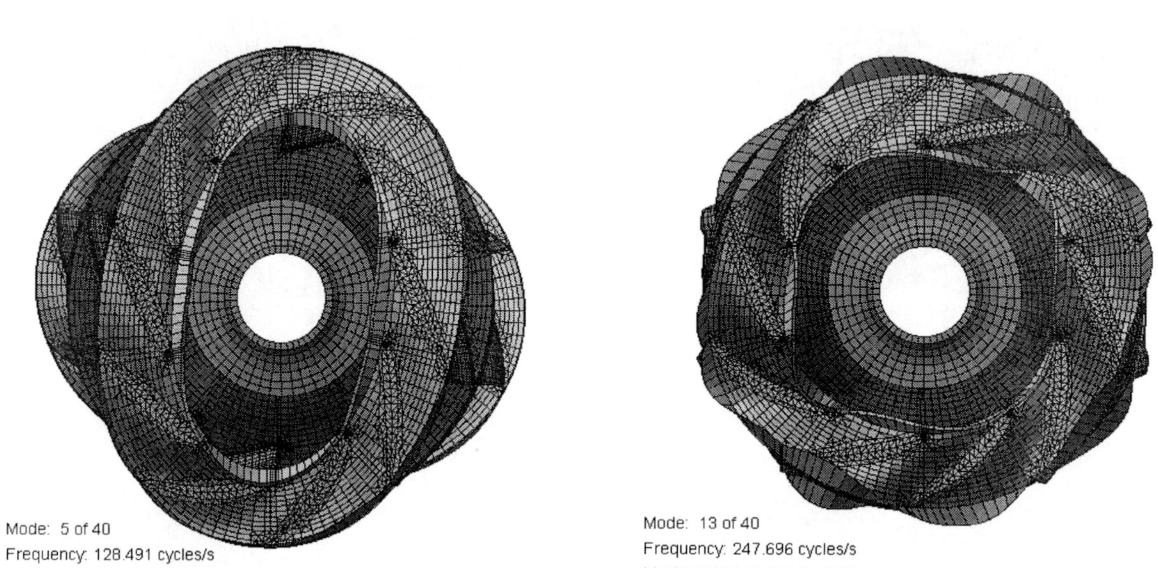

Figure 11: 2-Nodal and 4-Nodal Diameter Modes

EXAMPLE 1: The results of a finite element analysis are compared below to data acquired from a natural frequency test for an11-bladed DWDI fan wheel, designed to operate at 705 rpm.

2-Nodal Diameter Mode	FEA = 81.1 Hz	EMA = 91.0 Hz	FEA/EMA = 0.891
3-Nodal Diameter Mode	FEA = 120.4 Hz	EMA = 147.8 Hz	FEA/EMA = 0.815
4-Nodal Diameter Mode	FEA = 150.6 Hz	EMA = 185.0 Hz	FEA/EMA = 0.814

The FEA model contained a significant number of elements, which minimized the effect of the lumped mass formulation of the elements. Yet, the finite element analysis underestimated the natural frequencies by 11.9% - 18.6%. The amount of discrepancy between calculated values and actual data became greater as the complexity of the mode shape and the natural frequency (strain rate) increased.

EXAMPLE 2: The results of a finite element analysis are compared below to data acquired from a natural frequency test for a 10-bladed DWDI fan wheel, designed to operate at 1,200 rpm.

2-Nodal Diameter Mode	FEA = 92.0 Hz	EMA = 97.5 Hz	FEA/EMA = 0.944
3-Nodal Diameter Mode	FEA = 148.5 Hz	EMA = 165.0 Hz	FEA/EMA = 0.900
4-Nodal Diameter Mode	FEA = 175.3 Hz	EMA = 195.5 Hz	FEA/EMA = 0.897
5-Nodal Diameter Mode	FEA = 185.4 Hz	EMA = 207.0 Hz	FEA/EMA = 0.896

Once again, the FEA model contained enough elements to minimize the effect of the lumped mass formulation. The finite element analysis underestimated the natural frequencies by 5.6% - 10.4%. The discrepancy between calculated values and actual data became greater as the complexity of the mode shape and the natural frequency (strain rate) increased.

The finite element analyses listed for the above examples were performed by different analysts, using different FEA software, and employing different modeling techniques. Some of the variation between numerical analysis and experimental data is due to differences in modeling technique. However, the FEA results consistently underestimate the actual test results, which is contrary to the expected convergence of finite element analysis.

Many designers of fan wheels recognize the fact that finite element models of fan wheels, although appearing to be appropriate, consistently underestimate the natural frequencies associated with the n-nodal diameter modes. They apply factors to the FEA results to account for the discrepancy.

ASPECT RATIO & SKEWNESS

Another factor that is extremely important to the accuracy of the finite element solution is the variation of the elements used to model something relative to that of the parent element in the finite element library. This factor can be evaluated based upon the aspect ratio and skewness of each element in the model. The *aspect ratio* is the ratio of the length of one side of an element to the length of any other side.

Many FEA software codes contain isoparametric element formulations, where a parent element contained in the software library is mapped onto the actual physical model, using the shape functions as the means of co-ordinate transformation. This technique allows for a minimal amount of element formulations to be included in the library. These parent elements are typically perfectly square or cubic, depending upon the element type. Major deviations between parent elements and actual elements used in the model can lead to inaccuracies in the solutions provided by the model.

Correlation of FEA with Experimental Data:

For cases where equipment and structures exist and are in operation, experimental data can be used to correlate finite element models, providing insight into the nature of boundary conditions. Impact tests will identify natural frequencies and provide information relative to the amount of motion at bearings upon which a finite element model can be calibrated. The refinement of the model and the numerical value of boundary conditions can be modified until good correlation is achieved between the numerical calculations and the experimental results.

The appropriateness of the finite element type used to describe a structural or mechanical component can also affect the accuracy of the finite element solution. For instance, it may be inappropriate to use thin plate elements to model a thick concrete machine foundation. An assemblage of three dimensional solid elements would provide a more accurate solution.

Structural Dynamic Modification Studies:

Because a finite element model provides a more detailed mathematical description of a mechanical system than an experimental model, it is better suited for structural dynamic modification studies. An experimental model is typically a stick diagram that vaguely resembles the mechanical system being studied. The evaluation of structural modifications using experimental models is very limited and should be done only with an understanding of the approximations involved.

Figure 12 shows the mode shape for a natural frequency of an aluminum coil winder machine that was being excited during operation. The process line that the winder serviced operated at a constant speed. The rotational speed of the winder varied since the diameter of the coil on the winder was constantly changing. Several resonances were excited during the winding process as the frequency of the dynamic force produced by the rotating elements of the machine coincided with natural frequencies.

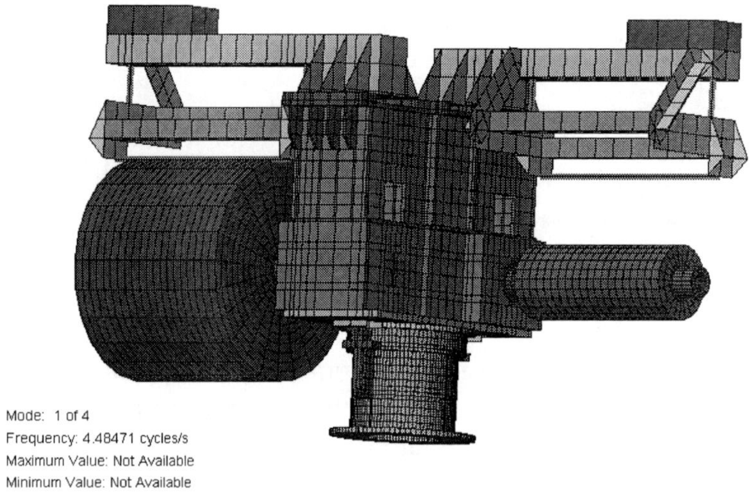

Figure 12: FEA Model of Winder

The finite element model was then used to formulate a modification to eliminate the resonance issues. Figure 13 shows the same winder, this time with a hydraulically actuated roll support that engages the machine during operation. The natural frequency was increased from 4.48 Hz to 8.68 Hz.

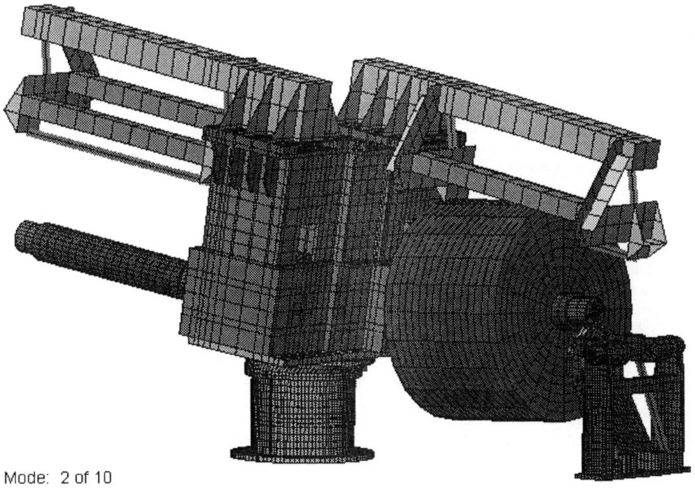

Figure 13: FEA Model of Modified Winder

Structural Dynamic Design Audits:

The finite element method provides the vibration analyst with a numerical technique to evaluate the structural dynamic characteristic of a machine and/or structure during the design stage, prior to fabrication. Natural frequencies and mode shapes can be predicted to insure that equipment does not operate at or near any resonant frequencies.

The stiffness of supporting structures/foundations or lack thereof can be evaluated. Areas of structural weakness or probable resonances can be addressed during the design stage where changes are easily accomplished. Once a structure or machine is installed, modifications can be limited by access restrictions, difficult to implement due to the nature of the problem (i.e. soil-foundation interface stiffness), and very expensive. An analysis that supports the structural dynamic design of a mechanical system can minimize potential problems in the field and reduce the cost of a project. The cost to modify a machine or structure in the field, after the installation thereof, can be very expensive. Vibration problems originating from the resonant excitation of equipment can result in delayed plant start-up and increased maintenance outages. The cost of lost production can be tremendous, evaluated in some cases as millions of dollars per day.

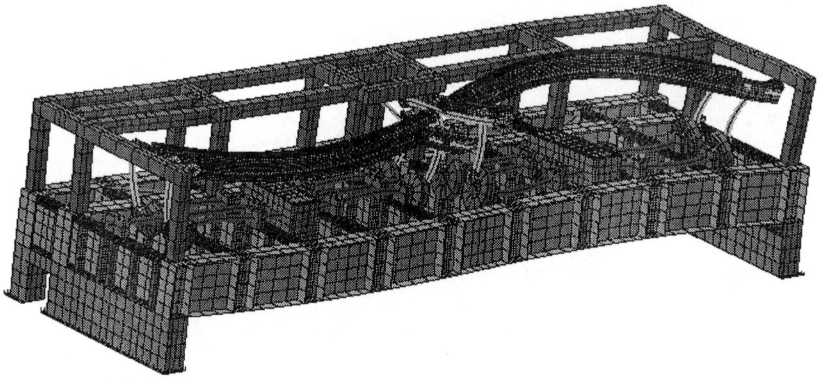

Figure 14: FEA Model of Straw Sewing Machine

Figure 14 contains the results of a finite element model developed to assist with the design of a proprietary machine. It is a machine that sews straw into sheets used for erosion control of seeded slopes, particularly in highway construction. The finite element model was used to evaluate the structural dynamic characteristics of the proposed machine design, identify potential problem areas, and provide corrective changes during the design stage. This is definitely a pro-active tool, providing an analysis with warnings of potential problems prior to fabrication and installation.

Finite element evaluation during the design stage is recommended for any machine that could have a significant impact on the productivity of an operation. The cost of a numerical study is insignificant compared to that of delayed project completion, lost production, and/or equipment repair or replacement after installation. Modifications can be made on paper instead of in the field.

FEA of Rotating Components:

Impact tests performed to identify the natural frequencies of rotating components of machines are done while the machine is "at rest". During operation, as the machine part rotates, the natural frequencies can change due to centrifugal stiffening and gyroscopic effects. The amount of increase due to centrifugal stiffening and gyroscopic effects can vary significantly and is dependent upon several factors including the rotational speed, stiffness, and mass of the rotor.

For cases where an impact test identifies the "at rest" natural frequency of a rotor to be less than the operating speed, it is important to evaluate the effect of centrifugal stiffening and gyroscopic effects to insure that during operation the rotor does not become resonant. Excitation of the natural frequencies of a rotor, such as a fan wheel and shaft, can lead to the sudden and catastrophic failure.

An impact test performed to identify the "at rest" natural frequencies of a centrifugal fan rotor, including shaft and fan wheel. According to the impact test, the principal natural frequency of the fan rotor was 15.25 Hz (915 rpm). The nominal operating speed of the fan was 20 Hz (1200 rpm). Since the "at rest" natural frequency was less than the operating speed of the fan, it became necessary to evaluate the effects of gyroscopic and centrifugal load stiffening.

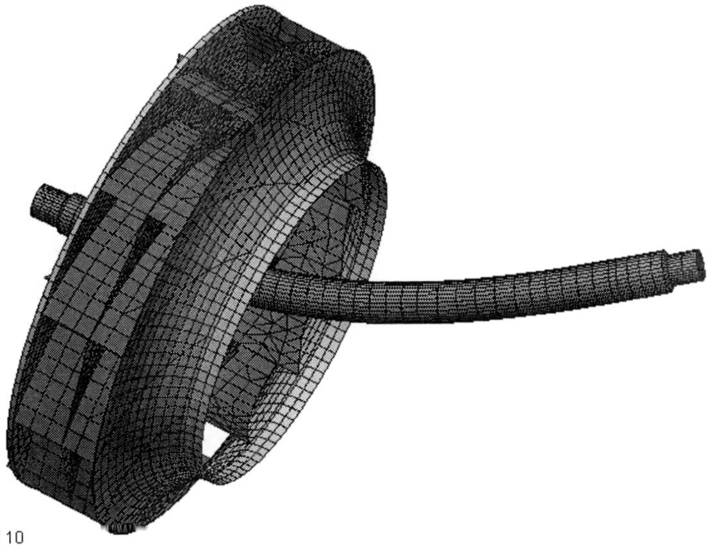

Figure 15: Mode Shape @ 15.6 Hz

A finite element model, shown in Figure 15, was developed to determine the modal parameters of the fan rotor. It was correlated to the experimental data, which provided information relative to the stiffness of the bearings that was used as boundary conditions for the numerical model. The number and type of elements (refinement of model) were modified until agreement between numerical predictions and experimental data was reached. The final correlated finite element model provided an estimate of 15.6 Hz (936 rpm) for the "at rest" natural frequency. The numerical estimation was very close to the experimental data. The subsequent stress stiffened finite element analysis indicated that the increase in natural frequency that could be expected during operation would not be sufficient to remove the fan from possible resonant excitation. Either the shaft or fan wheel would have to be modified. The finite element model was then used to develop an acceptable modification which included a thicker back plate for the wheel and a larger diameter shaft.

The finite element method can also be used to estimate stress levels in rotating equipment during operation. This includes both centrifugal stresses and stresses due to the application of dynamic loading, such as unbalanced forces. The stress results can then be input into a fatigue algorithm to predict the life of a mechanical component. Centrifugal stresses are generally associated with low cycle fatigue, being applied and removed once per start-up. Stresses due to dynamic forces produced by rotating equipment are typically associated with high cycle fatigue, being cycled once or more per revolution.

Conclusion:

Experimental methods such as spectral analysis, operating deflection shape testing, and experimental modal analysis can be used in any root-cause study to identify the source and nature of a vibration problem. However, using the experimental data to formulate solutions to the problems identified during the root-cause study can, in some cases, be very difficult. The finite element method provides the vibration analyst with a powerful numerical tool that can be used to evaluate the effectiveness of proposed solutions to vibration problems, regardless of the complexity of the modification. Because a finite element study uses numerical models that have a greater degree of resolution and refinement than experimental models, accuracy of results obtained from the finite element analysis will be superior to that of any experimental model. Finite element analysis is a valuable tool in evaluating the structural dynamic characteristics of machines and structures during the design stage, even prior to prototyping. The method can be used to estimate operating stress levels and fatigue life of mechanical components and to estimate the natural frequencies and mode shapes for equipment and supporting structures.

STUDY OF THE DYNAMIC BEHAVIOR OF A PLANETARY GEARBOX BY VIBRATION ANALYSIS, OPERATING DEFLECTION SHAPE AND FINITE ELEMENT METHOD

Santos, E. R. O.[1]
edmilson.santos@skf.com
Severino, R. H.[1]
henrique.severino@skf.com
Santana, S. S.[1]
silas.santana@skf.com
Correr, E. M.[1]
mario.e.correr@skf.com
Massoti, J. G.B.[1]
joao.massoti@skf.com
1-SKF, Cajamar, Brazil

Abstract:
Planetary gearbox presented failures of internal components. In order to solve this problem the following techniques were used: Vibration Analysis, Operating Deflection Shape and Numerical Simulation. Vibration Analysis was aimed to identify vibration levels and operational frequencies of the equipment. In this case dynamic misalignment of the first and second stages of planetary was observed. ODS technique allowed to evaluate the operational modes presented by the structure of gearbox, and so, regions with relative movements in the operational frequencies of the first and second stages were identified. Mode shapes calculated by the Modal Analysis are similar to operational modes obtained from ODS. Therefore, applying structural modifications proposal, a significant increase of system stiffness was observed: the first and second natural modes had its frequencies increased from 32.38 Hz and 46.94 Hz (current condition) to 66.82 Hz and 84.54 Hz (proposal), respectively. Also in numerical simulation, Component Modal synthesis technique was used, which enabled the best bearing arrangement for the gearbox. After application of structural modifications, the vibrational levels comply with ISO 10816-3, due to structure resonance elimination.

Keywords: Bearings; Mode shape, Natural frequency; Operating Deflection Shape; Planetary gearbox; Vibration

1. Introduction:

A planetary gearbox exemplified in Figure 1, presented failures on internal components (1st and input shaft). In order to identify the cause of these failures a Dynamic Analysis study of the planetary gearbox was performed, which provided proposals to solve this problem. Such proposals were implemented and new measurements were performed in this system, and thus, it was verified that the dynamic behavior was stable and vibrational levels comply with ISO 10816-3. In this study the following techniques were used: Vibration Analysis, Operating Deflection Shape - ODS Analysis and Numerical Simulation by Finite Element Method and Component Modal synthesis.

Figure 1 – Planetary Gearbox [1]

2. Background:

Solutions for constant vibration problems frequently have been performed by customers themselves from trial and error. Such actions can be more time consuming than necessary, increase cost and it do not reach the objective, which is trouble free operation. In order to obtain success in the vibration problems solving, a detailed study should be performed. In this article, numerical methods for structures and rotating machine modelling were correlated to experimental techniques, such as Vibration Analysis and Operating Deflection Shape – ODS, to obtain robust solutions for constant vibration problems. Thus, this robust approach provides the best solutions without creating new problems, and consequently, there will be a better planning to repair actions and also reduce downtime.

Vibration Analysis was applied to guide Operating Deflection Shape Analysis. This evaluation was performed by studying operational mode obtained by the ODS technique in the excitation frequencies source. If the source is close to natural frequencies, ODS mode would describe an approximate mode shape of the natural frequency [2].

Operating Deflection Shape – ODS is used to identify operational modes of structures or rotational machinery. ODS can be performed in time or frequency domain from vibration measurements of some structure points. These modes can be obtained by data measured simultaneously or measured data considering a reference signal. Measurements are performed in housing, metallic base, structure and frame machinery, and it can be in: displacement, velocity or acceleration. Such vibration measurements performed from ratio, which is defined by Frequency Response Functions - FRF, when excitation force is feasible to measure, is given by Eq. (1):

$$T_{ij}(\omega) = \frac{S_{ij}(\omega)}{S_{jj}(\omega)}, \quad (1)$$

where, S_{ij} is the cross power spectral density between the force and acceleration response and S_{jj} is the auto power spectral density. In case of ODS, when the excitation force cannot be measured, a response signal is assumed as reference [3]. Transmissibility measurements are often used among the ODS techniques performed from response signals as reference. It is calculated in the same way as the Frequency Response Functions, Eq. (1), but whereas the FRF is the ratio of the response divided by force, transmissibility is the ratio of the response divided by the reference response [3].

The Finite Element Method – FEM is a general method which can be used to simulate the dynamic behavior of structures and rotational machinery or part of them (components). In FEM the structure is subdivided into a large number of finite elements. For a good accuracy, a Finite Element model usually consists of thousands of degrees of freedom with associated equations [4]. The evaluation of dynamic behavior of systems allows studying root cause of failures due to high vibration levels. In this article FEM was used to calculate natural frequencies and their mode shapes of frame planetary gearbox, which were correlated to operational modes from ODS. Another numerical technique, Component Modal synthesis - CMS, was used to model the dynamic problem that involves rotational parts (bearings and gears), consists on dividing a complex structure among substructures, and solving problems in an optimized way [4].

3. Results and Discussion:

3.1. Vibration and ODS Analysis:

Instrumentation and software used in the measurement of vibration analysis, and in ODS measurement, as well as in the numerical simulation are described in the Table I.

Table I – Instrumentation and software used in the measurement of Vibration and ODS Analysis, as well as, in the Numerical Simulation

Data Collector/Analyzer:	Microlog GX 70;
Sensors:	Accelerometers CMSS2200
Analysis/Management Software:	@ptitude Analyst 8.0.168.0;
3D CAD Software:	Creo Parametric 2.0;
ODS, FEM e *CMS* Software:	ME'scopeVES 5.0; ANSYS® 15 and SimPro Expert 3.4

In Tables II and III technical data of Motor and Planetary gearbox is described. Such information is relative to operation conditions used during vibration measurement and ODS.

Table II – Motor data

Equipment:	Motor
Power:	1470.99 kW (2000 cv)
Rotation:	1329 RPM

Table III – Planetary Gearbox data

Equipment:	Planetary Gearbox
Input speed:	22.2 HZ (1329 RPM)
Output speed:	0.075 Hz (4.52 RPM)
Gear Mesh Frequency – 1 (GMF1):	212.17 Hz
Gear Mesh Frequency – 2 (GMF2):	29.56 Hz
Gear Mesh Frequency – 3 (GMF3):	4.82 Hz

Vibration and transmissibility measurements were performed using the following set up: 100 mV/EU; 6400 lines, end frequency 1500 Hz, Hanning window and operational condition was high load. Figures 2(a) and 2(b) present all points where vibration and ODS measurements were performed on planetary gearbox system.

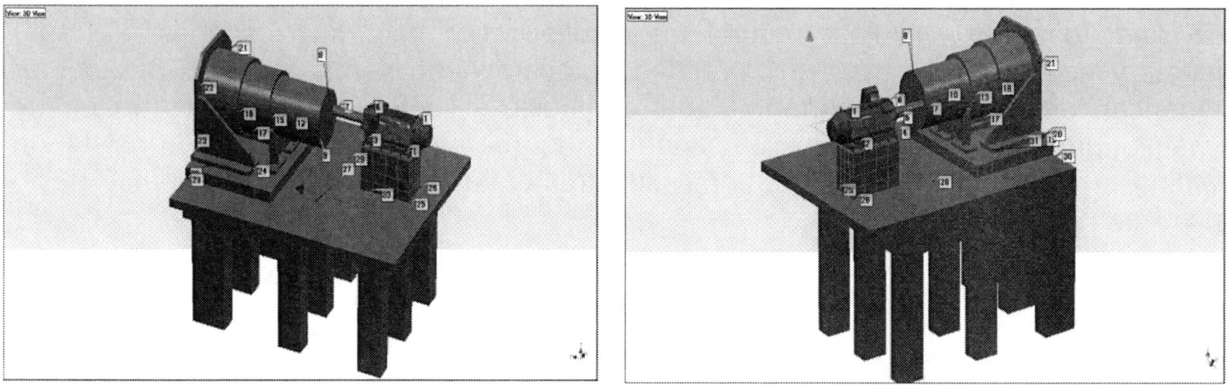

(a) (b)

Figure 2 – (a) and (b): Measurement points used in the vibration and ODS analysis

In the measured direction 7X spectra, which is horizontally placed on the input stage of planetary gearbox, it was verified the following frequencies: 24 Hz, 48 Hz, 29 Hz and 58 Hz. Such frequencies are highlighted in the range of interest. 48 Hz and 58 Hz frequencies represent 2nd harmonic of second stage gear meshing frequency. Therefore, this event characterizes dynamic misalignment due to 2nd harmonic highlight, Figures 3(a) and 3(b).

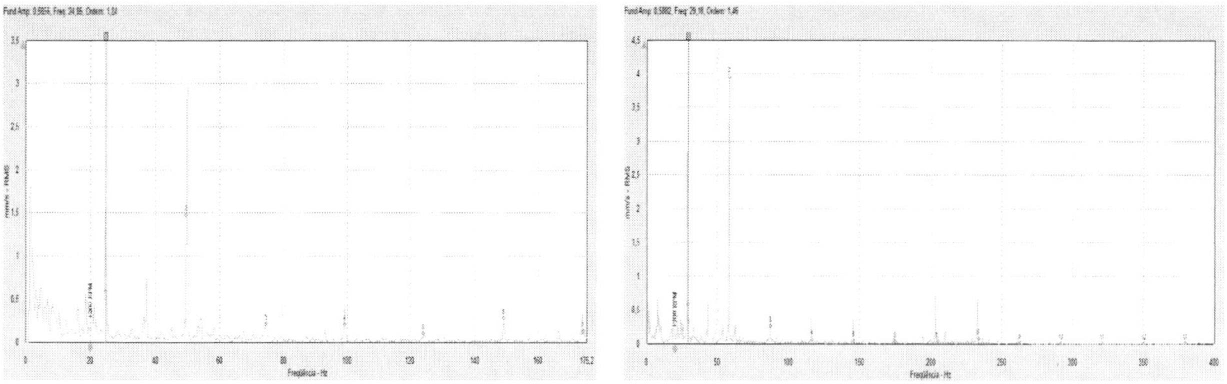

(a) (b)

Figure 3 – (a) and (b): Vibration Spectra measured in the 7X direction placed in the horizontal of the planetary gearbox input stage

Spectra measured in the 2X and 14Z directions show 30 Hz and 386 Hz highlighted frequencies, which represent GMF 2nd stage and the second harmonic of gear meshing frequency of first stage. In the 30 Hz region is possible to verify exciting characteristic of natural frequency in Figure 4(a), and in 2nd harmonic of the GMF1 there is a gear meshing frequency misalignment, Figure 4(b). According to [5] and also field experience [6] the base of energy spikes is very broad and may be made of many small energy spikes, which is on indication of natural frequency excitation.

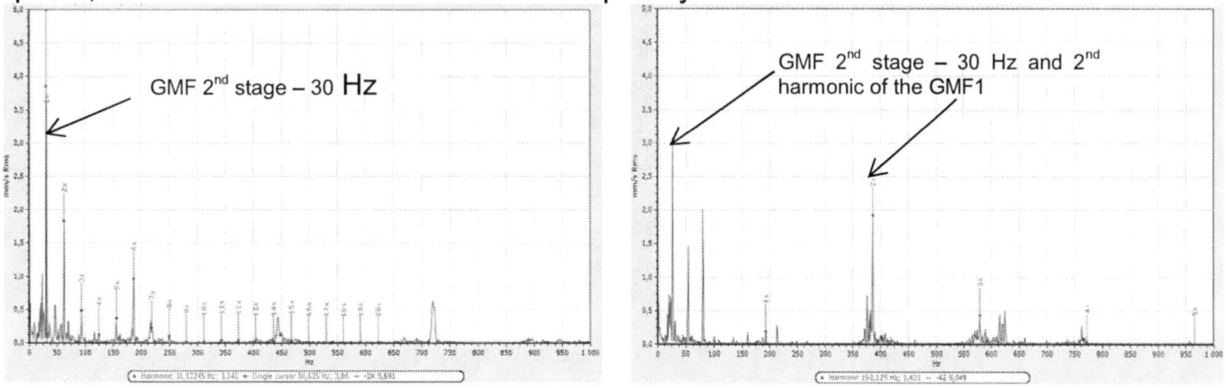

(a) (b)

Figure 4 – (a): 30 Hz region presents the characteristic of natural frequency excitation and (b): 2nd harmonic of the GMF1 misalignment

From ODS Analysis it was verified that the planetary gearbox presented operational mode that generated relative movement, which caused frame gearbox dynamic misalignment. Spectrum from Figure 4(a) measured in 2X direction shows GMF from 2nd stage with characteristic of natural excitation frequency, and the operational mode obtained by ODS in this region frequency (32.6 Hz) shows the planetary structure displaces on the vertical direction, Figure 5(a). This movement can cause amplitude of GMF sidebands increase, as verified in spectrum, Figure 4(b). The spectrum of Figure 3(b) shows that dynamic misalignment in 48 Hz also happened, thus operational mode in this frequency presents horizontal direction displace, Figure 5(b).

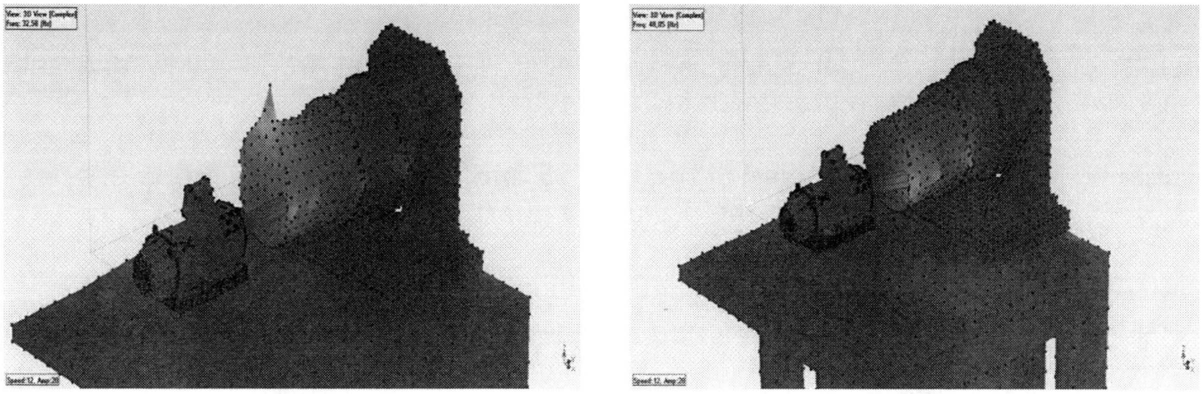

(a) (b)

Figure 5 – (a) and (b): Operational modes obtained by ODS in 32.6 Hz and 48 Hz frequencies highlighting the vertical and horizontal displacement, respectively

3.2. Numerical Simulation of Frame Planetary Gearbox By Finite Element Analysis – FEM:

Modal Analysis was used to calculate natural frequencies and its mode shapes respectively of gearbox frame, and then, assure numerically the possible existence of natural mode shape. The 3D CAD (Computer Aided Design) model was built based on the technical drawings, Figure 6, and used in the program of FEM, in which mass, material and boundary conditions were defined.

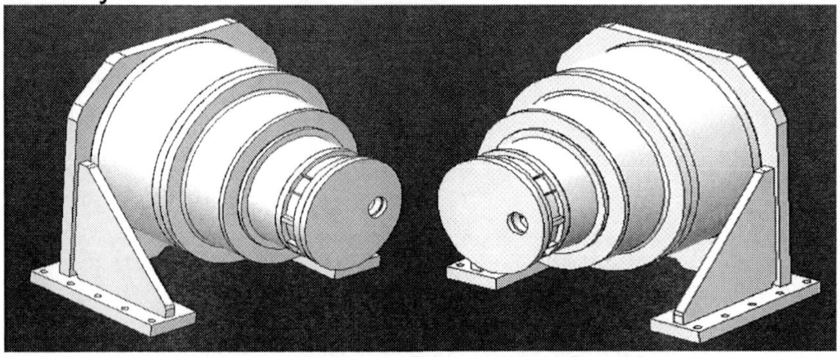

Figure 6 – Frame Planetary Gearbox

Boundary conditions considered in Modal Analysis are presented in the Figure 7, which are fixed support in foundation, gear, bearing and shafts masses.

Figure 7 – Boundary Conditions applied to frame planetary gearbox

Results obtained by Modal Analysis were first 15 natural frequencies and their respective mode shapes, represented in Table IV.

Table IV – Natural Frequencies of the first 15 and their respective mode shapes frame planetary gearbox

Mode	Frequency [Hz]
1	**32.385**
2	**46.947**
3	68.948
4	125.86
5	172.14
6	244.54
7	280.06

8	283.79
9	288.87
10	293.16
11	298.39
12	368.87
13	376.69
14	440.11
15	448.62

Results obtained by FEM, show that there is a mode shape in 32.385 Hz, which causes bending of frame planetary gearbox in vertical direction, and other mode shape in 46.947 Hz displace one horizontal direction. Thus, mode shape in 32.385 Hz (left side) was correlated with the operational mode in 32.6 Hz by ODS (right side), Figure 8(a), and also mode shape in 46.947 Hz (left side) was correlated to operational mode in 48 Hz (right side), Figure 8(b). Therefore, it was observed that mode shape by FEM and operational mode by ODS in each respective frequencies presented the same dynamic behavior. Then, Finite Element model is calibrated, and so will be presented the structural modification proposal to displace the natural frequencies in 32.385 Hz and 46.947 Hz, avoiding frame planetary gearbox resonance.

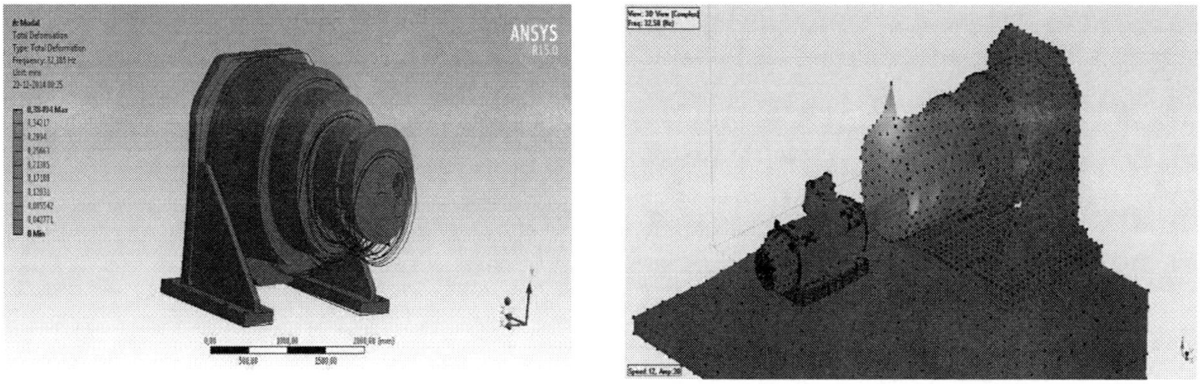

(a)

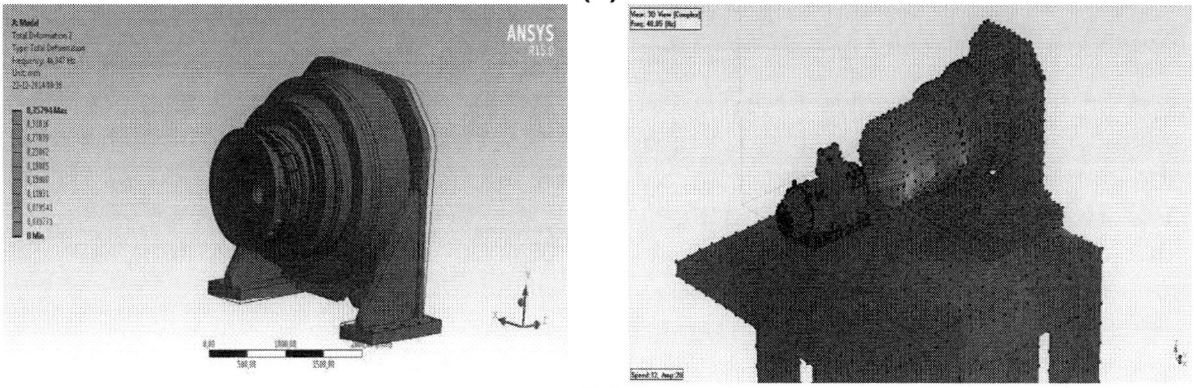

(b)

Figure 8 – (a): Simulated and Operational mode in 32.385 Hz and 32.6 Hz, (b): Simulated and Operational mode 46.947 Hz and 48 Hz, respectively

The proposal for the solution to avoid the resonance in 30 Hz and 48 Hz region is presented in Figure 9.

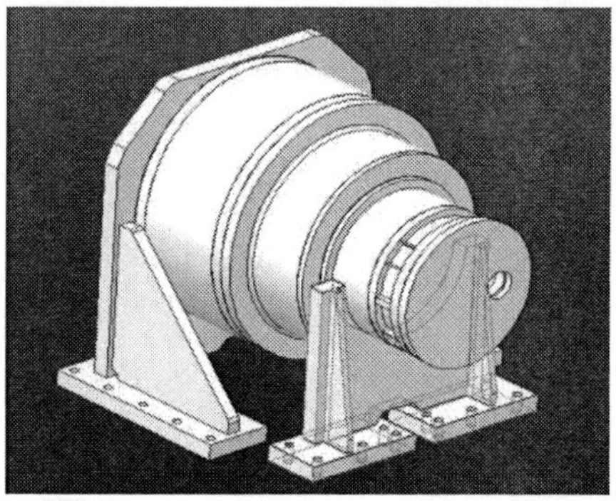

Figure 9 – Proposal frame planetary gearbox

Results of Modal Analysis obtained for the support proposal, presented in Table V.

Table V – Natural frequencies of the first 15 proposal frame planetary gearbox

Mode	Frequency [Hz]
1	66.823
2	84.546
3	150.04
4	154.09
5	183.00
6	272.46
7	288.22
8	299.34
9	307.23
10	314.25
11	347.36
12	376.85
13	440.58
14	468.73
15	481.55

Based on the Modal Analysis results could be verified that the structural modification proposal obtained increase stiffness of frame planetary gearbox, where the first mode shape had its frequency shifted from 32.385 Hz (current condition) to 66.823 Hz and 46.947 Hz to 84.546 Hz. This structural modification eliminated resonance risks. The Figure 10 highlights excitation frequencies obtained in the measurement after the proposal solution, thus its levels reduced.

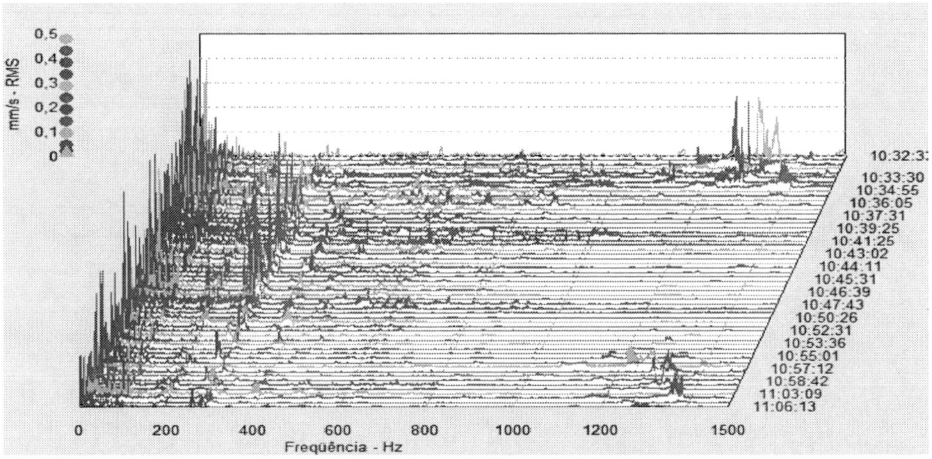

Figure 10 – Vibration spectra cascade of the planetary gearbox measured points after structural modification

Figure 11 presents overall global levels of measured vibration in each coordinates (x, y, z) for high load operational condition. These values were obtained from the measured spectra before and after structural modifications application. It was observed that the vibration levels after the implemented proposal solutions are compliant with ISO 10816-3, Figure 11.

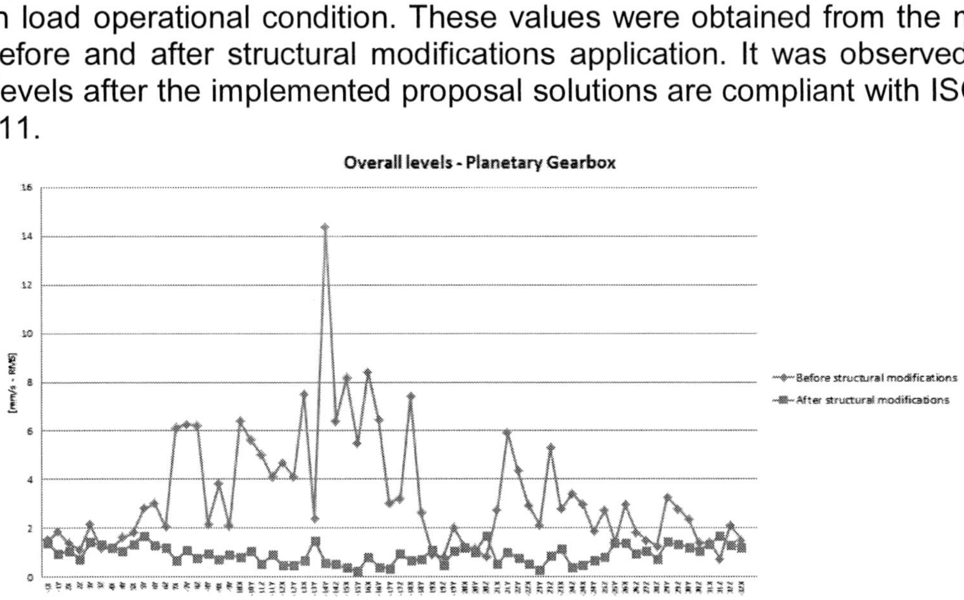

Figure 11 – Overall global levels to high load operational condition of measured spectra before and after implemented proposal solution

ODS analysis before the applied proposal solution presented operational modes that caused relative movements in the 32.6 Hz and 48 Hz frequencies of radial and vertical direction. Such movements, which caused dynamic misalignment, have not been observed after the applied proposal solution in all frame planetary gearbox. Figures 12 and 13 present operational modes in 32.6 Hz and 48 Hz frequencies, respectively, where on the right side there are operational modes measured before and on the left side, after proposal implementation.

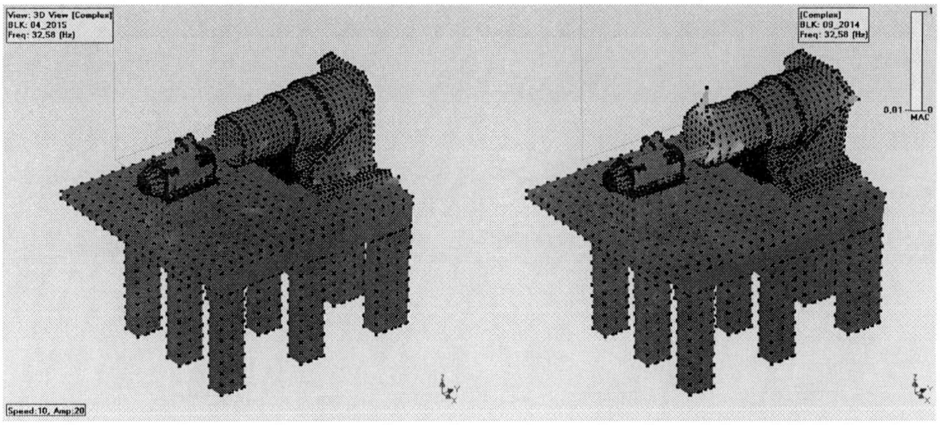

Figure 12 – Comparison of operational modes obtained by ODS in the 32.3 Hz. Before (right side) and after (left side) the proposal implementation

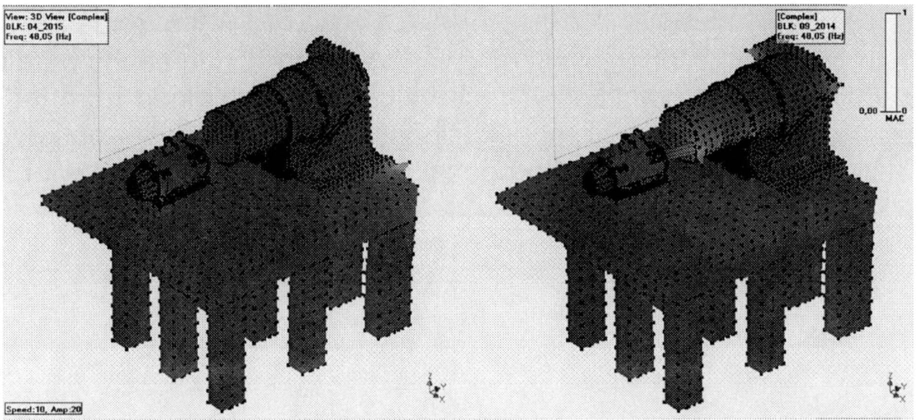

Figure 13 – Comparison of operational modes obtained by ODS in the 48 Hz. Before (right side) and after (left side) the proposal implementation

3.3. <u>Numerical Simulation Bearing Arrangement of Planetary Gearbox</u>:
Internal components were evaluated numerically by CMS technique, which is implemented in the software SimPro Expert 3.4. Figure 14 shows the model with boundary conditions, which are: 1470 kW, input speed 1329 RPM, ISO VG 460 lubricant, -/17/14 particles as ISO 4406.

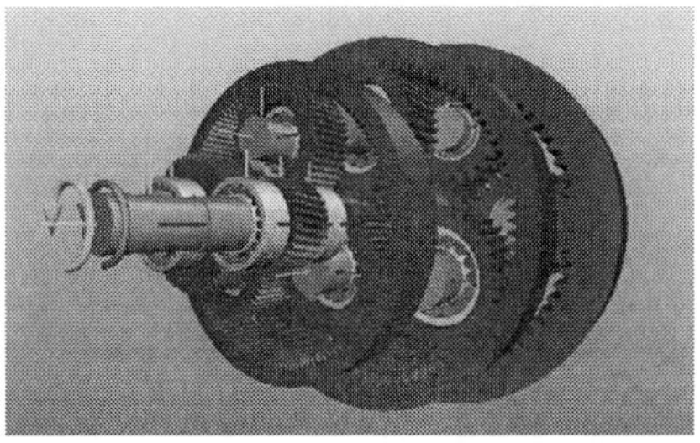

Figure 14 – Planetary gearbox model made in SimPro Expert 3.4

Results obtained by vibration and ODS analysis showed that the input shaft and first stage planetary gearbox need a detailed evaluation. Therefore, it was calculated fatigue life, and conclusion is a bearing minimum life of 79,000 hours, Table VI. Table VII shows acting forces in the bearings, which are considered average magnitude (C/P≅8.0).

Table VI – Fatigue life – Normal condition

Bearing	ISO 281/TS 16281:2008 [h]	SKF Advanced
Input Shaft - 23224 - LA	75851.98	79099.17
Input Shaft - 23224 - LOA	729487.16	696240.18
Sun 1st Stage - 23036 - LOA	288619.86	328275.83
Sun 1st Stage - 23130 - LA	2080871.72	1839560.78

Table VII– Acting forces – Normal Condition

Name	Type	F_x [kN]	F_y [kN]	F_z [kN]	M_{yz} [kNm]	M_{xz} [kNm]	M_{xy} [kNm]
Planet 1st St. 22324 - 1	SRB	-0.00	0.00	0.00	-0.00	0.00	-0.00
Planet 1stSt. 22324 - 2	SRB	0.00	-0.00	0.00	0.00	-0.00	0.00
Planet 1stSt. 22324 - 3	SRB	-0.00	-0.00	0.00	0.00	0.00	0.00
Input Shaft 23224 - LA	SRB	26.30	47.56	15.22	0.00	-0.00	-0.00
Input Shaft 23224 - LOA	SRB	11.67	47.62	0.01	-0.00	0.00	-0.00
Sun 1st St. 23036 - LOA	SRB	-30.13	-46.20	-15.22	0.02	-0.01	-0.00
Sun 1st St. 23130 - LA	SRB	-7.03	-49.30	0.01	0.00	0.00	-0.00

By evidenced misalignment in the Vibration and ODS analysis, it was generated fatigue life versus misalignment graphic. Therefore, a bearing minimum life of 140,000 hours (0.800 mm misalignment) was noticed, Figure 15(a). Figure 15(b) shows internal components of planetary gearbox operating in misalignment condition.

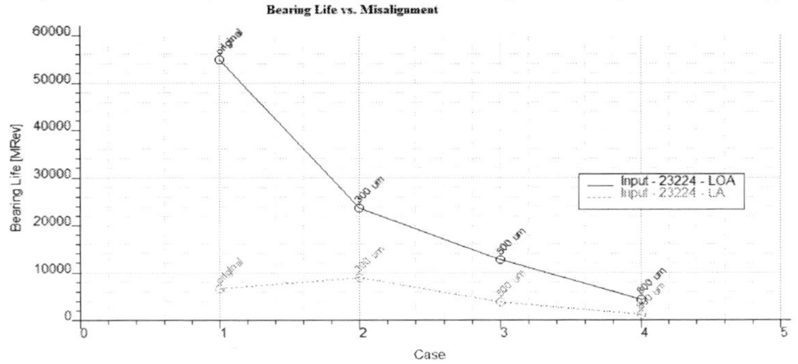

(a) (b)

Figure 15 – (a): Fatigue life versus misalignment and (b): internal components of planetary gearbox operating in misalignment condition

Through imposed misalignment condition, forces on the bearings become of high magnitude (C/P≅4.0). Dynamic misalignment affected input shaft bearings, sun shaft and also the 1st stage planet, mainly. Therefore, it was proposed a bearing arrangement less sensitive to misalignment, which is SRB - Spherical Roller Bearings and CARB - Toroidal Roller Bearing. Bearing arrangement should be applied in the input shaft and 1st stage sun shaft. SRB - Spherical Roller Bearing – 23224 CC/W33 should be applied in the input shaft (driver side) and CARB - Toroidal Roller Bearing - C 3224 on the opposite side. In the 1st stage sun shaft (driver side) bearing is C 3130 and in the opposite side 23036 CC/W33. The fatigue life versus misalignment was calculated to bearing arrangement, as can be observed in the Figure 16(a), where a bearing minimum life is 16,000 hours (0.800 mm misalignment). The dynamic behavior also leads to fritting corrosion on the outer diameter contributing to premature failures. SRB and CARB arrangement is less sensitive to this failure mode, Figure 16(b).

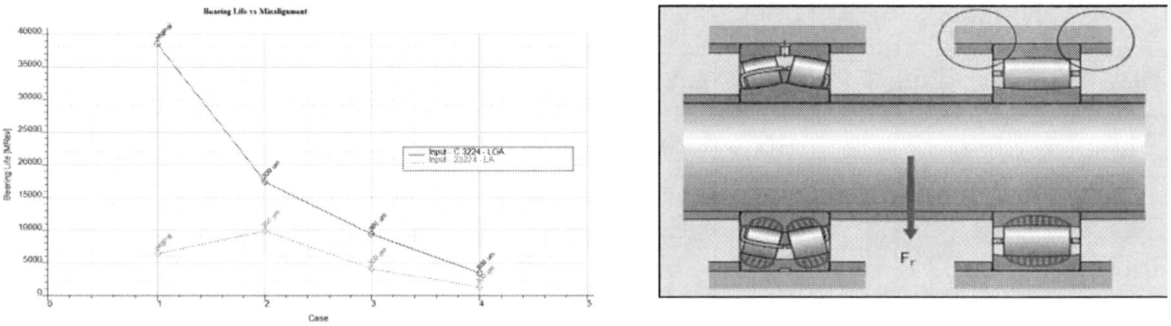

(a) **(b)**

Figure 16 – (a): Fatigue life versus misalignment condition – with CARB; (b): SRB+CARB Arrangement

4. <u>Conclusion</u>:

The planetary gearbox mainly presented internal components failures (bearings and gears) in the 1st and input shaft. A study was performed in this machinery through the following techniques: Vibration and ODS Analysis, Numerical Simulation (FEM and CMS). Such techniques highlighted excitation frequencies with natural characteristic, operational modes, which in these frequencies caused relative movements. Results obtained by Vibration and ODS were correlated to numerical simulation, which allowed calibrate FEM model. Afterwards, a proposal solution was given: structure (support) and bearings arrangement, avoiding system resonance.

5. <u>References</u>:

[1] Celulose Online, <http://celuloseonline.com.br/> 05/15/2015;

[2] Onari, M. M., Boyadjis, P.A., Solving Structural Vibration Problems Using Operating Deflection Shape and Finite Element Analysis, Proceedings of the Twenty-Fifth International Pump Users Symposium, 2009, pages 85 to 102

[3] Schwarz, Brian J., Richardson, Mark H. Introduction to Operating Deflection Shapes. In: CSI Reliability Week, CSI Reliability Week, Orlando, FL: Vibrant technology, October 1999, 7 pages;

[4] Tillema, H.G., Noise Reduction of Rotating Machinery by Viscoelastic Bearing Supports, PhD thesis, University of Twente, Enschede, The Netherlands, 2003, 187 pages;

[5] Technical Associates of Charlotte, Use of Vibration Signature Analysis to Diagnose Machine Problems, 1997, Chapter 6;
[6] @ptitude Exchange - SKF, A Guide to the Interpretation of Vibration Frequency and Time Spectrums, sept. 2011, 118 pages.

Biography:

Edmilson Santos holds a Bachelor's and Master's at Mechanical Engineering, Solid Mechanic São Paulo State University, UNESP (2000) and Ph.D, Mechanical Engineering, Solid Mechanic and Design State University of Campinas, UNICAMP (2006). Professor in Mechanical Engineering Course, Department of Mechanical Engineering, Maringá,/PR, Brasil until July/2009. Has experience in Mechanical Engineering, focusing on Mechanical Engineering, acting on the following subjects: Spectral Element Method, Energy Finite Element Method, Energy Spectral Element Method. Since 2009 has been working at SKF do Brasil with ODS – Operating Deflection Shape, Numerical Simulation by Finite Element Method – FEA; Conventional Vibration Analysis and CAD 3D Solid Modeling.

VIBRATION SYMPTOM OF ELECTRICAL MALFUNCTION IN A SYNCHRONOUS VERTICAL HYDRO GENERATOR

Sherub Tharchen and Dawa
Center of Excellence for Vibration and Thermographic Analysis
Research and Development Center
Chhukha Hydropower Plant
DrukGreen Power Corporation Limited
sherubtcoe@gmail.com; namdawa2020@gmail.com

Abstract: Vibration monitoring is the most beneficial test for identifying mechanical and electrical problems of rotating machines. Similar test was carried out in four identical generators of Chhukha Hydropower Plant, and the generator of unit no.1 observed to be behaving differently from rest of the generators. To identify the problem, frequency spectrum relationship between running speed and electrical vibration was studied. The vibration spectra's were collected at different operating modes which is the detrimental factor to differentiate between electrical and mechanical problem. At load, steep dropped in vibration amplitude close to 70% with dominating 2LF vibration components was observed and such behavior is noticeable only in lower guide bearing. The cause of such behavior is discussed in this article using machines vibration spectra analysis and other historical background. Also, static air gap is plotted to support the electrical problem analysis. Further, the article also recommends checking the condition of generator stator winding and rotor.

Keywords: Mechanical; Electrical; 2LF; frequency spectra; orbit; static air gap and AGMS

Introduction: One of the most important information sources to detect problems in the machine today is through vibration measurements. Vibration is a natural product of the mechanical dynamic, electrical phenomenon and hydraulic interaction and this can be seen for all rotating machines. Undoubtedly, each malfunction generates unique vibration signals that can be seen and analyzed in the frequency spectrum of the signals.

Similarly, arrays of vibration signals were captured from the four identical generators of Chhukha Hydropower Plant using displacement transducer and accelerometers. The generators are identical in terms of manufacturer, size, containment structure, load rating, operating cycle, installation, consecutive serial number, RPM and commissioned from 1986 to 1988. Despite the alike and several modifications to the main bracket, unit-1 generator was oscillating as high as 420 microns peak-peak and exhibiting unique behavior on load and no load. Accordingly, the inherent high vibration since commissioning and the aftermath modification led to the replacement of upper and lower bracket of this unit. A decreased in vibration amplitude was observed after replacement of the brackets, however, machines inherent dynamic instability continues to persist even after replacement of the brackets. Breakdown or forced shutdown is an expensive criterion for this machine since it fuels millions of revenue to the nation.

Machine Operating History: Generator no.1 was commissioned from 1986 to 1988 and since then the machine was oscillating as high as 420 microns peak-peak. Several modifications in the bracket and rotor balancing were carried out to reduce the high vibration. All such activities couldn't resolve the problem fully. Finally, the OEM, Bharat Heavy Electricals Limited was consulted for the rectification. The OEM [1] re-designed, manufactured and installed the new brackets. Even after changing the brackets not much improvement was seen and the machine dynamic instability remained same. By referring the machines operational history [2] and using the present vibration signature trend, more reasoned analysis was carried out to differentiate the nature of problem.

Measurement Setup: The machine under investigation is a 3 bearing vertical turbine generator propelled by 21 buckets pelton turbine with rated power of 84 MW and speed of 300 rpm (5 Hz). The bearing gaps at UGB, LGB and TGB were set at 0.44 mm, 0.34 mm and 0.46 mm diametrically. As the bearings are journal type, eddy current probes 90^0 (Figure 1) apart were installed at each bearing positions. In addition, high frequency accelerometers were installed at upper and middle guide bearing casing to monitor the casing vibration. A temporary optical keyphasor was installed on the turbine cover to measure rotational speed and vibration phase. The signals from individual sensor were input into a portable analyzer. Portable analyzer was set at 800 numbers of lines, 1000 Hz Fmax, 2560 sample rate, 2048 number of samples and the hanning window was used to collect the vibration signals. The vibration signals were recorded at constant speed no load, and with excitation and load. All the measurements were recorded in micron meter peak to peak and mm/s.

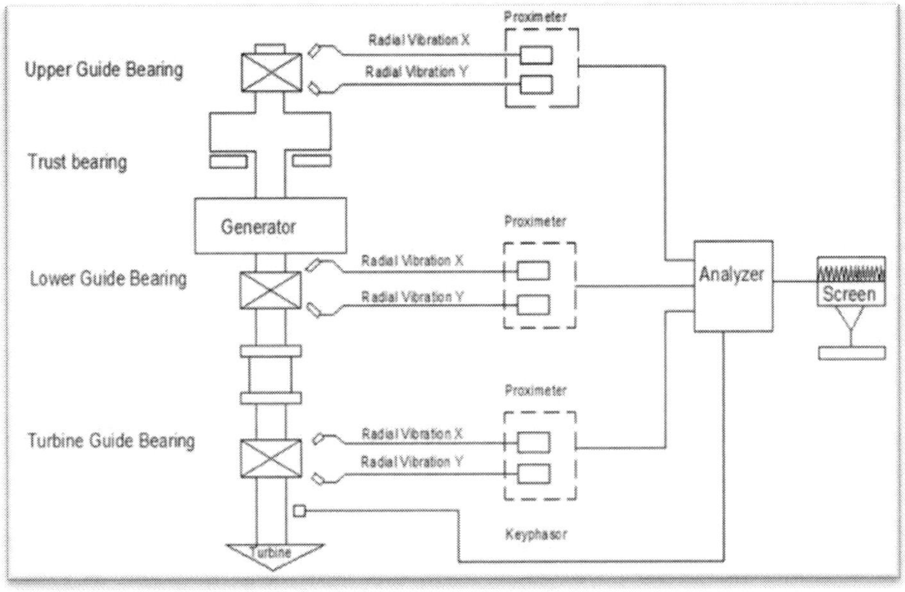

Figure 1. Typical 3 Bearing Vertical Hydro Turbine Generator with Measurement Setup

Results and Discussion: Figure 2 and 3 indicates vibration trend at different operating modes. Relative shaft displacement levels of 232 and 225 micron meter peak-to-peak were observed at upper and lower guide bearing on mechanical spin. However, there is an indication of downward trend on excitation and on load. The case was more severe at the lower guide bearing with 70% drop in shaft displacement (Figure 1b). Whereas, upward trend for casing vibration (Figure 2b).

This behavior is an evidence of possible electrical problems. For the ideal machine the vibration levels should not differ irrespective of the operating pattern. Machines behavioral change with change in operating mode is one way of considering possible electrical problem.

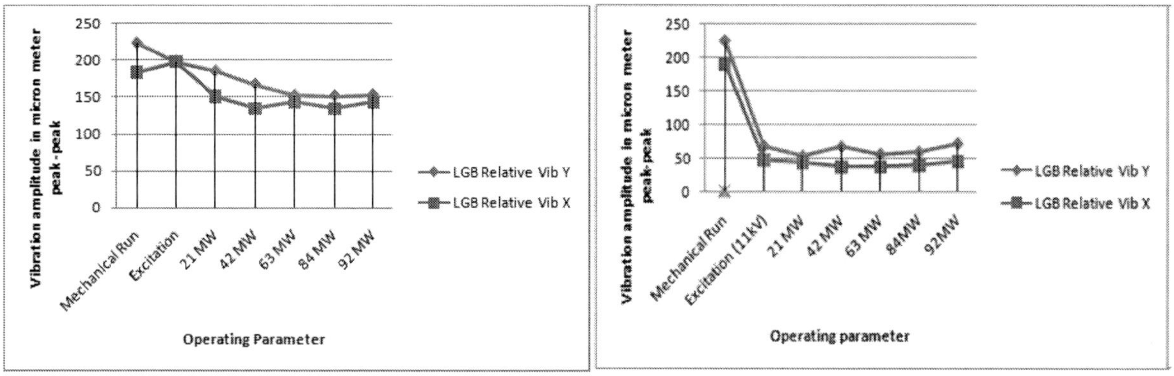

(a) (b)

Figure 2. Shaft Relative Displacement levels a) At Upper Guide bearing and b) At Lower Guide Bearing

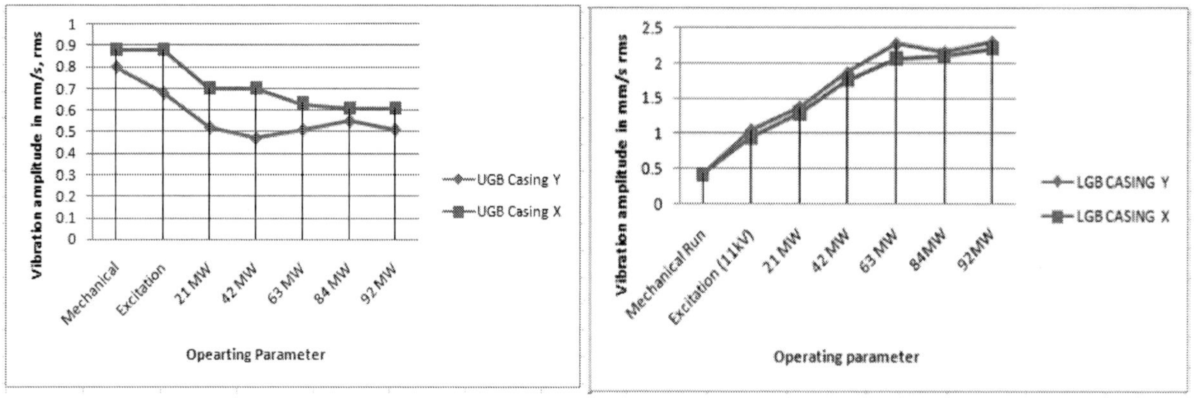

(a) (b)

Figure 3. Generator Casing Vibration Levels a) At Upper Guide Bearing Casing and b) At Lower Guide Bearing Casing

Waterfall plot in Figure 4 & 7 and spectrum plot in Figure 5, 6 & 8 further supports the possibility of electrical problem. Twice the line frequency (2LF) vibration component is indicated at lower guide bearing generator casing. Such behavior was noticeable only during electrical mode as shown in Figure 6 and pulls back to normal indication as shown in Figure 5. The complete spectral indication is shown in Figure 7. The case was different for upper guide bearing generator casing and the spectral signatures can be compared for any ideal machine as shown in Figure 8. Possibility of loose stator winding is suspected and is more significant at the bottom side of the windings.

If the electrical windings of the stator are loose, the vibration level at twice the line frequency will increase. This condition is very destructive because it abrades the insulation on the conductor, leading to shorted turns and eventual short circuits to ground and stator failure [3].

Excitation of twice the line frequency (100 Hz) is always measurable vibration components in an electric machine, but the concern is surging of 2LF when machine changes to electrical mode.

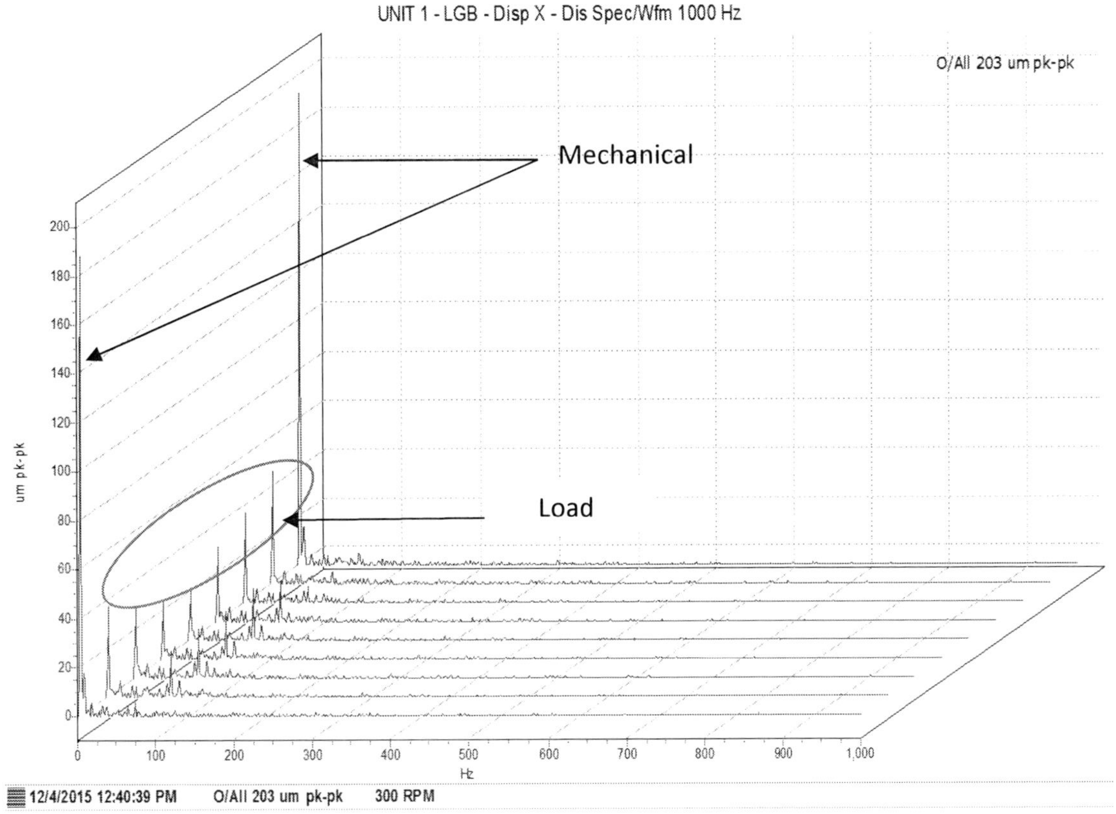

Figure 4. Waterfall Plots of Relative Shaft Displacement at Lower Guide Bearing

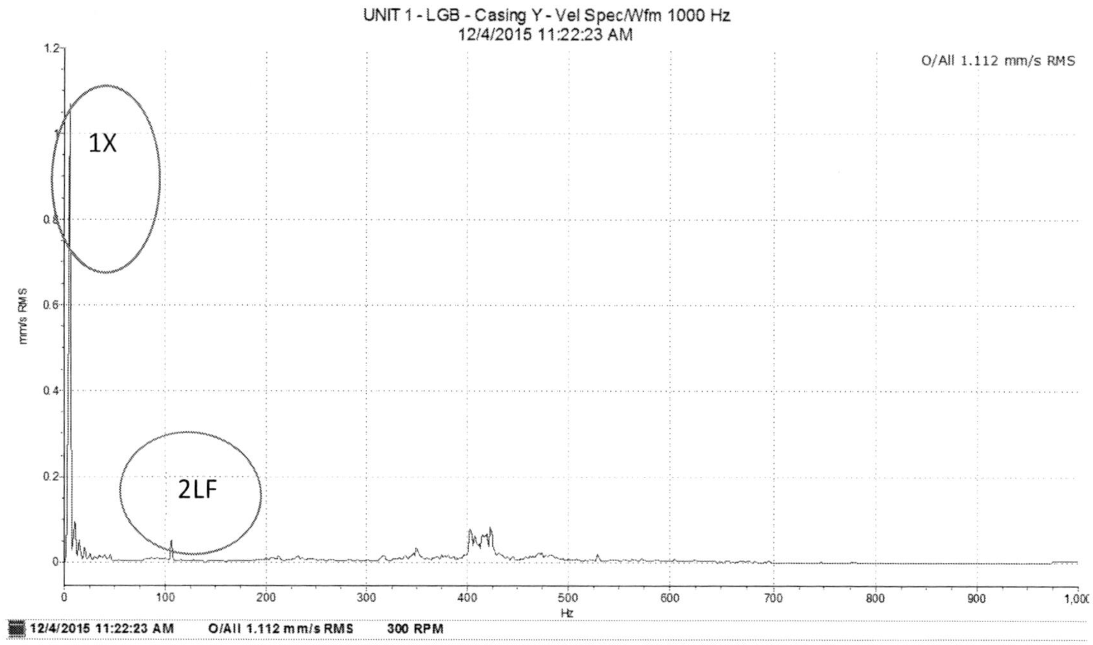

Figure 5. Spectrum Plot of the Lower Guide Bearing Generator Casing on No Load

Figure 6. Spectrum Plot of the lower Guide Bearing Generator Casing on Load

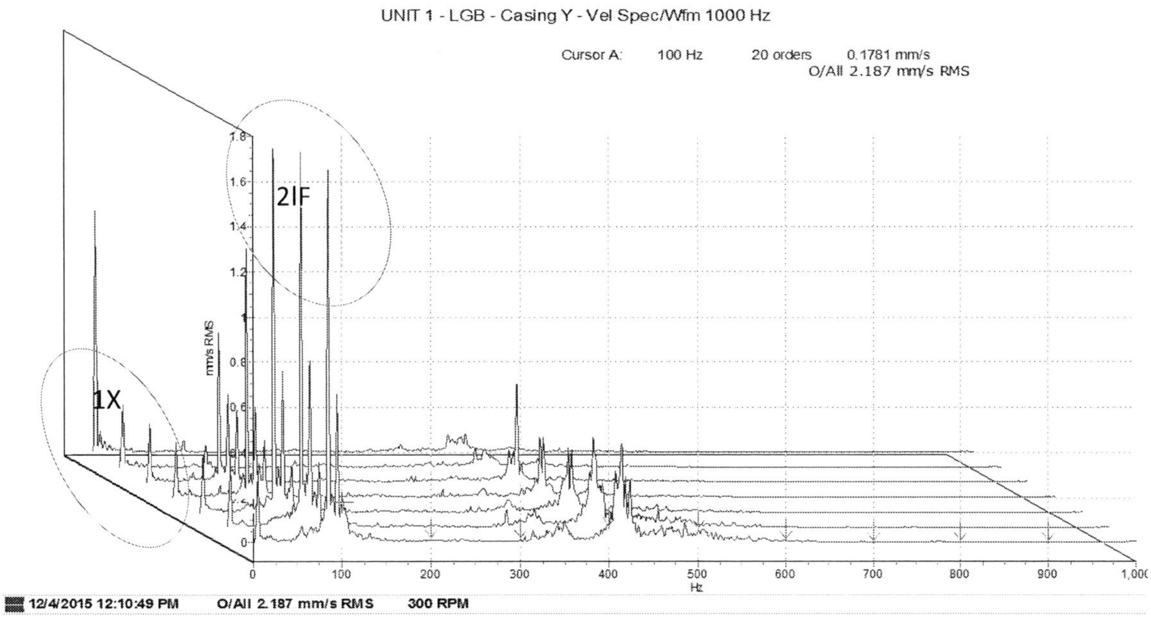

Figure 7. Waterfall Plots of Lower Guide Bearing Generator Casing on Load

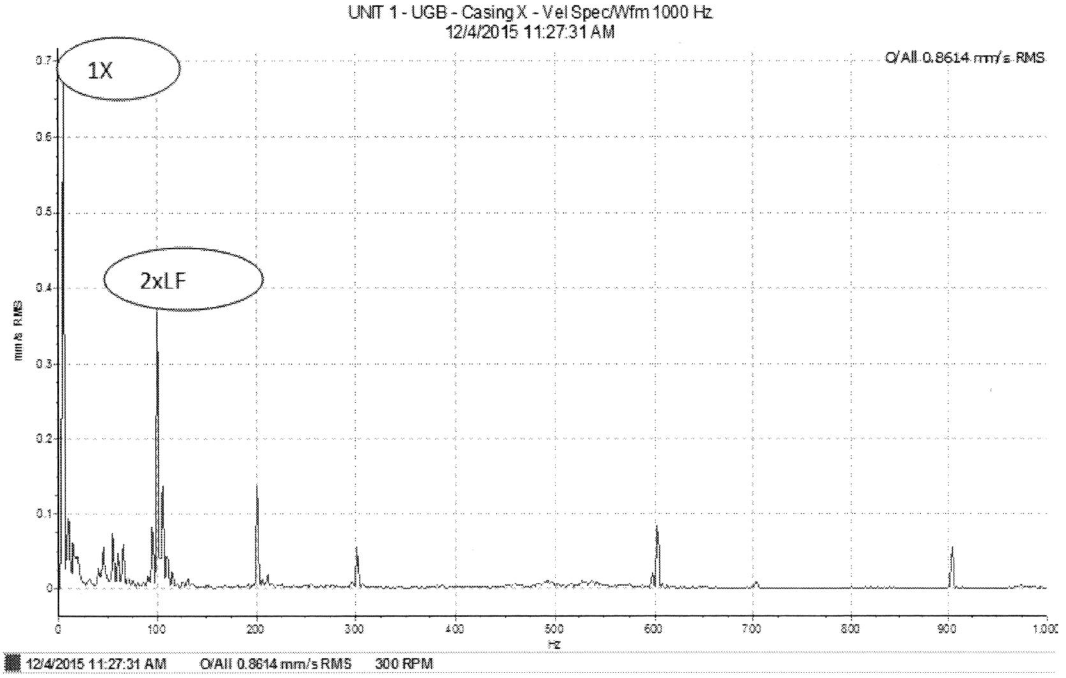

Figure 8. Spectrum Plot of the Upper Guide Bearing Generator Casing on Load

Further, the shafts orbit plots for lower guide bearings were compared between no load in Figure 9(a) and load in Figure 9(b). A significant difference in orbit shape was observed. Similarly, the shafts orbits shape for upper guide bearings were also compared between no load in Figure 10(a) and load in Figure 10(b). Not much difference in orbit was observed. Possibility of unbalance magnetic flux distribution due to disturbances in air gap is also suspected.

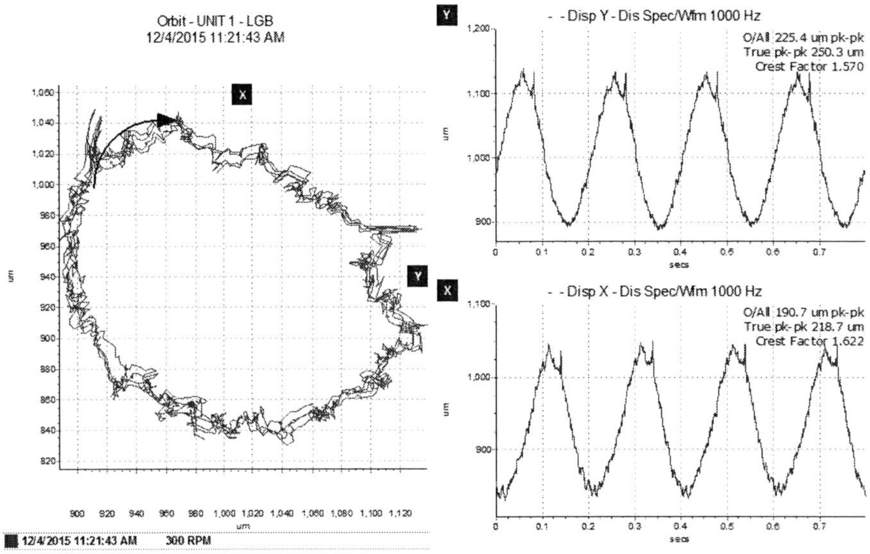

(a)

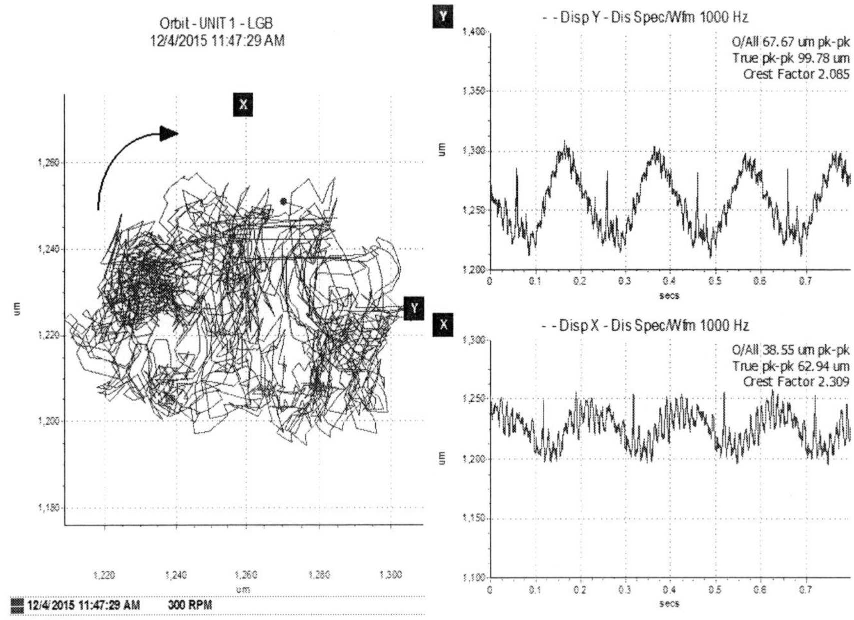

(b)

Figure 9. Shaft Orbit Plot Shape at Lower Guide Bearing a) Mechanical Spin and b) Excitation and load

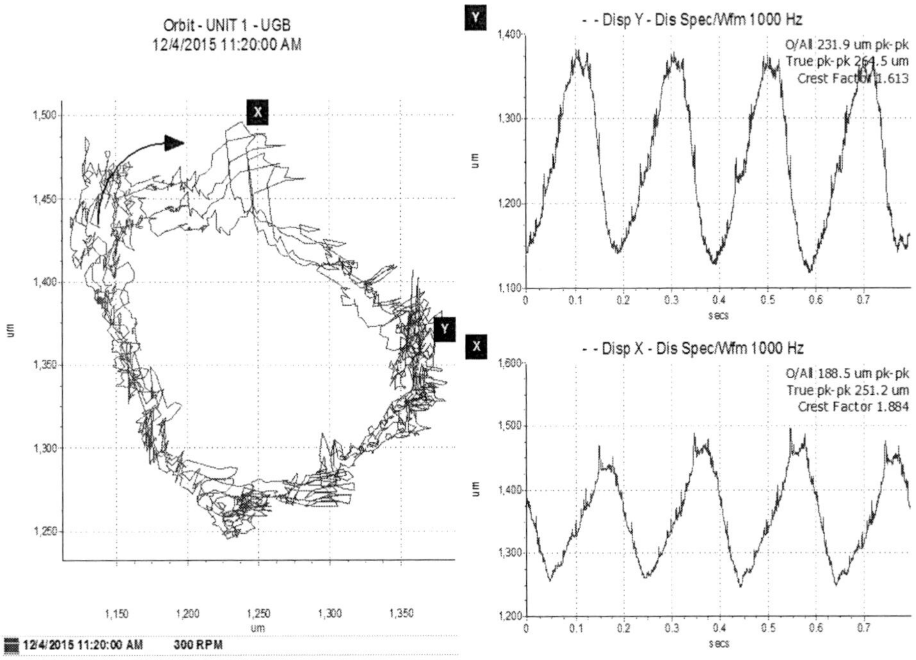

(a)

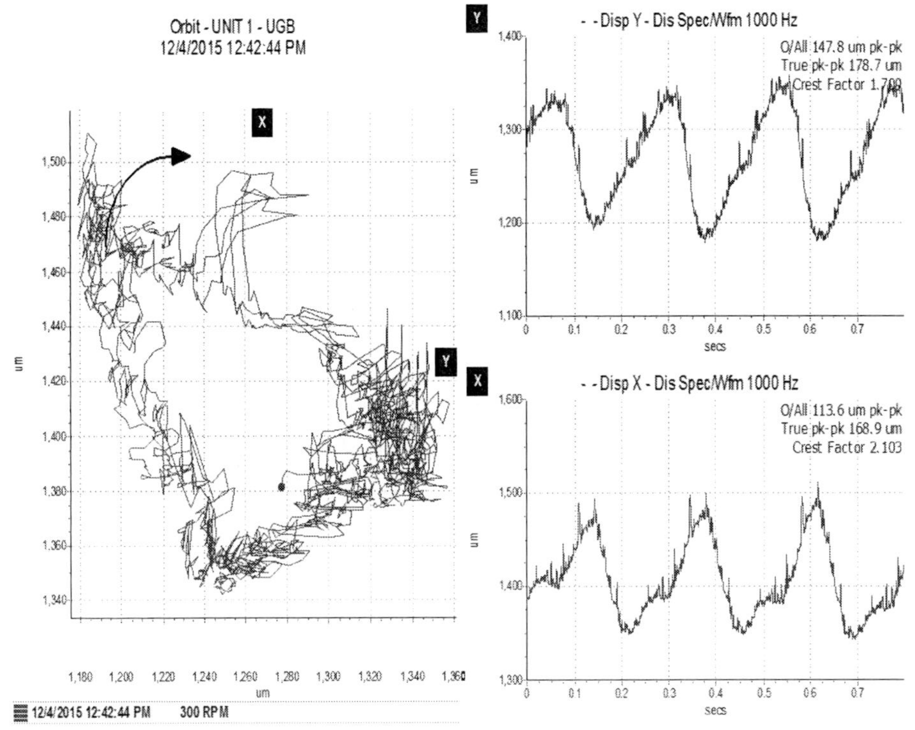

(b)

Figure 10. Shaft Orbit Plot Shape at Upper Guide Bearing a) Mechanical Spin b) Excitation and Load

Another reasoned judgment was by plotting the static air gap. Non uniformity of the air gap in rotating electric machinery leads to a non uniform permeance of the air gap. This unsymmetrical air gap and unbalanced magnetic pull (UMP) creates unsymmetrical forces on the rotor [4]. Inconsistent air gap as shown in Figure 11(b) could be the possible reason for machines unique behavior. The case was more significant at the bottom side of the rotor. The machine under investigation is not equipped with Online Air Gap Monitoring System (AGMS) and has limited access to data interpretation. An Online Air Gap Monitoring System could detect dynamic air gap and possibility of elevated explanation.

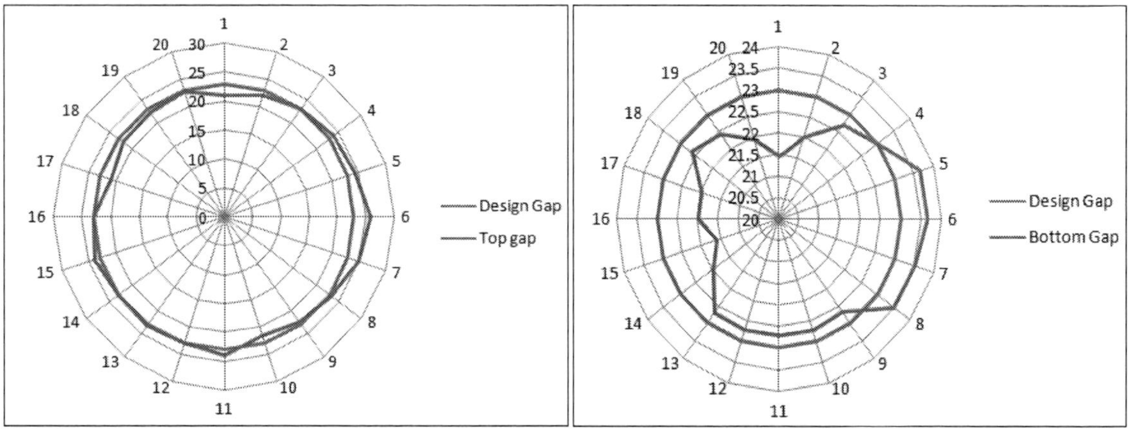

a **b**

Figure 11. Static Air Gap a) Measured from Top and b) Measured from Bottom

Another explanation was using phase analysis. Spiky fall and rise in vibration amplitude on no load and load with 180^0 phase shift as shown in Figure 12. One of the judgments for indentifying the electrical problem is phase shift analysis. From the plot it is obvious that the machine under study is electrically behaving abnormal.

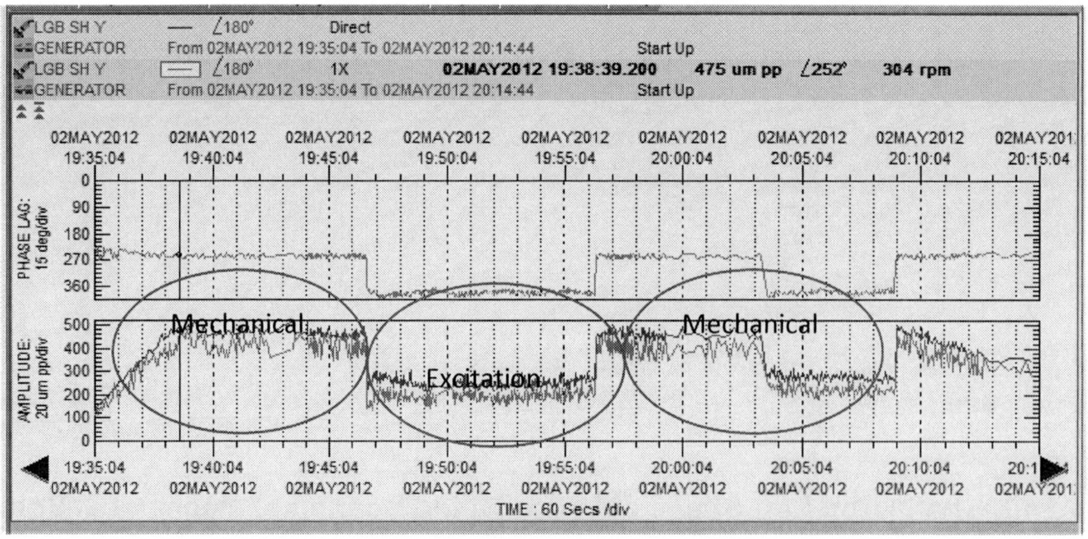

Figure 12. Trend plot indicating spiky fall and rise in vibration amplitude on excitation with 180^0 phase shift.

Conclusions: The vibration signatures were collected at different operating modes to identify the nature of the problem. Through various critical graph analyses, it was observed that the machine under investigation is predominantly suffering from electrical problem. Significant dynamic behavior change is observed when machine rides from mechanical to electrical and vice-versa. Predominant electrical vibration (2LF) was noticeable on excitation and load, and fades away on mechanical spin. Another unique character was abrupt drop in vibration amplitude followed by 180^0 phase shift on excitation.

The signal characters and other behaviors easily distinguish electrical problem from that of mechanical problem. From the vibration signatures, loose stator winding and stator soft foot is suspected. One of the recommendations through the findings was consider checking the looseness of the stator winding, stator soft foot and installation of online air gap monitoring system. Shifting and centering of rotor with respect to stator was also recommended.

The machine under investigation will be taken up for major rehabilitation with stator winding rebuilding. More detail and reasoned judgment is expected after rebuilding of the winding. Also, it is expected to reshape the content of this paper with the completion of the rehabilitation.

Acknowledgements: The authors would like to thank Druk Green management for their unconditional support and motivation throughout. Above all, to all the colleagues and site supervisors who undeniably put their effort to install vibration sensors repeatedly for the vibration test. It was only because of their faith and support in the Center (CoEVaTA) that the completion of this paper has been possible.

References:

[1] Bharat Heavy Electricals Limited, September, 2006, Phase II Rectification and Modernization of Chhukha Hydropower Plant, Druk Green Power Corporation Limited.

[2] Tenzin and P. Rabi. 2013, Unbalanced Magnetic Pull in a Vertical Hydro Generator, 25 Years of Learning, Technical Conference 2013, Druk Green Power Corporation Limited, pp. 469-490.

[3] Mobius Institute, 2014, Diagnostic Guide, Useful Charts and Tables, pp. 33-34.

[4] F. Jawad and T. Iman, Extension of Winding Function Theory for Nonuniform Air Gap in Electric Machinery, IEEE Transactions On Magnetics, Vol. 38, No. 6, November 2002.

BIOGRAPHIES:

Sherub Tharchen started his professional carrier in July of 2000 as electrical engineer with Chhukha Hydropower Power Corporation Limited and closely associated for the maintenance and operation of generating units and its auxiliaries. For the last six years he has been fully engaged in Non Destructive Groups. He is a certified Vibration Analyst Cat III; Certified Infrared Thermography Level 1 (ISO 9001, itc); Certified Magnetic Particle Inspection Level II (SNT-TC-1A of ASNT) and Certified Laser Alignment. Till date, he has authored technical paper titled "Asymmetrical Vibration of Shunt Reactor: Tala Hydropower Plant"; "Root Cause Analysis of Fire Outbreak from 11 MVA Furnace Transformer No. II at SD Eastern Bhutan Ferro Silicon (P) Limited"; "Analysis of Changing Vibration and Temperature in a Vertical Hydro Generator"; "Recognition of Noise as Occupational Health Hazard and Machine Condition Monitoring Technologies"; "Detecting Abnormal Heating Of Earth Rod by Infrared Technology and "A Case of Reappearing Thermal Anomaly in an Electrical in an Electrical Connection at Kurichhu Hydropower Plant".

Dawa: He has a MScE in Mechanical Engineering from the University of New Brunswick, Fredericton, NB, Canada and B.Tech in Electrical Engineering from Malaviya National Institute of Technology, Jaipur, Rajasthan, India. He has authored technical papers titled "Machining Process Characterization using FFT analysis of Acoustic Signal" and published in Proceeding of the 24th Canadian Congress of Applied Mechanics (CANCAM), Saskatoon, Canada conducted from June2-6, 2013.

Presently, he is working as the Head of the Center of Excellence for Vibration and Thermographic Analysis, under Research and Development Division of Druk Green Power Corporation limited. He is a certified Vibration Analyst Cat II. He was closely associated with the work of complete assembly, erection, testing and commissioning of two numbers of 200/40/20T EOT cranes, all the six Generating Units of 170MW each including auxiliaries i.e. Main inlet Valves, Hybrid cooling water system, Drainage and De-watering System, G-100 Governing Systems etc... four Circuits of 400kV XLPE Insulated Power Cables and Mechanical Workshop Equipment of Tala Hydropower Plant from 2003 to 2014.

MACHINERY VIBRATION RELATED TO POOR BOLTING PRACTICES

Ken Singleton, Manager
KSC Consulting L.L.C.
Bristol VA
ksingleton@vibrationconsulting.com

Bob McGinnis P.E.
McGinnis Mech Engineering L.L.C
Kingsport TN
mmengineering@charter.net

Abstract: Poor bolting practices are often identified as a source of machinery excessive vibration. The design and selection of foundation anchor bolts, skid hold down bolts, bearing housing and machine feet mounting bolts often do not receive the attention necessary to achieve reliable bolted connections. Problems such as bolt failure in fatigue, loose bolting, changes in machine alignment, etc., are common. Various techniques are discussed to identify bolting issues which include visual inspection, vibration data, and operating deflection shape analysis (ODS). Examples of machinery with poor bolting practices are provided. Definitions pertaining to bolting and general guidelines are provided.

Keywords: Alignment; anchor bolts; fatigue; Junker machine; ODS; vibration.

Introduction: The authors have spent many years analyzing machinery failures, most exhibiting high vibration or rapidly failing components. An on-site analysis typically begins with a bottom to top visual inspection of the machine or machines in question capturing areas of concern with notes, photos, and video. One issue seems to almost always be present when this in-depth inspection of the machine is done. Inadequate and downright poor bolting practices are identified in virtually every case. Examples of a broken anchor bolt is shown in **Figure 1**, missing shims, see **Figure 2**, thin flat washer yielded into a large hole, see **Figure 3**, and only the corners of a bolt bearing against a washer, see **Figure 4**. These findings are common. This may not always be the primary problem prompting the investigation, but it is one that can be prevented with knowledge and application of proper bolting techniques. Attention to this overlooked issue will also be a valuable tool in improving machinery reliability.

Let's begin by saying that you will find little to no specific bolting information provided in your installation and operating documents for the machine. Very little practical information is provided in machine design, maintenance reliability, or other technical reference books and publications. It is our goal, that at the end of this presentation, you will leave with new knowledge and an approach to solve some of the machinery reliability problems related to bolting you are so often confronted with or perhaps were even unaware that you have.

Examples of Problems Found

- **a.** Thin flat washers.
- **b.** Washers with over-sized inside diameters
- **c.** Bolts and nuts installed in oversize holes with improper flat washers or no washers
- **d.** Missing fasteners
- **e.** Broken fasteners
- **f.** Inadequate quantity of anchor and/or hold-down bolts
- **g.** Wrong grade of bolt and/or nut

h. Improper design to allow for bolt "free stretch"
i. Improper bolt/nut/washer design to prevent self-loosening
j. Improper design of attachment points for bolts
k. Poorly designed flexible machine and bearing frame assemblies
l. Improper size/length/installation foundation anchor bolts

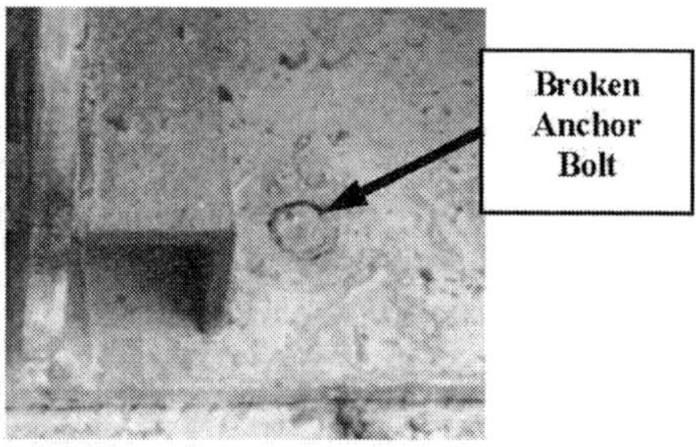

Figure 1. Broken Anchor Bolts on Fan Base.

Figure 2. No Grout Under Frame. Relative Movement Between Frame

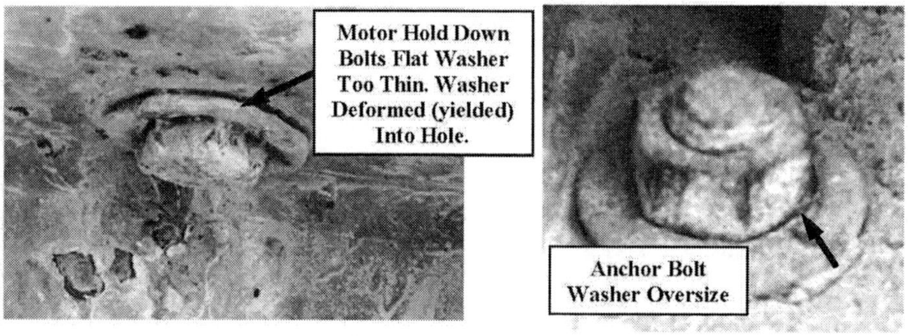

Figure 3. Washers Too Thin and the Washer ID Too Large (Only Corners of Nut Contacting the Washer).

Wrong bolt or nuts for the application - A bolted joint may load the fastener in tension or shear or both. The correct selection of the bolt, washers, nut or preloading device must be properly specified.

- All-thread rod should never be used for any foundation, baseplate, or soleplate fastener application.
- Do not use stud bolts as defined by ASME B16.5 standard for mounting or hold-down bolts. These are specifically designed for pipe flanges.
- Do use S.A.E. Grade 5, 8 or Metric Class 8.8, 10.9 bolts and nuts, unless otherwise specified. Bolts and nuts should be the same grade. Never use grade 2 or unmarked fasteners.
- Grade 5 and Grade 8 bolts should not be exposed to temperatures above 450°F or below -50°F. Expect a reduction in yield strength at elevated temperatures.
- Do not substitute ASTM A325, A490, F568M, or F1852 for S.A. E. fasteners. Fasteners made to these ASTM specs are specifically for structural applications.

- Fully threaded fasteners should not be used except on thin frame assemblies as shown in **Figure 4**.
- Threaded holes in base and sole plates for hold-down bolting should be used only as a last resort. An example is a 3,950 HP Motor bolted to the base plate using 1" bolts in tapped holes, see **Figure 5**. Cast, drawn, and rolled plates and machine components typically have but a fraction of the yield strength of grade 5 and 8 fasteners. Stripped threads are common.
- Coarse threads should typically be selected over fine threads. Fine threads have a tendency to strip.
- The largest bolt which will fit the assembly holes should be used. Where motor alignment is required the motor bolt holes should be larger than the base plate holes.

Figure 4. Full Tread Bolt for Thin Mounting Structures

- Thick hardened washers should always be used under the fastener head or nut. Place the washer under both or at minimum under the one which will rotate during tightening.
- Bolt threads should not extend into both the sole or baseplate and the machine mounting foot.
- The shear plane is the plane between two or more pieces under load where the pieces may move parallel to each other, but in opposite directions.
- The bolt threads may be included in the shear plane or excluded as shown in **Figure 6**.

Figure 5. 3950 HP Motor, 1" Bolts in Base Tapped Holes.

- The capacity of a bolt is greater with excluded threads
- In addition to the clamping force to hold machines in position, shifting due to thermal growth or loading may impose axial and shear loads.

Fastening design fails to allow for bolt "free stretch" - It is important that bolted joints be clamped with enough force to prevent relative movement which can cause loosening of the joint as well as fretting corrosion. A bolt must also have adequate preload (stretched) to effectively clamp the joint. In order for the bolt to sustain the proper preload without failure in tension a minimum free stretch length is required. For anchor bolts, reference [9 pg 19], states minimum free stretch length of 7 bolt diameters for rotating machinery and 12 bolt diameters for reciprocating machinery. Adequate free stretch length is also important for hold down bolts on motors, bearing

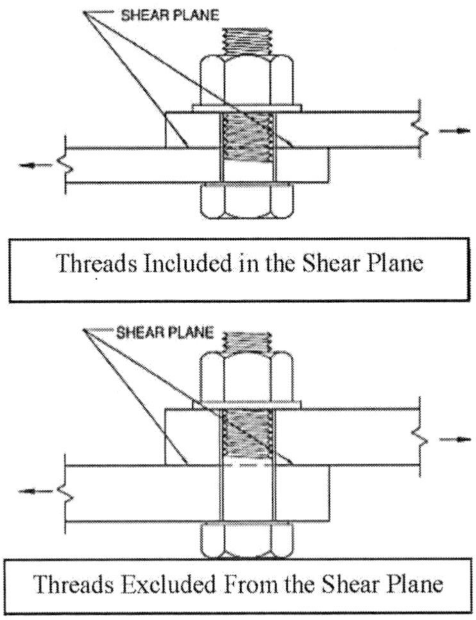

Figure 6. Bolt Threads May Be Included In Shear Plane or Excluded.

housings, machine cases, etc. If the bolt does not have adequate free strength length, the bolt may fail in tension or fatigue; see examples in **Figures 1 & 7**.

Figure 7. Machine Frames Bolted Together. Bolt and Frame Fretting Corrosion Indicated Looseness (Arrows). No Flat Washer. Bolt Failed in Fatigue Which Was Indicated by Beach Marks and Final Rupture Zone.

A bolt's free-stretch length is the distance between the nut (actually several threads into the nut or tapped hole) and the head, see the illustrations in **Figures 8 & 9**. The length of the thread engagement in the nut or tapped hole should be sufficient to break the bolt in tension, reference [10 pg 52]. The nut wall thickness and length must also be taken into consideration.

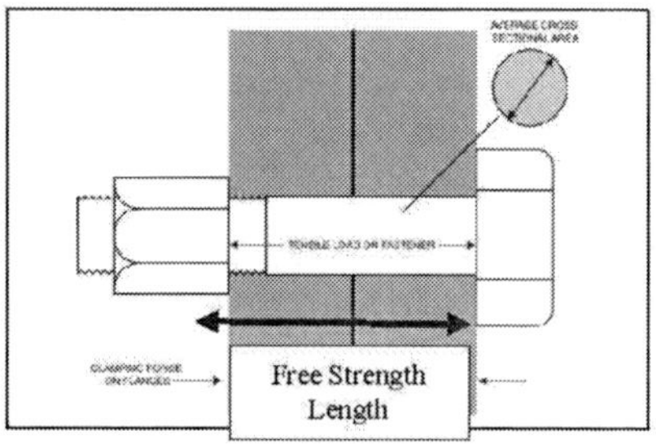

Figure 8. Free Stretch Length of Bolt, Reference [16 pg 3].

Tensile strength is the mechanical property most widely associated with standard threaded fasteners, reference [15]. The tensile strength is the maximum load in tension that the bolt can support prior to fracture. The load in tension that a bolt can withstand is given by equation **(1)**.

$$P = St\ X\ As \qquad (1)$$

Where:

P = Tensile load or Clamp Load lb_f
St = Tensile Strength psi
As = Tensile Stress Area in^2

For example, for a 3/4" SAE J429 Grade 5 HCS bolt:

St = 120,000 psi
As = 0.3340 in^2

P = 120,000 psi x 0.3340 in^2 = 40,080 lb_f

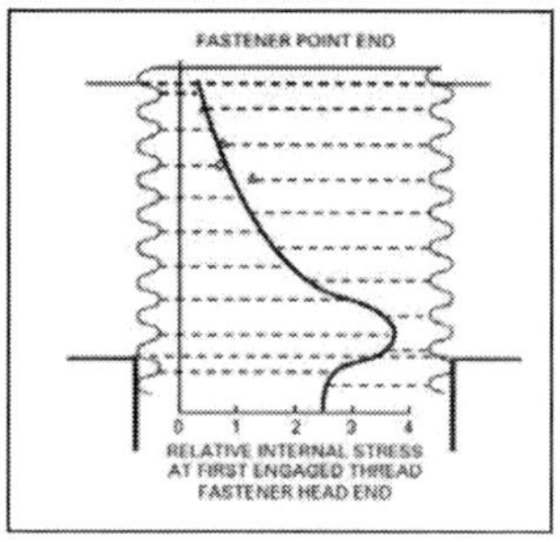

Figure 9. Relative Internal Stress Of Fastener Threads, Reference [10 pg 55].

A tensile stress-strain diagram is shown in **Figure 10**, [15 & 16]. If the bolt is deformed or stretched by applying load in tension (stress), as long as the load does not exceed the Yield Point, the bolt will return to its original shape after the load is removed. The Typical Clamp Load is taken as 75% of the Proof Load. The Proof Load is typically 85-95% of the material Yield Point. If the bolt is subjected to a load exceeding the material Yield Point (in threaded section of bolt), it will neck down and elongate further with a reduction in stress. It will not return to its original shape once the load is removed. Strain is the amount of elongation divided by the original length of the stressed section.

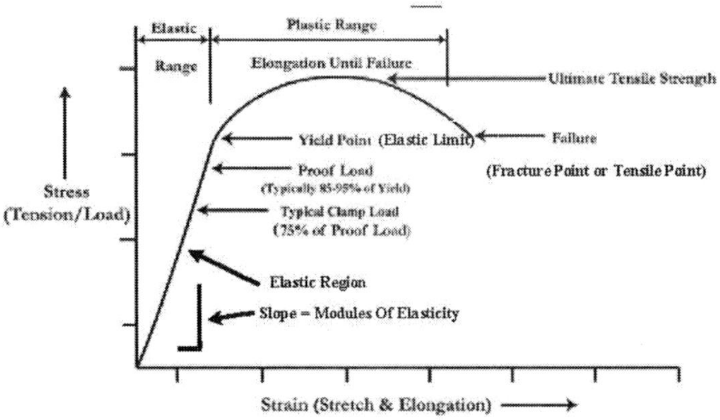

Figure 10. Tensile Stress-Strain Diagram, Ref [15, 16].

Stress/strain curves for some materials used for bolts are shown in **Figure 11**, [15]. Note that harder, higher tensile strength fasteners such as ASTM A574 Socket Head Cap Screw (SHCS) and SAE J429 Grade 8 are less ductile than lower strength SAE J429 Grade 5 and ASTM A307 Grade A. Although the ASTM A574 and SAE J429 materials have high tensile strength, the overall length of the strain curve is decreased when compared to materials J429 and A307.

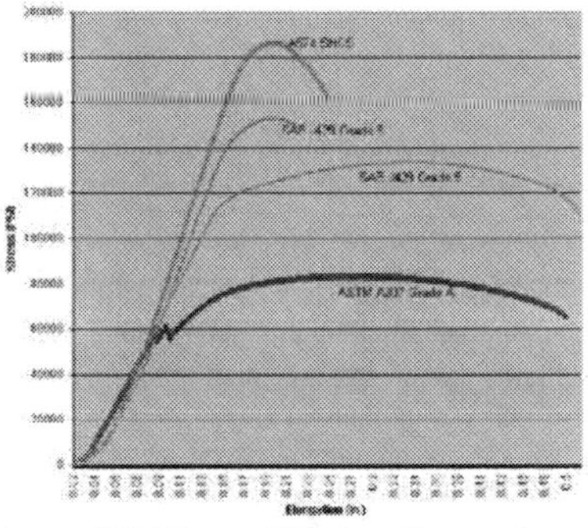

Figure 11. Stress/Strain Curve for Different Materials, Reference [15]

Washers, thick wall pipe with ends machined, etc., may be used to provide the specified bolt free-stretch length. Some examples are shown in **Figures 12 & 13**.

Figure 12. Recip Compressor Crankcase Hold-Down Bolts Using Pipe to Provide Adequate Free-Stretch Length.

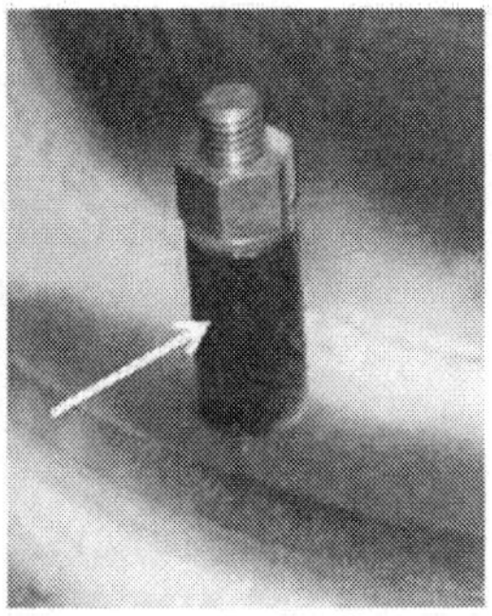

Figure 13. Pressure Vessel Hold-down Bolt on Recip Compressor Skid, Use of Pipe to Provide Adequate Bolt Free-Stretch Length.

Very high local pressures in the contact areas on the threads and the nut and bolt head contact surfaces occur when the bolt is tightened. Local plastic deformation at these interfaces results in flattening of surfaces. The flattening is also called embedding. The amount of embedding influences the reduction of bolt stretch or preload, [14].

Improper torque applied to bolt or nut - Torque is probably the most often used method of tightening a bolted joint after the "tighten by feel" method. The common torque wrench is quick and easily implemented. However, according to [Ref. 31, 32 & 33] a torque wrench may only be accurate to +/-35%. Torque generates parasite torsion stress, see **Figure 14** [33 pg 9], in the bolt or stud which may result in failure of the bolt before adquate preload occurs. Only about 10% of the torque typically is used to tension or stretch the bolt, see **Figure 15**. Most of the torque is expended overcoming friction, 50% under head and 40% threads according to reference [36 pg 5].

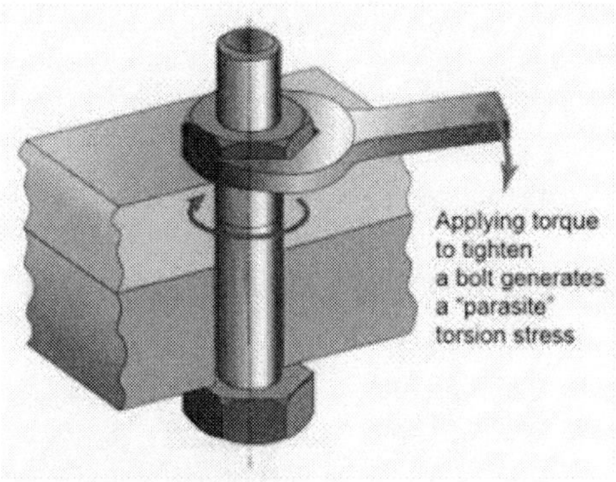

Figure 14. Applying Torque Generates a Parasite Torsion Stress [33 pg 9]

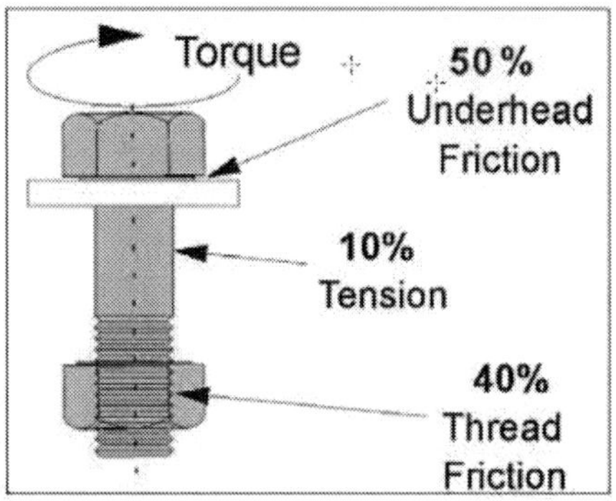

Figure 15. Where Does The Applied Torque Go? [36 pg 5]

There are many portable torque wrenches available which include hydraulic, impact, tools with pneumatic or electric adjustment, impact wrenches with stored energy via a torsion bar or other means. Many of the portable torque wrenches can be calibrated using a Skidmore-Wilhelm gage, reference [34] or Torque-Tension Research Head (load cell), reference [36 pg 9]. The Skidmore-Wilhelm gage and Torque-Tension Research Head measure the tension or pre-load of the bolted assembly. For critical applications, testing several bolt-nut-washer assembles provides a method of calibrating the torque wrench or tool to significantly reduce bolt tightening error.

Another method of achieving bolt preload is Torque-Angle Monitoring With Control, references [31, 36], see **Figure 16**. Torque is applied until a specified threshold level is obtained then the joint is tightened by rotating the nut a specified angle. This method is reportedly accurate <=5% in the joint elastic clamping zone per reference [36, pg 12-13]. Calculations to determine the nut rotation angle in the elastic clamping zone for a ¾-10 bolt, see **Figure 17**, showed that if the bolt free-stretch length was only 1 bolt diameter, the nut rotation angle was less than 2 degrees. If bolt preload relaxation was 0.002 inch, the bolted assembly would become loose (all preload lost) for

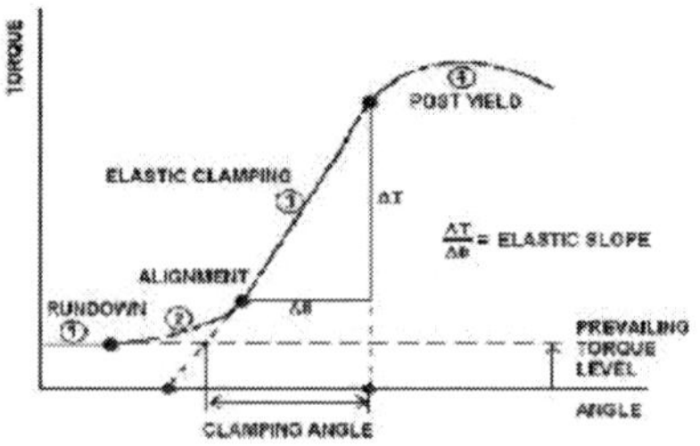

Figure 16. Torque-Angle Monitoring With Control, [31, 36 pg 15].

bolt diameters 1 thru 3. Therefore, an advantage to long free-stretch length of 7 or more bolt diameters is that the nut angle of rotation is greater and thus easier to measure and the potential error decreases. The longer bolt also retains more preload if the joint relaxes.

3/4-10 Bolt	Bolt Grip Length Inches	Preload psi	Bolt Stretch Inch	Remaining Pre-Load lb_f Relaxation= 0.002 inch	Nut Advance in Degrees (In The Elastic Clamping Zone)
1 Bolt Diameter	0.75	60,000.00	0.0005	None	1.6
3 Bolt Diameters	2.25	60,000.00	0.0014	None	4.9
7 Bolt Diameters	5.25	60,000.00	0.0032	6,832	11.5
12 Bolt Diameters	9	60,000.00	0.0055	11,594	19.7

Figure 17. Calculations for Nut Rotation Angle to Achieve 60,000 PSI

Absence of flat washers where oversize holes exist – Although this may seem improbable, we have found bearing housing and motor hold down bolts installed with no flat washers. In a few instances, only the corners of the nut were actually contacting the base plate as shown in **Figure 18**.

Figure 18. Motor Hold down Bolt, No Washer.

Wrong type of flat washers – Flat washers are typically used as the contact surface for the bolt head or nut. As the nut is tightened, galling may occur so the washer serves as a sacrificial surface. The washer also spans the hole through which the bolt is inserted. One problem we have documented is that thin flat washers often yield into the hole (dish), as shown in **Figure 19**, which results in loss of bolt pre-load. The dishing also has a centering effect (side load on the bolt) which has a tendency to cause the motor, bearing housing, etc., to move. This movement can result in the machine moving out of alignment. Consider using hardened extra thick flat washers, reference ASTM [45] see table in **Figure 20**. Up through 1 ½ in, the washers are through hardened. A stiff washer with large diameter increases joint stiffness. The strength of the washer must be at least equal to the strength of the bolt. These are available from many sources such as Grainger and Fastenal.

Figure 19. Motor Hold Down Bolts, Oversize Flat Washers Deformed Into the Base Plate Holes.

Nominal Size	Nominal OD	Nominal ID	Min Thickness (in)	Max Thickness (in)
½	1.063	0.531	0.305	0.375
5/8	1.313	0.688	0.305	0.375
3/5	1.468	0.813	0.305	0.375
1	2.000	1.063	0.305	0.375
1.125	2.250	1.188	0.305	0.375
1.250	2.500	1.375	0.305	0.375

Figure 20. Extra Thick Hardened Flat Washers, Ref ASTM [45].

These are available from many sources such as Grainger, Fastenal, etc.

Figure 21 Split Lock Washer Reference [12].

Use of split or lock washers –According to reference [12], work completed during the 1960's in Germany found that the effectiveness of locking devices such a split lock washers actually aided loosening. As a result, the conventional spring lock washer, shown in **Figure 21**, is no longer specified. Consider removing split lock washers from your stores inventory.

Expecting bolting to remain properly tensioned when high vibration is normal – In a paper published by Junker reference [4], it was discovered that lateral vibration applied to a bolted assembly would cause self-loosening. The Junker Machine (named after the inventor) was developed in Germany in the 1960s and is still used today to test various fasteners. Under high vibration, most fasteners will self-loosen; see reference [22] which has a link to a YouTube video. To address self-loosening, thread locking devices were developed and can be grouped into three categories: 1) Free Spinning, 2) Friction Locking and 3) Chemical Locking.

a. Free Spinning type uses a plain bolt with a circumferential row of teeth under the washer head. These teeth are ramped and allow the bolt to rotate in the clamping direction but lock into the bearing surface when rotated in the loosening direction. Brand names include Nord-Lock [21], Oglaend, Disc-Lock, etc.
b. Friction Locking is divided into two groups; metallic friction locking and distorted thread. The Philidas nut and Nyloc nut are included in this group. The distorted thread includes Stanley Spiralock, reference [24].
c. Chemical Locking includes adhesives which fill the gaps between the male and female threads bonding them together. Chemical Locking provides the greatest resistance to vibration loosening, reference [12 pg 3]. Loctite, reference 42, is a well-known brand.

Excessive quantity of shims – Stainless steel shims are now commonly used (since the 1980's) when aligning machines. Occasionally, primarily in older plants, we still find carbon steel shims in use. The carbon steel shims are typically heavily corroded. Shims act as springs and thus tend to soften the bolted joint especially a stack of shims like that shown in **Figure 22**. The bolt should always be the weaker element of a joint in order to maintain bolt preload. Consider using as few stainless steel shims as possible under machine feet (3 to 5) and a single solid shim under cast bearing housings.

Figure 22. Too Many Shims at Corner of Gearbox.

Poor grouting practice related to anchor bolts – This remains a serious issue even today after many years of seminars and published papers providing detailed information on proper grouting. There are primarily two types of grout, cementitious and epoxy. Cementitious is typically used as filler which holds the shims in place upon which the soleplate or skid rests. The soleplate or skid is clamped by anchor bolts to the concrete foundation with shims at each anchor bolt. Epoxy grout, reference [9] (when properly used), glues or bonds the soleplate or skid to the concrete foundation. Where epoxy grout is suitable, (not for use in high temperature area such as exhaust end of gas turbine) it is considered superior to cementitious grout due to compressive strength of 14,000 psi and 2,100 psi tensile strength per reference [38]. But, whichever type grout is used, the anchor bolt must have adequate free stretch length, reference [9, pg 19]. The grout, shims and soleplates should handle the compressive stresses imposed on them, reference [39, pg 24].

Thin base plates or soleplates - For most non-API equipment, the soleplates and base plates were deficient in size and design. Charlie Jackson, reference [41] stated that steam turbine soleplate thickness should be minimum 3" thick, epoxy grouted to a monolithic poured re-bar re-enforced concrete mezzanine. The soleplate should be larger than the turbine base by 6" or more on the outer perimeters. Coat tops of soleplates with 0.5 mil electroless nickel, reference [43], against corrosion. .

Bolt Tightening Methods – Generally there are two methods, torque or tension. Torque is used to rotate a nut, stretch the bolt and clamp the joint. Tension is implemented using mechanical devices, reference [20], or hydraulically actuated tools, reference [33], to stretch the bolt without rotation. A nut is typically spun down and snugged while the bolt is under tension and stretched. When the force applied to the bolt is removed, the nut maintains the bolt stretch. There are many problems with torque because the majority of the torque is expended in friction. Tension is more accurate but also more expensive due to the tooling or the type of nut used

Mechanical Chocks: Vibracon reference [13], see **Figure 22**, and ExactAlign reference [44]. There are adjustable supports for use under diesel engines, natural gas engines, motors, etc. The advantage is easier alignment changes since shimming is not required

Figure 23. Vibracon Chocks, Reference [13].

Wedge Locking and Friction Based: Ref [17]: The principle of friction locking is based on the increased friction in the thread and the surface of the bolt head/nut. The primary disadvantage of assembling dry or unlubricated bolts is the increased torsion stress in the joint. The high torsion may cause the fastener to yield at a lower preload than expected. Since friction conditions are uneven, the bolt may not reach the necessary preload. Insufficient preload is the most common cause of bolting failure due to fatigue.

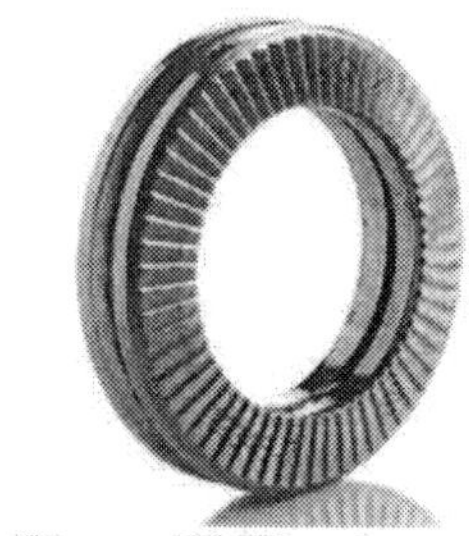

Figure 24. Nord-Lock X-Series Washers Ref [25].

Thread lubricants are often used to minimize the friction and to obtain more uniform clamp load. However, lubrication will significantly reduce the locking ability of any friction based method.

The wedge locking method offered by Nord-Loc, see **Figure 24 & 25**, Disc-Lock, etc., is based on tension instead of friction. The most common example of the wedge locking system is a pair of washers which have cams with a rise greater than the thread pitch of the bolt. The washer pair is installed cam-face to cam-face. When the bolt/nut is tightened, teeth grip and lock the mating surfaces, allowing movement only across the cam faces.

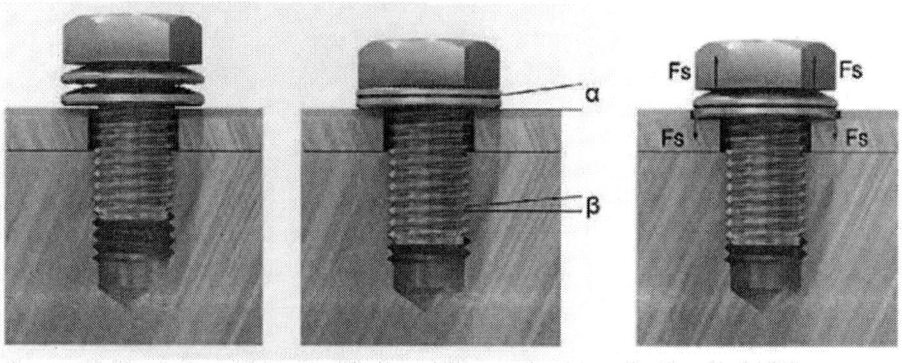

Figure 25. Nord-Lock X Series Washers Illustrating Installation, Ref [25].

Any rotation of the bolt/nut is blocked by the wedge effect of the cams. The wedge locking ability is not affected by lubrication.

The Spiralock by Stanley Ref [24] uses a wedge ramp at the root of the internal thread which only engages when the joint starts to build clamp load during tightening, see **Figure 26**. The crest of the standard external threads draws tightly against the wedge ramp which is reported to eliminate all radial clearance and create a continuous spiral line of contact between the internal and external threads. One of KSC's clients has reported successfully using the Spiralock in assembly of underground coal mining equipment.

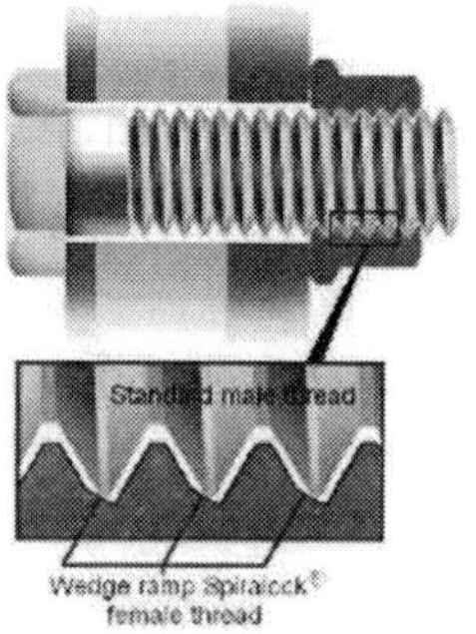

Figure 26. Stanley Spiralock, Ref [24].

Since the Spiralock has special internal thread geometry, special taps and gages are required. The locking action is in one direction only. Identifying holes tapped with Spiralock taps is difficult so a means of tracking assembles using this type fastener should be implemented.

Bolting Criteria

a. Washers: (Flat, Flat Extra Thick Hardened, Split Lock, Bellville, Wedge Locking, Direct Tension Indicating, etc.)
b. Thread Lubrication Effect:
c. Secure Bolting
    i. Double Nuts, Serrated, Ribbed (Rip-Lock, Nord Lock, DISC-LOCK WASHER
    ii. Chemical Thread Lockers, Polymer Patches, Adhesive Patches
    iii. Nuts and Bolts with Thread Locking Features

Proper Bolting and Potential Solutions Summary

a. Develop and implement standards and specifications regarding hold-down and anchor bolting
b. Bolt Grades: grade 5 vs grade 8 (8.8 vs 10.9 metric) Grade 1, 2 and unmarked don't use!!
c. Verify proper materials, design and installation at initial installation
d. Always check fasteners for proper "Grade", especially after maintenance is performed. Be aware of imported counterfeit fasteners.
e. Insist on exta thick, hardened washers with proper inside diameter be installed on motor, bearing housing and machine frame hold-down bolts.
f. Eliminate split lock washers from your bolt assemblies and inventory.
g. Determine proper bolt torque and insist that proper torqueing procedures be followed and document in procedures.
h. Using "Torque-Angle Monitoring with Control" as previously described will provide more accurate tensioning provided the bolt diameter to length is 1 to 7 or greater.
i. Use specialty type nut locking devices where needed, especially where high vibration is present.

Example 1: AMCA Arrangement 8 Fan. High Amplitude Vibration, Anchor Bolts Breaking:

Vibration Analysis: Vibration data showed high amplitude 1X and orders or harmonics. Spectra and time domain data measured on the fan end bearing housing with loose hold down bolts are shown in **Figure 27**. Multiples of the run speed frequency in the spectrum and relative motion of the bearing housing are indicated in the time domain data.

The 3D model of the fan and foundation, see **Figure 28**, was developed in ME'scopeVES, reference [30]. Cross channel vibration data were measured at various locations on the motor, fan base and concrete foundation for the ODS. The vibration amplitudes at various points are shown in **Figure 29**. The vibrating shape or pattern of the fan was rocking side-to-side, pivoting about

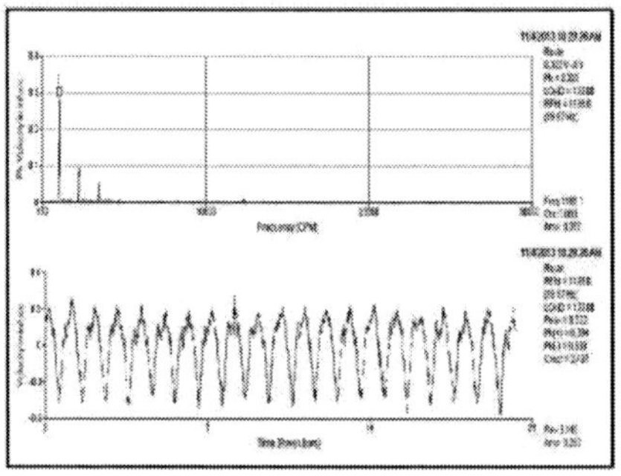

Figure 27. Fan End Brg Housing Ver, Time Data in Velocity in/sec pk.

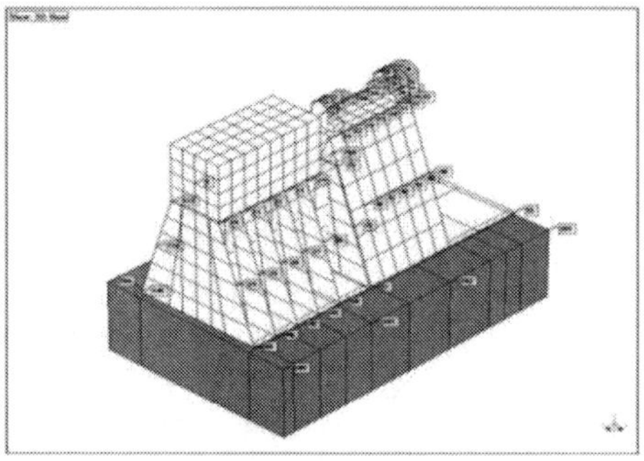

Figure 28. Fan Model Developed in ME'scopeVES, Reference [30].

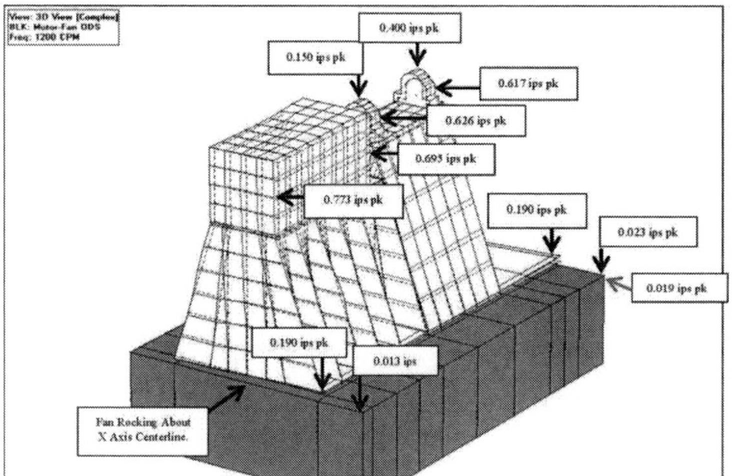

Figure 29. Animated ODS at Fan Run Speed Frequency.

the X Axis just above the surface of the concrete base parallel to the shaft's centerline. The severe rocking motion of the entire fan assembly was a result of fan wheel unbalance (forcing function) amplified by a rigid body mode natural frequency near running speed and fan rotor 1st critical. The rigid body mode was caused by loose mounting of the fan base to the concrete foundation (failed anchor bolts). The failed anchor bolts also lowered support stiffness which lowered the rotor 1st critical down to run speed frequency.

Fan frame anchor bolts had only ½ inch free stretch length. Some of the anchor bolts were loose and other bolts had experienced fatigue failure. The anchor bolt washers were too thin resulting in deformation of the washers and inability of the bolted assembly to remain properly tensioned. Typically a sleeve is installed around the bolts to prevent concrete and grout from contacting the bolts. The sleeve allows the bolt to stretch over an adequate length of at least 7 diameters to prevent over stressing during tightening. This long stretch length also allows the bolt to be more elastic than the other components in the joint.

The motor hold down bolts had flat washers which were too thin and had deformed into the base holes. This created a loose mounting and allowed relative motion or vibration of the motor to the base. The fan bearing housing hold down bolts also used flat washers. Both bearing housings exhibited looseness but the fan end bearing had significant relative motion to the bearing support plate. There was very low amplitude vibration of the concrete pad indicating it had very little participation in the fan vibration.

Several recommendations were made to address the bolting issues that included installing anchor bolts, as shown in **Figure 30**, full grouting and stiffening of the fan base, see **Figure 31**, and installing Grade 8 bolts with thick washers under the motor and bearing housing hold-down bolts and nuts. In the end, the decision was made to replace the fan.

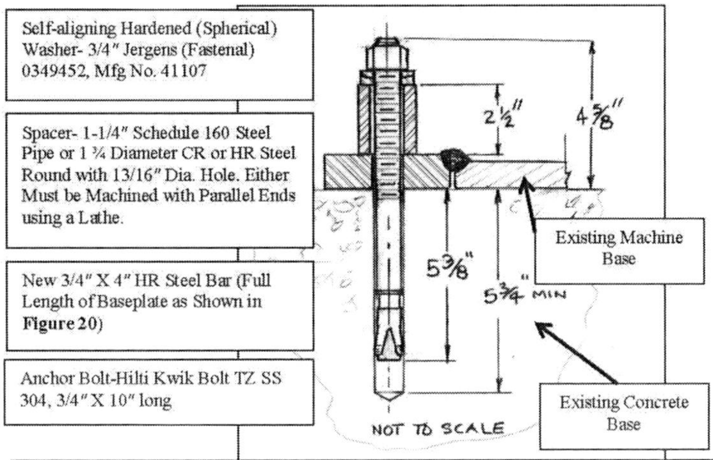

Figure 30. Anchor Bolt Installation Detail.

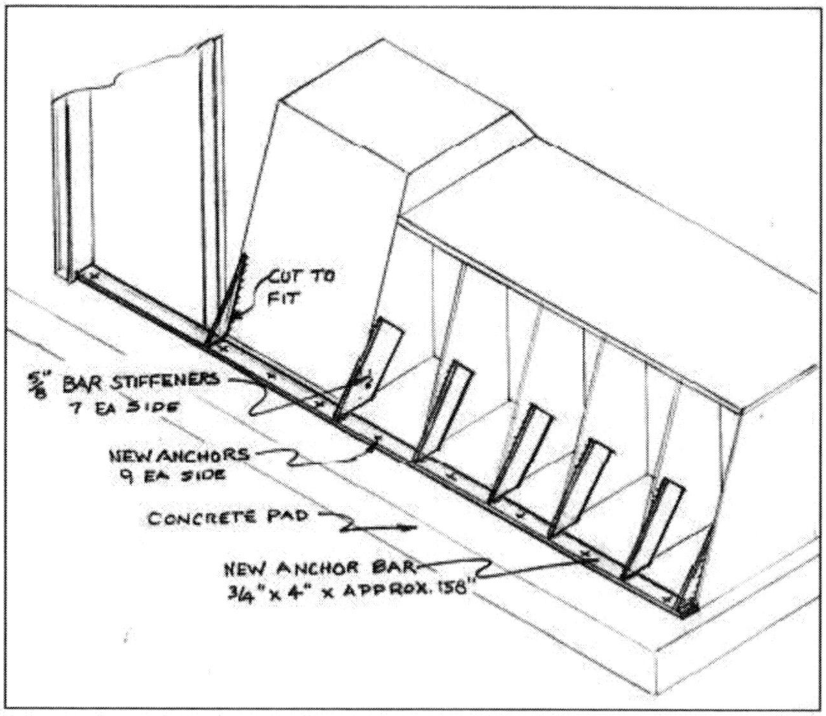

Figure 31. Anchor Bar and Bracing Isometric Sketch.

Example 2 AMCA Two Arrangement 8 Process Fans

The reported reliability problems for two process fans (A & B) were short bearing life, high amplitude vibration and previous failure that resulted in a bent fan shaft. The fans were 500 HP, 56,433 ACFM, AMCA Arrangement 8. Our initial findings verified elevated vibration amplitudes plus loose fan bearing housing and motor hold down bolts and bearing defect frequencies. Vibration data measured during coastdown indicated fan rotor unbalance as the primary forcing function. An ODS clearly showed motor relative motion of the bearing housings and motors to the support plates and foundation rocking of Fan A.

The vibration data at Fan A inboard bearing housing, see **Figure 32**, showed 1X, 2X and 3X as the primary frequencies. The overall amplitude measured 0.152 in/sec pk. Time waveform data measured 2 g's.

Fan bearings were spherical roller type with four Grade 5 hold down bolts. The hold down bolt washers were dished with only the corners of the bolt heads contacting the washers, see **Figure 33**. Split shims had been inserted under the bearing housings instead of solid shims resulting in the center of the bearing cast housing being unsupported. The bearing housing support surfaces had been machined to 2 to 3 mils/ft flatness.

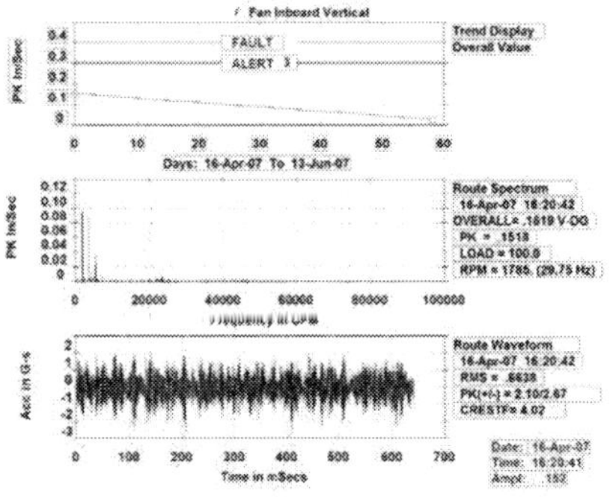

Figure 32 Initial Vibration Data at Fan Inboard Bearing Housing Vertical Direction.

Both fan's bearing housing hold down bolts (under the base plate), see **Figure 34**, had a stack of washers that included a flat washer, split lock washer and square washer burned from bar stock by cutting torch. All bolt assemblies were found loose.

Some of the motor hold down bolts had low carbon steel flat washers which had dished into a large hole. Some bolts had no washer with only the nut corners contacting the support plate, see **Figures 35-37**. The holes in the base plate supporting the motor were much too large.

Figure 33. Bearing Housing Hold Down Bolt Washers Dished, Full Shim not Used Under Housing.

Figure 34. Fan A, Bearing Housing Hold Down Bolts.

Figure 35. Motor Hold Down Bolt, Corners of Nut

Figure 36. Motor Hold Down Bolt, Oversize Hole, No Washer.

Figure 37. Fan A, Motor Hold Down Bolt With Deformed Flat Washer.

Coastdown data on one fan is plotted in Bode format is shown in **Figure 38**. The vibration response was typical of unbalance with no indication of structural resonance.

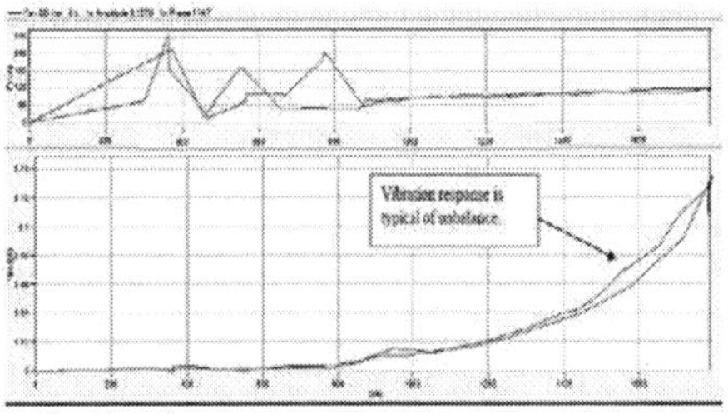

Figure 38. Fan A, Coastdown Data, Bode Plot, Fan OB Brg Hor.

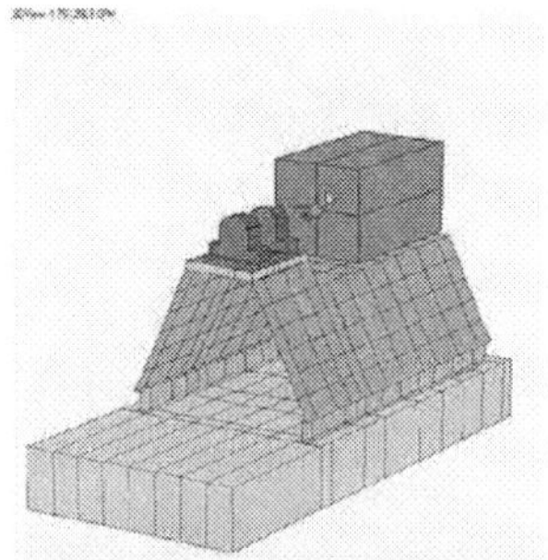

Figure 39. Fan Foundation, Base, Motor & Bearings ME'scopeVES [30] Model.

ODS data was acquired using one reference accelerometer and five roving accelerometers. The model of the fan foundation, fabricated steel base, bearing housings and motor was developed in ME'scopeVES, [30] see **Figure 39**

The ODS animations at running speed showed bearing housing and motor relative movement to the base plate. Vibration data measured at the motor right-front foot of Fan A at 1X was 0.050 in/sec pk and 0.039 in/sec pk on the support plate, see **Figure 40**. Amplitudes were lower at Fan A Wheel End Bearing Housing measuring 0.025 in/sec pk on the top of the housing and 0.016 in/sec pk on the base plate, see **Figure 41**.

Corrective actions were as follows:

1. The fan rotor was balanced per ISO 1941-1 for <u>over hung rotor</u> to G2.5.
2. Wheel hub advance on the tapered fit was calculated to maintain interference fit allowing for differential thermal growth of the shaft, fan hub and tapered sleeve.
3. The fan bearings were replaced.
4. Bearing housing hold down bolts with hardened washers torqued per specification.

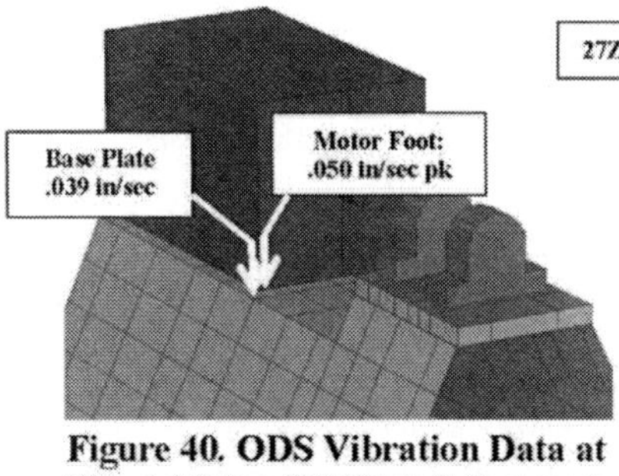

Figure 40. ODS Vibration Data at Fan A Motor RF Foot & Support Plate 1X Running Speed.

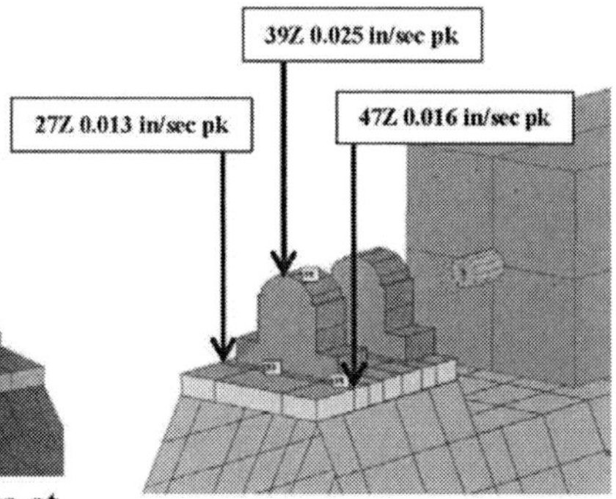

Figure 41. ODS Vibration Data at Fan A, Wheel End Bearing Housing and Base Plate.

Vibration was reduced 88% at 1X and 90% at 2X.

Example 3: New Installation of Slurry Pumps, Excessive Vibration

Three pumps, 500 Hp, 1170 RPM , 5508 GPM, 212 ft head, had been installed at a mining facility. [Ref 8] The units were direct coupled, variable frequency drive in a waste disposal system. One of the units is shown in **Figure 42**. The client was concerned about the very high amplitude vibration of the motors and the skids reported by the consulting company monitoring vibration periodically. Resonance was suspected.

Figure 42. Pump Assembly Naming Convention.

The motors and pumps were mounted on fabricated skids which were bolted and epoxy grouted to individual concrete bases. Visual inspection, vibration data, ODS and Experimental Modal Analysis (EMA) were conducted and identified the following:

- Motor hold down bolts – flat washers yielded into base plate holes.
- The pump base mounting foot underneath the pump housing did not have a hold down bolt to the skid.
- The six (6) anchor bolts on the bottom of the skid were fastened through a very flexible flange or "wing". The thin flanges of this design allow excessive flexure.
- The anchor bolt washer's inside diameters were too large and the washers were too thin. This resulted in a small clamping area for the nuts, and deformation of the washers.
- There were an inadequate number of anchor bolts on the lower section of the skid frame.

- There was lack of bonding of the epoxy grout to the concrete and the structural steel frame at some locations.
- Lack of bonding of the concrete mounting base to the concrete mat The mass of the concrete mounting base was undersized according to the information provided for the weight of the motor, pump, fabricated frame and the dimensions of the concrete. Typically, the mass of the concrete should be a minimum of three times the mass of the rotating machinery mounted to the concrete base.
- Resonant frequencies of the fabricated skid and motor pedestal near 1X the motor run speed frequency and pump blade passing frequency (4X).
    - The resonant frequency at 1X the motor run speed was side-to-side rocking mode shape of the motor and flexure of the motor pedestal.
    - The resonant frequency at 4X the motor run speed was a twisting motion about the Z axis of the motor and motor pedestal.

The skid was fabricated of wide flange beams and flat plate. The flanges of the wide flange beams were used to install the skid hold down bolting see **Figure 43 & 44**. This attachment design lacked adequate rigidity.

The skid hold-down bolts were carbon steel and had already started corroding. The washer holes were oversize with only the corners of the nuts make contact. The motor hold-down bolt washers were low carbon steel with oversized holes which had deformed into the holes in the motor support base, see **Figure 45**. All bolts were found loose.

Figure 43. Motor End of Skid, Caron Steel Anchor Bolts, Over

Carbon steel shims had been inserted under some motor feet and had started corroding.

Figure 44. Flanges of the Skid Wide Flange Beams Were Used For Anchor Bolt Hold Down.

Figure 45. Motor Hold Down Bolts Had Oversize Washers Which Had Deformed Into the Holes.

The ODS Model developed in ME'scopeVES Ref [30] is shown in **Figure 46**. Vibration amplitudes in/sec pk for 1X run speed are labeled for some locations. The motor outboard bearing housing had the highest amplitude at 0.826 in/sec pk.

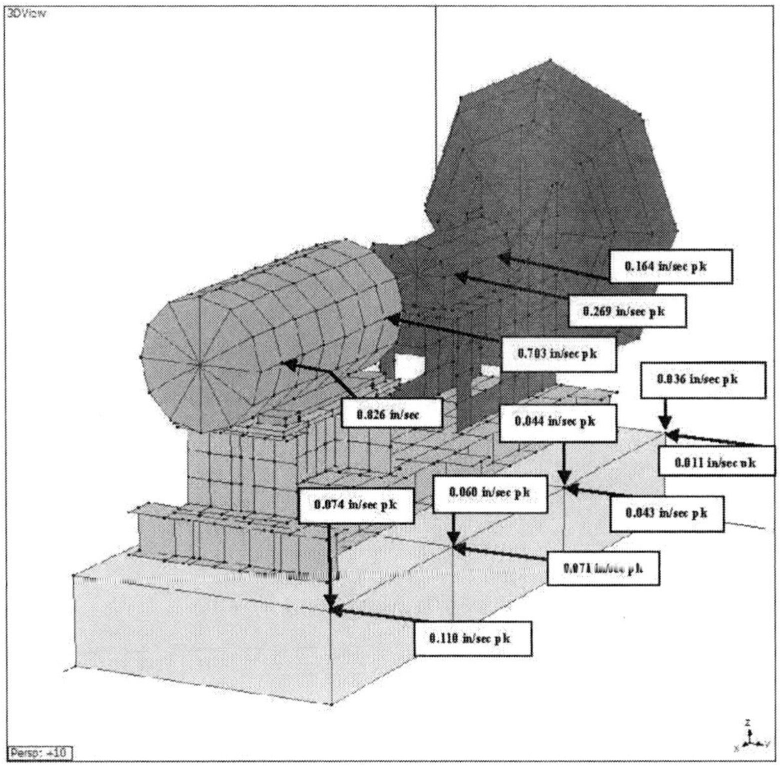

Figure 46. Average Vibration Amplitudes for Operating Deflection Shape Model At 1X Run Speed Frequency.

ODS spectra data measured at the motor right-front foot and base vertically are shown in **Figure 47**. The data is plotted with log magnitude scaling to more clearly show the indications of resonance near 1X and 4X run speed (pump vane passing).

The vertical vibration was much lower amplitude than horizontal. As shown in **Figure 47**, the Motor Right-Front foot measured 0.157 in/sec pk while the base plate under the foot measured 0.085 in/sec pk.

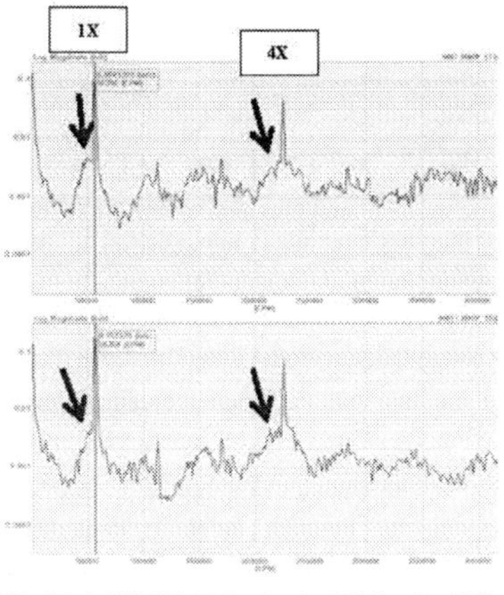

Figure 47. Spectrum at Motor IB Right Front Foot and Support, Log Magnitude Scaling Is Better Showing Indication of Resonance.

Corrective Recommendations: There were many problems identified with the installation of the pumps but this article focuses on bolting problems. A drawing, see **Figure 48**, was provided Ref 9 to show the proper installation of anchor bolts, Ref [9].

Recommended modifications to the skid and motor support are shown in **Figure 49**. This drawing showed the proper method of skid design for the hold down bolts. Stiffeners were recommended to significantly increase the rigidly of the motor support.

Thicker washers (1/2" thick) and Grade 5 bolts were recommended for the motor hold down bolts.

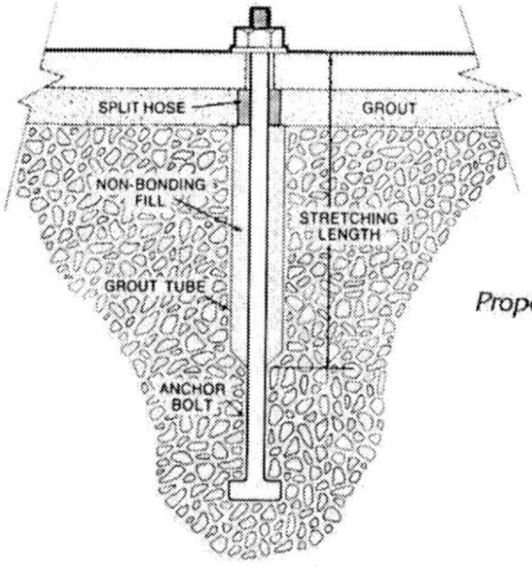

Proper method of installing anchor bolts.

Figure 48. Anchor Bolt Installation Ref [9 & 38] Escoweld 7505E/7530 Machinery Grout.

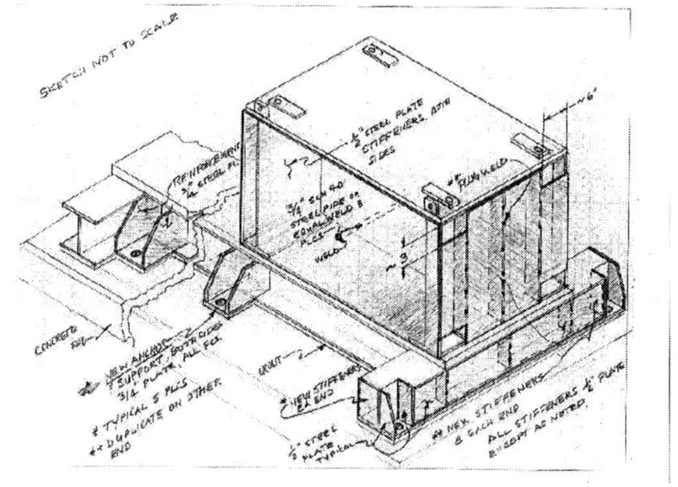

Figure 49. Drawing Showing Modification of Skid.

Example 4: Turbine-Generator Train Bearing Housing & Pedestal Excessive Vibration

Following an outage of a turbine-generator train, vibration of some of the bearing housings and pedestals was found to be very high amplitude. There were other issues identified during the investigation but the focues of this article is on bolting. The unit consisted of three turbines, see **Figures 50-51**, high pressure turbine (H1P), two low pressure turbines (LP1) & (LP2), Generator and Exciter. The bearing housings and pedestals of LPA & LPB, bearings 4 and 5, see **Figure 52**, had very high amplitude vibration primarly in the axial direction. Note that the pedetal hold down bolts has structural steel angle welded to the nut and top of the bolts, see **Figures 53-54**.

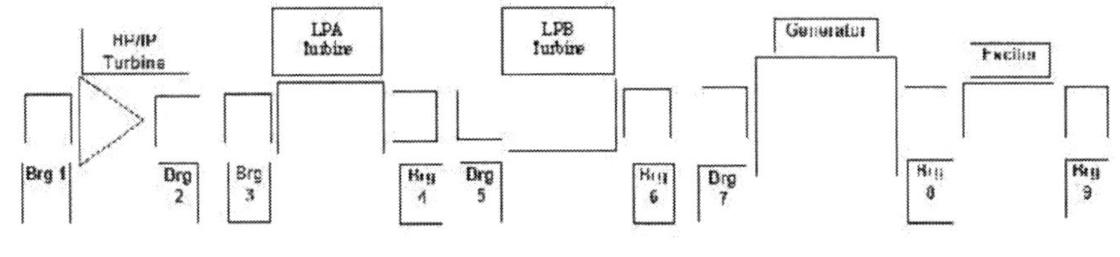

Figure 50. Layout of Turbine-Generator Train.

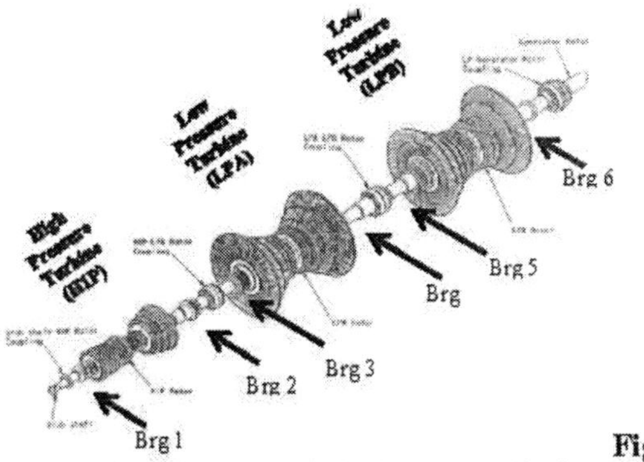

Figure 51. High and Low Pressure Turbines.

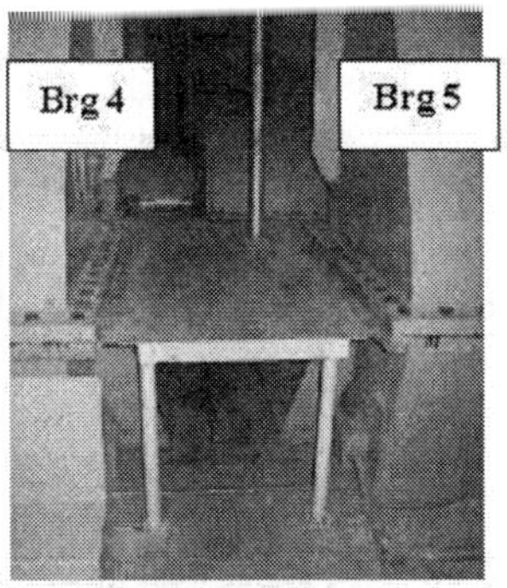

Figure 52. LP1 & LP2 Bearings 4 and 5 Pedestals Right Side.

Figure 53. Bearing 4 Pedestal & Soleplate LPA Right Side.

Figure 54. Bearing 5 Pedestal & Soleplate LPB Right Side.

The vibration spectral data showed indication of structural resonance near running speed, see **Figures 55-56**. The spectra are plotted with log magnitude scaling to aid identification of natural frequencies.

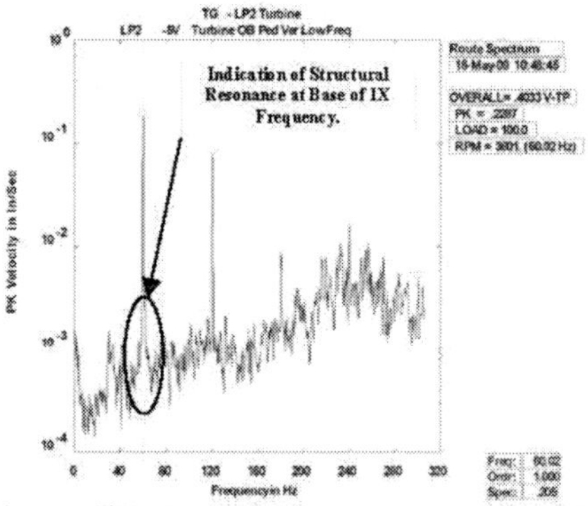

Figure 55. Frequency Spectrum, Brg 5 Casing, Vertical.

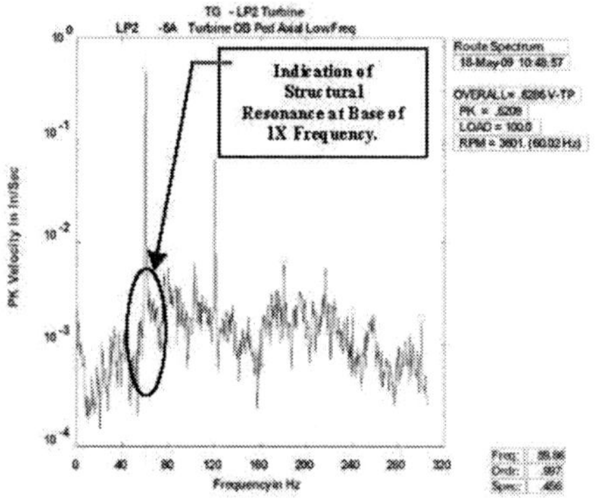

Figure 56. Frequency Spectrum at Brg 5 Casing, Axial.

The bearing housings and pedestals of the LP Turbines were modeled using ME'scopeVES, see **Figures 57-58**. ODS data were measured and imported to ME'scopeVES for animation. The ODS data measured on Brg 4 & Brg 5 pedestals showed flexure of the pedestals and relative movement of the pedestal base plates and soleplates. Vibration was highest amplitude near the central points of the casing split lines and bearing housings. Inadequate clamping by the hold down bolts (no bolt pre-load) was indicated.

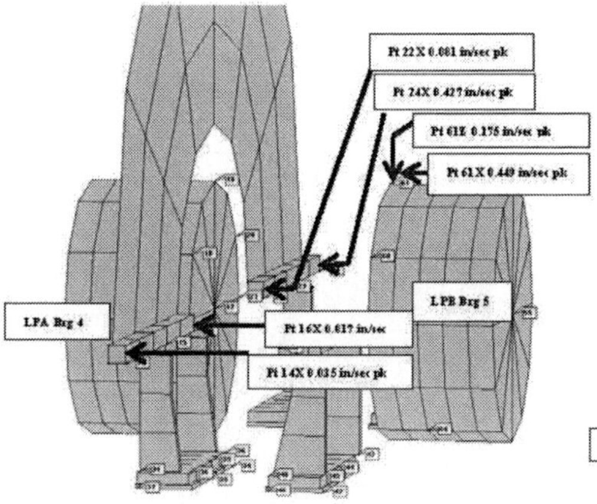

Figure 57. ME'scopeVES Model, LPA Brg 4, LPB Brg 5 Bearing Housing Pedestals & Soleplates.

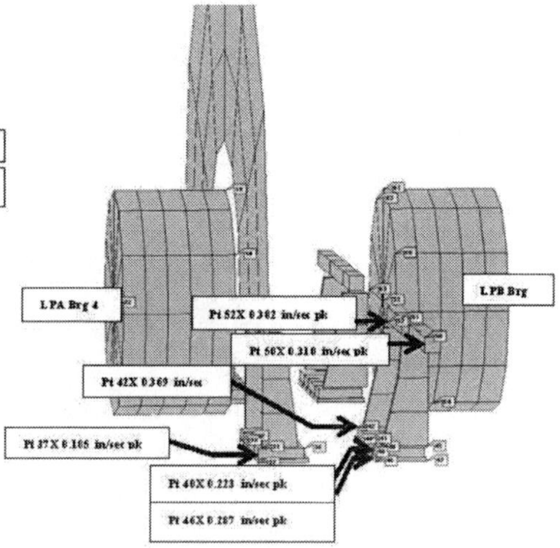

Figure 58. ME'scopeVES Model, LPA Brg 4, LPB Brg 5 Bearing Housings, Pedestals & Soleplates.

Additional testing to include checking tightness (adequate pre-load) of hold down bolts, modal testing to identify resonance, coastdown data collection from the proximity probes and casing accelerometers was recommended. After these issues and other problems were identified, the client's decision was to bring the outage contractor back on site to resolve the identified problems and collect the recommended vibration data.

References:

1. Barbee, A, Bracher, B, McGinnis, B, Singleton, K, **Case Study Analysis of Process Fan Failure And Bearing Housing/Shaft Design**, Vibration Institute National Training Seminar 2013.
2. McGinnis R, Singleton K, **Power House FD & ID Fans Analyzing Concrete Foundation Resonance,** Vibration Institute, Piedmont Chapter, May 8, 2015.
3. Monroe, Todd R., Monroe, Perry C. Jr, **"The Road To Reliable Pumps"**, 23rd International Pump Users Symposium (2007).
4. Junker, G., **New Criteria for Self-Loosening of Fasteners Under Vibration**. SAE Paper 690055, 1969.
5. Fastener transverse vibration test machine, Junker vibration test,http://www.boltscience.com/pages/junkerstest.htm.
6. Temitope, Stephen J. **Condition Monitoring of Bolted Joints**, Department of Mechanical Engineering, The University of Sheffield, June 2015.
7. Kaas, Michael, **Stop Loosening of Fasteners**, Fixing Magazine, www.fastenerandfixing.com
8. McGinnis R., Singleton, K, **New Pump Installation, Analysis of Excessive Vibration Root Cause**, Vibration Institute Piedmont Chapter, May 2008.
9. Escoweld 7505E/7530, **Anchor Bolt Installation Reference**, Escoweld 7505E/7530 Machinery Grout.
10. Unbrako Engineering Guide, http://unbrako.com/docs/engguide.pdf
11. StressTel, **Guide To Ultrasonic Inspection of Fasteners**, http://www.ge-mcs.com/download/ultrasound/bolt-extensometer/stresstel-fasteners.pdf
12. Bolt Science – **Vibration Loosening of Bolts,** http://www.boltscience.com/pages/vibloose.htm
13. SKF **Vibracon Application Guide**, https://skf-solution-factory-marine-services-674702.c.cdn77.org/downloads/SKF-Vibracon-Application-guide.pdf
14. Bolt-Science, **Tutorial on the Basics of Bolted Joints,** http://www.boltscience.com/pages/basics1.htm
15. Bolted Joint Design, Fastenal Engineering & Design Support, https://www.fastenal.com/content/feds/pdf/Article%20-%20Bolted%20Joint%20Design.pdf
16. StressTel Ultraxonic Testing Equipment, **Guide to Ultrasonic Inspection of Fasteners**, http://www.ge-mcs.com/download/ultrasound/bolt-extensometer/stresstel-fasteners.pdf
17. NORD-LOCK, http://www.nord-lock.com/wp-content/uploads/2014/03/BOL1_14_en.pdf BOLI_14_en.pdf
18. **Fastenal Technical Reference Guide**.pdf, https://www.fastenal.com/
19. NORD-LOCK, http://www.nord-lock.com/wp-content/uploads/2015/03boll_15_3n.pdf
20. NORD-LOCK, http://www.nord-lock.com/superbolt/multi-jackbolt-tensioners/flexnuts/introduction/

21. NORD-LOCK, http://www.nord-lock.com/nord-lock/multifunctional-wedge-locking/x-series-washers/joint-guide/
22. NORD-LOCK, **Junker Machine Test Video**, https://www.youtube.com/watch?v=IKwWu2w1gGk
23. Applied Bolting Technology, **Applied Bolting Brochure**, http://www.appliedbolting.com/our-brochure.html
24. Stanley, **Spiralock Fasteners & Threading Tools Product Specifications,** http://www.stanleyengineeredfastening.com/sites/www.emhartamericas.com/files/downloads/Spiralock%20Products%20Catalog.pdf
25. NORD-LOCK, **X Series Washers 2012**, http://www.nord-lock.com/2012/09/28/nord-lock-x-series-bolt-security-without-compromise/
26. SmartBolts, **Visual Indication System Product Catalog,** www.smartbolts.com
27. Hilti North America Product Technical Guide, www.us.hilti.com/anchors
28. Jergens Workholding Components.pdf pg 260, pg 268, http://www.jergensinc.com/Spherical-Flange-Assemblies
29. RBW Manufacturing, **Helpful Hints,** www.rbwmfg.com/wp-content/uploads/2012/02/HelpfulHints2.pdf
30. Vibrant Technology, www. http://vibetech.com/
31. Oberg, E., Jones. F., Horton, H., and Ryffel, H: "**Machinery's Handbook**", 27th Edition, Industrial Press Inc., New York, 2004
32. "Criteria for Preloaded Bolts", NSTS–08307, Rev. A, NASA, 1998
33. SKF, "**Bolt-Tightening Handbook**", https://www.google.com/?gws_rd=ssl#q=skf+bolt+tightening+handbook
34. Skidmorewilhelm, "**User Manual Model MZ-100 Bolt Tension Calibrator**", http://www.skidmore-wilhelm.com/
35. Raynor-Keck, Lisa, "**Torque Vs. Tension: Is There a Clear Winner?"**, Fastorq, www.fastorq.com
36. Shobert, Ralph S. P.E., "**Engineering Fundamentals of Threaded Fastener Design and Analysis**", http://www.hexagon.de/rs/engineering%20fundamentals.pdf.
37. Grainger, https://www.grainger.com/product/TE-CO-Heavy-Duty-Washer-2YJG5?s_pp=false&picUrl=//static.grainger.com/rp/s/is/image/Grainger/2YJF8_AS01?$smthumb$
38. Escoweld 705E/7530, **Technical Bulletin # 1612D**, http://na.itwengineeredpolymers.com/downloads/tds/Escoweld-7505E-tds.pdf
39. Lee, James Pl, Golod, Yelena S., **Foundations for Dynamic Equipment**, Reported by ACIS Committee 350, http://www.inti.gob.ar/cirsoc/pdf/fundaciones/ACI-351-3R-04.pdf
40. RotoBolt Load Monitors, http://rlrowan.com/products/rotobolt-load-monitors
41. Jackson, Charlies, **Reliability Treatise Part 1**, CompressorTech, January-February 2000.
42. Loctite, **Application Wall Chart V10**, http://www.loctite.com.au/aue/content_data/311083_7129_Application_Wall_Chart_V10.pdf
43. Micro Plating Inc., http://www.microplating.com/?gclid=CKmZoY2igcsCFQhkhgod91cE4w
44. Rowan, Robert .L. & Associates, Inc., http://rlrowan.com/products/exactalign-adjustable-motor-supports-for-skids/

Academic Paper Submissions

The following papers are being published following double-blind review.

NOISE REDUCTION IN MEDICAL DIAPHRAGM PUMP USING VIBRATIONAL AND ACOUSTICAL ANALYSIS

Andrew B. Paustian and Aidan F. Browne
William Lee College of Engineering
University of North Carolina at Charlotte
Charlotte, North Carolina 28223
{ apaustia, aidanbrowne } @uncc.edu

Abstract: A medical device company sought to identify, quantify, and develop remediation strategies for noise inherent in the design of a two-sided diaphragm pump. The pump's application requires patients to use the device daily for extended periods of time; the focus of this study was to decrease the noise production of the pump. Analysis and testing was performed on multiple specimens. The initial phase of the study consisted of comparative sound pressure level tests, vibration level tests, forensic disassembly, and identification of noise generation and transmission mechanisms. The second phase made various modifications to the flow path and analyzed their effects compared to the baseline. The final phase of the study incorporated various design changes into the moldings of the pump and the comparative analysis of their effects; the focus of the modifications varied the geometry of the inlets and outlets in order to reduce turbulent flow. The study produced favorable results, attenuating the pump noise on the order of 10 dB at the more energetic frequencies. The study was unable to reduce the sound energy at the 2x fundamental frequency; this energy was attributed to the diaphragm, which was expressly excluded from modification within the scope of the study.

Keywords: acoustical; pump; noise; vibration; FFT; reduction; CFD;

<u>Overview:</u> A medical device company sought to identify, quantify, and develop remediation strategies for the sources of noise inherent in the design and construction of a two-sided diaphragm pump. The pump is used in applications where a patient has exposure to the acoustic output of the device for extended periods of time in close proximity. It was desirable to find ways to mitigate the amplitude of the acoustic signature of the pump, without significantly changing the design of the pump.

To perform the acoustic and vibrational analysis, a chamber was constructed and instrumented; pumps were suspended elastically from the ceiling of the chamber, and fitted with multiple high-sensitivity ICP accelerometers. The bandwidth of measurement for the sound data was 1 Hz-20 kHz; for the vibration data it was 0.5 Hz to 10 kHz, based on instrumentation capabilities. Fast Fourier transforms (FFTs) were performed on all data. Video capture and analysis was also performed on a cross-sectioned specimen. Finally, computational fluid analysis was performed on the production flow path and used to compare potential improvements of various geometry changes. The majority of the acoustic data was in the range of 20-2000 Hz, classifying the pump noise as low pitched. The motor fundamental rotational frequency was approximately 50 Hz; the greatest amount of acoustic energy appeared at multiples of the fundamental of 2, 4, 6, and 10 times. Across the tested bandwidth, the longitudinal vibrations (in-line with diaphragm pistons) were consistently higher magnitude than the lateral vibrations (orthogonal to diaphragm pistons); significant vibrations occurred at 1, 2 and 3 times the fundamental frequency. The only large vibrational spike that correlated with a large acoustic signature frequency spike was at 2x the fundamental frequency. The 2x fundamental vibration and acoustic spikes were attributed to the diaphragm actuating twice per rotation.

Equipment: All acoustic and vibrational testing was conducted in a one cubic foot noise reduction chamber constructed from packing foam. The pump specimen was suspended inside the chamber by a suspension cord running through the top of the chamber. Power, speed control, tachometer output cable, air plumbing and accelerometer cables were negotiated through a side of the noise reduction chamber for attachment to the pump specimen. A sound level meter was placed through a port on the opposite side of the chamber from the wiring and plumbing. Due to the fact that the pump design is proprietary, we are unable to include photographs or schematics.

The sound level meter used for testing was a SoundPro SE/DL Series SP-SE-1-1/1 fitted with a pre-polarized condenser microphone made by Bruel & Kjaer. The acoustic data was captured and an A-weighting was applied. The time series data was collected from an AC output on the unit and connected via minijack-to-BNC cables to an NI cDAQ-9174 equipped with a 9234 analog input module unit. Two identical accelerometers were used for each trial; one in the horizontal position and a second in the vertical position. The accelerometers used were PCB Piezotronics high sensitivity ICP accelerometers with a bandwidth of 10,000Hz. The accelerometers were connected by coaxial cables with BNC connections. Petroleum wax was used to secure the mounting platform to the pump specimen to which the accelerometers were connected with a mounting stud. Acceleration and acoustic time series data was captured using a compact data acquisition system or cDAQ. The cDAQ was equipped with a four port 9234 analogue input module that collected data from two accelerometers, the SLM and the pump tach output. The cDAQ read the 4 channels simultaneously at a rate of 51,200 samples per second. The data acquisition occurred well over the required Nyquist Rate to capture the full audible spectrum.

During the data collection, the pump under test was connected to a back pressure regulator with 5/16 inch outer-diameter vinyl tubing. The pressure was maintained at 12 psi for all tests. To drive the motor at different rates a Tektronix AFG3102 function generator was connected to an open collector lead on the pump lead wires. By varying the duty cycle of a 15 kHz 5 Vpp pulse train the motor was able to increase and decrease operating speed and flow rate.

The time series data collected was processed in a LabVIEW 2014 virtual instrument (VI); both sets of acceleration data and the acoustic data were acquired simultaneously. Data was collected continuously at 51,200 samples per second but processed discretely into separate intervals that were dictated by the sampling rate and sampling period. 200k samples were collected for each FFT calculation over a time period of 3.90625 seconds. FFT averaging was not used; a Hanning window was applied to the time series data to eliminate discontinuities that appear in the data collection time interval and decrease the spectral leakage, allowing for a more accurate portrayal of the power contained in specific frequency spikes [1]. The resolution of the frequency was therefore 0.26 Hz, decreasing the chance that two separate acoustic phenomena are combined with a greater magnitude of acoustic power. The raw acoustic and vibrational data sets were exported to Microsoft Excel for plotting and labeling.

Acoustic analysis: In order to forensically determine the noise generation mechanisms and transmission paths, the acoustic signature of the pump under test was obtained. An example spectrum (obtained from pump specimen 2 operating at 45.06 Hz) is displayed in Figure 1. The majority of the acoustic power manifests in lower frequencies between 20 and 2000Hz. For clearer view of the majority of the acoustic signature the same pump and trial are replotted over a lower frequency band in Figure 2.

Figure 1: Acoustic signature of pump under test

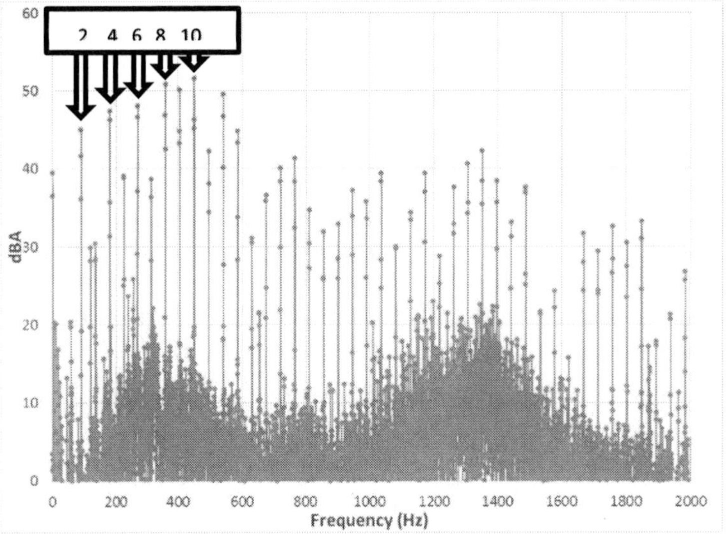

Figure 2: Acoustic signature zoomed to 0-2000 Hz

The larger magnitude acoustic spikes displayed in Figure 38 are evenly spaced as multiples of the rotational frequency of the pump motor. The 2nd, 4th, 6th, 8th and 10th multiples are noted on the plot to demonstrate the frequency spacing phenomena of the acoustic signature. The pump's acoustic output is classified as a low pitched noise with a maximum amplitude between 50 and 55 dBA. The majority of the acoustic power was transmitted through the inlet of the pump against the mean pulsating flow. Due to the inevitability of differences between pump specimens some display slight differences in acoustics such as slightly more power in certain fundamental multiple peaks, however, the general acoustic signature shape remained constant between pump tests.

Attempts were made to decrease the generated acoustic power or increase the transmission loss through the inlet plumbing. Reactive and absorptive acoustic attenuation devices (acoustic filters) were applied to the inlet of the pump to increase the transmission loss. Some filters did decrease the amount of transmitted energy, however, others did not behave as expected. Tweten, Nored and Brun point out that classical acoustic theory is limited by assumptions such as the fact that acoustic waves are not traveling in a medium with a mean fluid flow [2]. The unconventional effect of many of the filters can be attributed to the shortcomings of the theories that they were derived from. Many filters are supposed to cancel out a frequency via destructive interference of an oncoming wave [3]; Atypically, in the case of this pump, the main transmission path is against the pulsating mean flow; we believe that the induced pulses that are reflected back towards the source are not in phase with the acoustic pulses and thus do not have the desired destructive function.

Since the use of relative filters was not effective, absorptive filters and other devices to condition the acoustic waves were investigated. One of the most successful filters used was an inline restriction filter that altered the pulsating flow characteristics of the acoustics making the pulse of a longer period but of a lower magnitude [2]. Absorptive filters also proved successful on increasing the transmission loss thought the inlet, functioning to dissipate acoustic pressure energy into heat [4]. Though the absorption materials are less successful at dissipating energy at lower frequencies, the absorption filter noticeably decreased the overall noise output.

Vibration Testing: The accelerometer data for the same pump specimen was recorded and displayed in Figure 3. The majority of the acceleration energy appears below 200Hz which is expected due to the relatively low rotational speed of the pump motor.

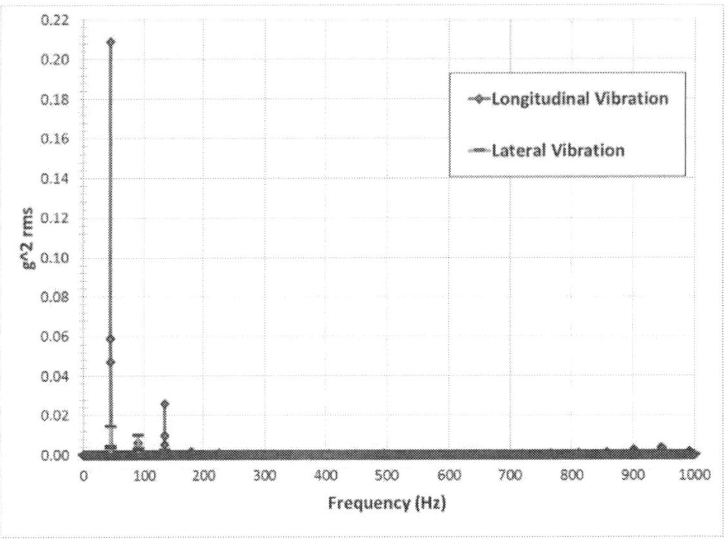

Figure 3: Longitudinal and lateral acceleration data

The large vertical spike at 45.06 Hz corresponds to the rotational speed of the motor driving the piston arms and eccentric. The other relatively large spikes occur at 2x and 3x the fundamental frequency, occurring at 90.11 and 135.17 Hz, respectively.

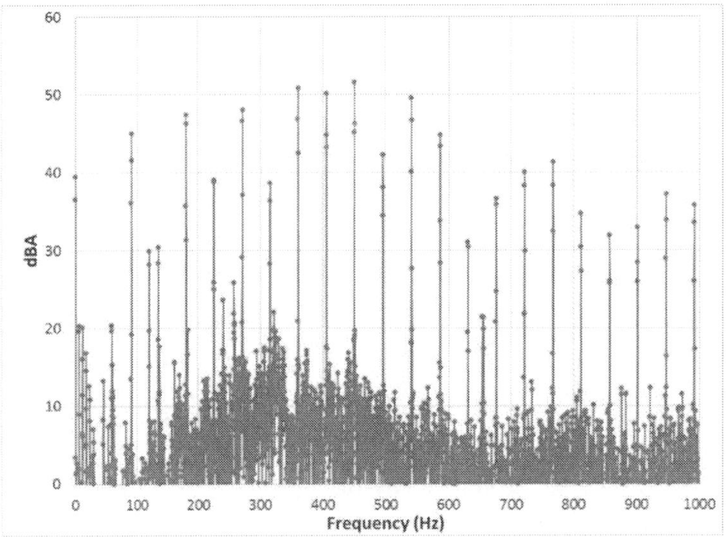

Figure 4: Acoustical data for pump 2

A 0-1000 Hz range power spectrum of the acceleration and acoustic data can be viewed in Figures 3 and 4. For the most part, the acceleration energy and the acoustic energy appear at different frequencies. The large vibrational spike that is present at the rotational frequency (45.06Hz) does not have a significant corresponding acoustic spike. Though the majority of the vibrational energy occurs as the first multiple of the rotational frequency, the acoustic signature becomes significant at the second multiple of the fundamental. The 2x fundamental acoustic spike contains a large portion of acoustic energy because the diaphragm is actuating twice per rotation. The diaphragm is inherently similar to a loudspeaker making for good acoustic coupling facilitating the transfer of mechanical energy to acoustic energy.

Acoustic signature forensics: The acoustic energy observed at lower frequencies is attributed to the diaphragm and mechanical components of the pump. The other low pitched frequency multiple peaks are bolstered by the harmonic effects of the 2x the fundamental spike and are also accredited to the mechanically induced acoustics. The diaphragm and other moving parts are a likely source of the acoustic spikes in the 0-600 Hz range because they are occurring as multiples of the rotating machinery speed. The mechanically induced sound also occurs at relatively low frequencies compared to other audible pulses which is common for mechanical noise [5]. It is also documented that low pitched sounds often can escape form the inner workings of machines through the exhaust plumbing [6]. This also supports the generation of the lower frequency content by mechanical components because the lower pitched noises are transmitted thought the intake, inherently similar to exhaust piping.

In piping applications four major acoustical sources are conventionally accepted: fully developed turbulent flow noise, acoustics created by turbulence induced from flow path geometry, noise from standing waves or mechanical resonance, and vibrations of induced by pipe walls or other vibrating/pulsating planes [5]. The other range of interest of the acoustic signature contains energy of lesser magnitude at frequencies above 600 Hz. The acoustic spikes are not induced by a standing wave or excited resonance due to the lack of vibrational data at these higher frequencies. The frequency range of 600-2000 Hz is also out of the range of mechanical excitation suggesting that the noise is turbulently generated.

Though little can be done to decrease the acoustic generation of a fully developed flow, steps can be taken to decrease additional turbulence introduced by flow path geometry. The relation of turbulence and aeroacoustics is well documented in [5, 7]. To decrease the flow path induced turbulence the pump head was investigated. Figure 5 shows that the standard pump head incorporates a 90 degree turn. The cross section of the flow path also reveals a cavity in the base of the flow path where air pockets may develop.

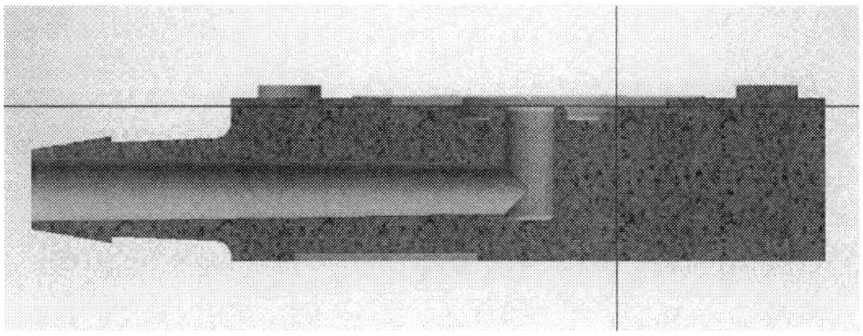

Figure 5: Standard pump head flow path

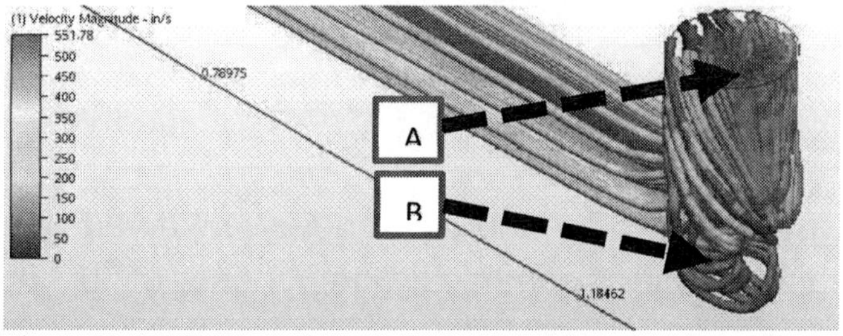

Figure 6: CFD analysis for standard pump head

Computational fluid dynamics is utilized to investigate the air flow interactions around the sharp turn. Figure 6 contains the computed air particle paths and speeds for the standard pump head under nominal operating conditions. The red color indicated high speed flow path while blue flow path indicates a slowed fluid particle flow. The CFD analysis illustrates that air flow is impacted by the reservoir in the turn. Having air flow into the reservoir introduces turbulence when the air reenters the normal flow path. The maximum velocity obtained during the analysis for the standard pump head was 551.78 in/s.

A new geometry was envisioned to quell the aeroacoustical noise introduced by the reservoir and sharp turn. Figure 7 is a cross section of pump had prototype 1 and Figure 8 is the CFD analysis of the flow path.

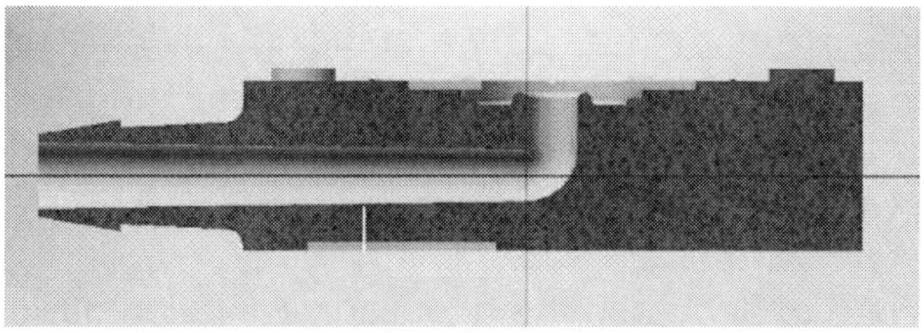

Figure 7: Smooth flow path (Prototype 1)

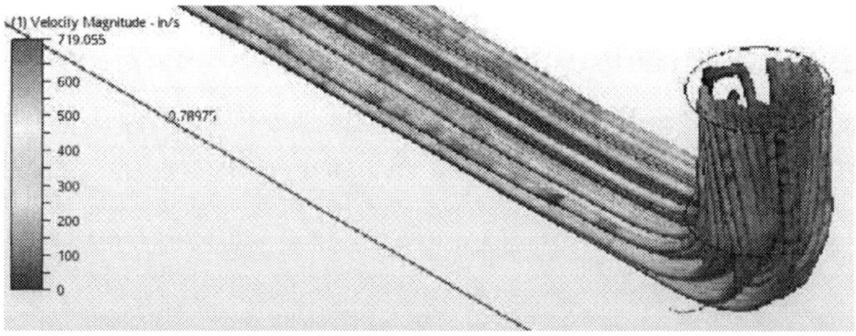

Figure 8: CFD of flow path for prototype pump head 1

The prototype head has the same port diameters but the sharp edge around the turn is removed and replaced with a smooth transition. The highest air particle velocity reported from this analysis was 719.055 in/s. The pocket of air that appears in the reservoir also is not apparent during CFD of prototype 1.

A second prototype head was created to incorporate filtering geometry into the construction of the pump. With the known success of an inline restriction filter prototype 2 sought to incorporate a flow path restriction into the internal molding of the pump specimen head. Figure 9 is a cross section of prototype pump head 2, and Figure 10 is a CFD analysis of the associated flow path of the same geometry.

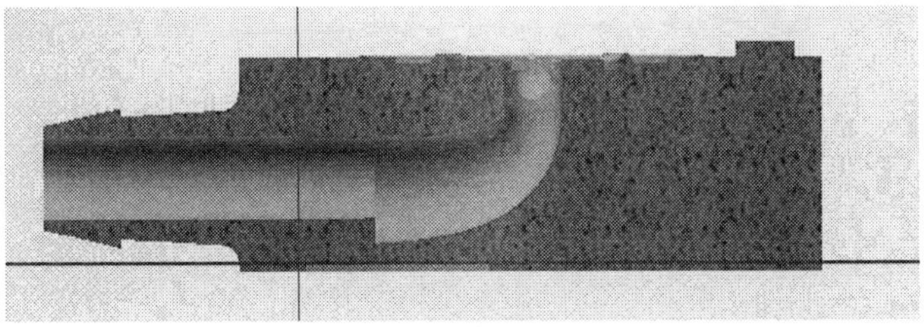

Figure 9: Curved inlet restriction (prototype 2)

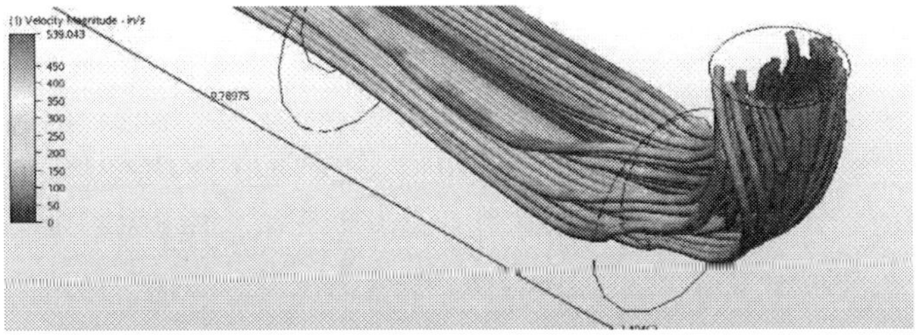

Figure 10: CFD of prototype 2

The expansion of the flow path on the curve is employed to have a minimal effect on the flow of the air while acting as a restriction of for the acoustics that are flowing in the opposite direction of the air flow. The maximum air flow velocity obtained during the CFD analysis is 539.043 in/s. The addition of the inline restriction decremented the flow rate by approximately 10 in/s.

An acoustic analysis was performed on both prototype heads. The comparison was with a pump outfitted with a 3D printed head on both sides of the dual diaphragm pump. Both of the prototypes were tested against a 3D printed head that maintained the original pump geometry.

Figure 11 is a comparison of the printed standard inlet arrangement and the printed prototype 1. The printed heads increased the variability of rotational speed of the pump making trial to trial comparisons more difficult. Figure 11 was created by averaging three trials of each printed head and comparing the overall results. The lower diaphragm acoustics were not affected by the altered flow path, however, in the 1000- 2,000Hz range the acoustics of prototype 1 display a reduction. A qualitative blind test was conducted and prototype 1 was chosen as a smoother and less irritating acoustic signature.

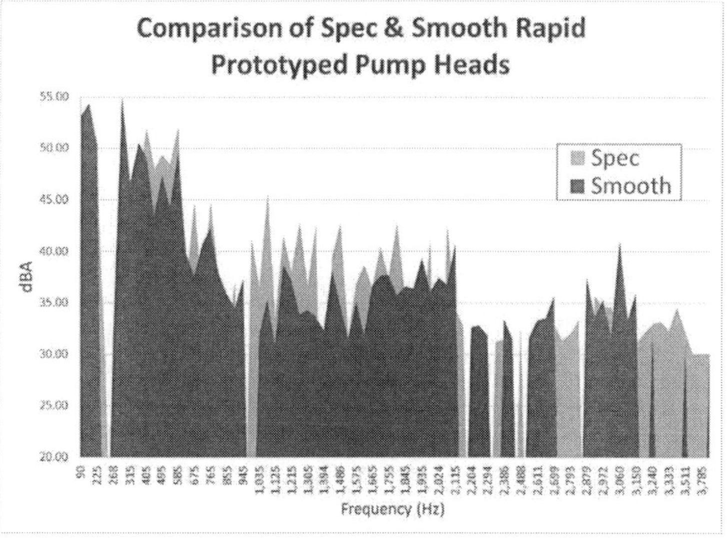

Figure 11: Acoustic comparison between printed standard and printed prototype head 1

Prototype 2 was not as successful at reducing the pump acoustic signature. A plot of the acoustic signatures of prototype 2 compared to a printed standard pump head is displayed in Figure 12. Though the prototype head did still increase the variability of the acoustic signature, the same analysis was not needed to compare similar peaks because prototype 2 increased the signature across the band. The addition of the inline restriction for the acoustic pressure pulses added additional sharp angles to the flow path geometry. The additional angles and pockets introduced in the restriction may be a cause of the increased acoustic pressure pules magnitudes.

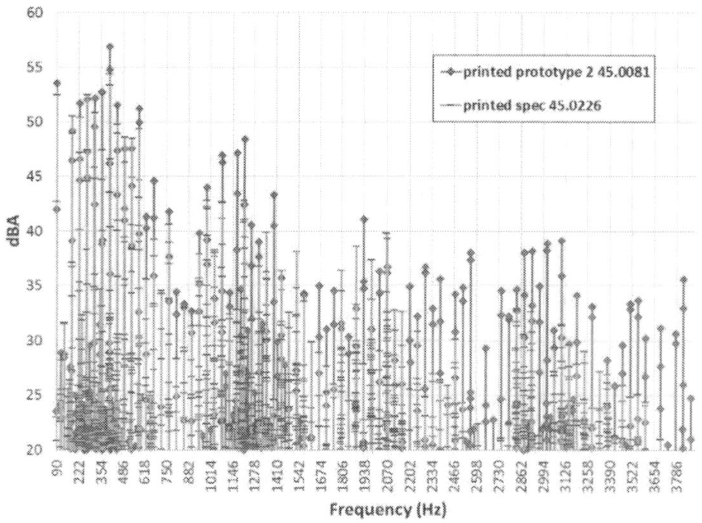

Figure 12: Acoustic comparison between printed standard and printed prototype head 2

Conclusions and Future Work: An acoustic and vibrational analysis was conducted on a dual-sided 12 volt DC diaphragm pump. The resulting acoustic signature displayed the pump operating noise as low pitched with the highest acoustic spikes below 1000Hz and around 50 to 55 dBA. The vibrational energy also appeared at relatively low frequencies. Both the major portions of the vibrational and acoustical power appear as multiples of the rotational frequency of the driving DC motor.

Because of the low frequencies that the majority of the noise is occurring at, the diaphragm and other moving mechanical components are accredited for the 0-600HZ noise. The shape and function of the diaphragm being similar to a loud speaker and actuating at twice the rotational frequency also support the assertion of the low frequency noise is mechanically produced.

The relatively higher pitched noise, occurring at frequencies in the 600-2000Hz range are attributed to the turbulent effects of the air flowing through the inner workings of the pump. Noise occurring at this frequency range is commonly found to be aeroacoustics. Resonant effects have been determined not to be the cause because of the lack of vibrational energy occurring at these frequencies.

Attempts were made to reduce the acoustic output of the pump both by increasing the transmission loss of the acoustic propagation path and by decreasing the acoustic generation of the pump. Prototype 1 was developed to decrease the generation of the pump via reducing the sharp and unnecessary geometry in the pump head. With good success using an inline restriction filter, prototype 2 was conceived to incorporate a restriction inside the pump head.

The prototypes were tested against standard pump heads that were also printed form the same manufacturer with the same material properties. Prototype 1 displayed good results in reducing the acoustics that were recognized to be turbulently induced. The CFD showed that the smooth transition around the inlet curve increased the overall flow through the inlet while reducing turbulence. Prototype 1 did not have a noticeable difference in the mechanically excited acoustics. When comparing prototype 2 to the acoustics of the standard printed head geometry it was found that the acoustic emissions have increased over the frequencies of concern. The addition of a restriction for the acoustic pulses acted as an expansion chamber for the pulsating mean flow. Though the CFD did not display a large difference with the addition of the inlet geometry, the additional sharp curves may have increased the turbulence of the air stream. Such interactions may be investigated further using newer models incorporating fully compressible Navier-Stokes Equations such as the ones described by Singh and Rubini [8]. The newer models account for pressure interactions as opposed to focusing on the fluid flow.

Additional sound level reductions may be possible with additional alterations to the geometry of the air flow paths within the pump. There are some proposed modifications that went unexamined due to limited resources to produce the 3D printing parts, but may be worth pursuing. Additional acoustical analysis of the updated design may likely lead to further reductions in flow path turbulence, and therefore acoustic signature.

References

[1] Harris, Fredric J. "On the Use of Windows for Harmonic Analysis with the Discrete Fourier Transform." *Proceedings of the IEEE* 66, no. 1 (1978): 51-83.

[2] Tweten, Dennis, Marybeth Nored, and Klaus Brun, Ph.D. "The Physics of Pulsations." In Gas Machinery Conference, 2008.

[3] Kinsler, Lawrence, Austin Frey, Alan Coppens, and james Sanders. Fundamentals of Acoustics. fourth ed: Publication Services, Inc, 2000.

[4] Atalla, N, R Panneton, FC Sgard, and X Olny. "Acoustic Absorption of Macro-Perforated Porous Materials." Journal of sound and vibration 243, no. 4 (2001): 659-78.

[5] Nored, Marybeth., Dennis. Tweten, and Klaus Ph.D. Brun. "Compressor Station Piping Noise: Noise Mechanisms and Prediction Methods ". Gas Machinery Research Council Southwest Research Institute®, 2011.

[6] Cuesta, María, and Pedro Cobo. "Active Control of the Exhaust Noise Radiated by an Enclosed Generator." Applied Acoustics 61, no. 1 (2000): 83-94.

[7] Carley, Michael. "Turbulence and Noise." Online: University of Bath, 2011.

[8] Singh, Nishant K., and Philip A. Rubini. "Large Eddy Simulation of Acoustic Pulse Propagation and Turbulent Flow Interaction in Expansion Mufflers." *Applied Acoustics* 98 (2015): 6-19.

BOND GRAPH MODEL OF RECIPROCATING COMPRESSOR FOR MACHINE CONDITION MONITORING

Enaiyat Ghani Ovy*, Mario Forcinito** and Qiao Sun*

* Department of Mechanical and Manufacturing Engineering,
University of Calgary,
2500 University DR NW, Calgary, AB, Canada.

** AP Dynamics Inc.
703 6 AVE SW 360, Calgary, AB, Canada.

Abstract: Reciprocating compressors have great implementations in oil and gas industries. Thus, there are strong demands for condition monitoring and machine maintenance techniques. Most common models are data driven that take a tremendous amount of time and budget to improve. Very few of them based on physics which are somewhat rudimentary and fail to capture essential physical phenomena. More complex models such as those based on Finite Element analysis are too abstruse to actualize and carry through its features in real world. Realizing challenging scenario of model based machine condition assessment, this research is dedicated to develop an analytical physics based model of reciprocating compressor. The mathematical model is first developed for small size one cylinder reciprocating compressor (OCRC) based on multi-body dynamics. Subsequently, it is extended for small size two cylinder reciprocating compressor (TCRC). All the dynamics equations are represented by bond graph which simulates a complete system. Effects of gravitational, frictional and constraint forces are investigated thoroughly on different parts of compressor. Responses of model are further analyzed by SOLIDWORKS motion analyses manager to examine model validity.

Keywords: Bond Graph; CAMPG; Condition Monitoring; Constraint Force; SOLIDWORKS.

1. Introduction:

Compressors are indispensable components in numerous industrial operations. Among these, reciprocating compressors have significant implementations in industries such as oil, gas and petrochemical industries because of its relatively high compression ratio. Due to its prevalent usage in industries, condition monitoring and fault diagnosis are now become fundamental after installation. There have been many complications reported in literature after reciprocating compressor installation [1] that could even lead to complete shutdown of the process. Besides, these abrupt and unexpected disruptions mean tremendous loss to industries. As a result, there is strong interest and effort geared towards condition monitoring and machine maintenance processes to closely detect any problem prior to severe damage. For decades, industries are using traditional data driven models that requires tons of previous records to give satisfactory predictions. Although data driven model may seem to be robust but this eventually means the machine learning method which is extremely time consuming and expensive. This leads to the physics based mathematical model development that gives deeper insight into the system [2]. There are some works regarding physics based model of reciprocating compressor. Viscomi and Ayre [3] worked on elastic slider-crank mechanism and concluded that the responses depend on five parameters: length, mass, damping, external piston force and frequency. Lin et al. [4] and Ha et al. [5] derived mathematical model of simple slider-crank mechanism based on classical Hamilton's principle and Lagrange multiplier methods. There are very few works on modelling reciprocating compressor by bond graph method. It is very common that the physics based model becomes too complicated which could have been easily avoided by application of bond graph method [6]. Engja [7] developed first bond graph model of a reciprocating compressor which focused extensively on thermodynamic aspects of the system rather than structural dynamics. In [2, 8] structural analyses was main concern on the bond graph model development of reciprocating compressor, but the dynamic model has many compliance elements of no physical significance and is also limited to simple analyses. In this paper, we develop mathematical models for OCRC and TCRC based on multi-body dynamics particularly considering planar motion of rigid bodies. The dynamics equations of motions are linked altogether by bond graph. Effects of gravitational, frictional and constraint forces are investigated on different parts of the compressor. Computer Aided Modelling Program with Graphical Input (CAMPG) [9] is used for bond graph modelling that is interfaced with MATLAB to solve differential equations. Model responses are further analyzed by SOLIDWORKS motion manager to examine the validity.

2. Some Insights on Bond Graph

Bond graph is a graphical representation of systems that is simply based on conservation of energy principle. The concept of bond graph was originated by H.M. Paynter [10] back in 1961. The distinguishing feature of bond graph is it has only nine elements by which all the systems from different domains (e.g. mechanical, electrical, hydraulic, acoustic etc.) or combination of these domains can be modelled unequivocally. The fundamental elements of bond graph are source of effort (SE), source of flow (SF), inductive element (I), capacitive element (C), resistive element (R), transformer (TR), gyrator (GY), 1-junction, and 0-junction. Besides, these elements are connected to each other by means of power bonds. These bonds represent the energy flow between two elements and thereby form a network of many sub models [11]. Causal analysis ascertains the direction of energy flow and attributes constitutive relations of different elements. The natural causality of inductive and capacitive elements is to take effort for flow generation and is to take flow for effort generation accordingly. These two elements are in fact the conserved quantities that signify the states of the system [11-12]. Resistors are energy dissipating elements and constitutive equation for this element is an algebraic relation between effort and flow.

3. Kinematics of the Reciprocating Compressor

We begin our model formulation with the kinematics of reciprocating compressor or simply slider-crank mechanism. Figure 1 shows a schematic of the OCRC. The length of the crank and connecting rod are denoted by l_1 and l_2 respectively. At any instant, the angular displacement of the crank and connecting rod are represented by θ and φ accordingly. Center of gravity (C.G.) of crank is located at half of its length and C.G. of connecting rod is located at one third of its length from crank-connecting rod junction point. From the geometry of figure 1, the following equations of positions and velocities of slider-crank mechanism may be derived. The horizontal and vertical components of C.G.'s linear displacement of crank are,

$$x_1 = \frac{l_1}{2}\cos\theta \text{ and } y_1 = \frac{l_1}{2}\sin\theta \quad (1)$$

The horizontal and vertical components of the C.G.'s linear velocity of the crank are,

$$\dot{x}_1 = -\frac{l_1}{2}\sin\theta\,\dot{\theta} \text{ and } \dot{y}_1 = \frac{l_1}{2}\cos\theta\,\dot{\theta} \quad (2)$$

The horizontal and vertical components of the C.G.'s linear displacement of the connecting rod are,

$$x_2 = l_1\cos\theta + \frac{l_2}{3}\cos\varphi \text{ and } y_2 = l_1\sin\theta - \frac{l_2}{3}\sin\varphi \quad (3)$$

The horizontal and vertical components of the C.G.'s linear velocity of the connecting rod are,

$$\dot{x}_2 = -l_1\sin\theta\,\dot{\theta} - \frac{l_2}{3}\sin\varphi\,\dot{\varphi} \text{ and } \dot{y}_2 = l_1\cos\theta\,\dot{\theta} - \frac{l_2}{3}\cos\varphi\,\dot{\varphi} \quad (4)$$

The horizontal component of slider's C.G. displacement is,

$$x_3 = l_1\cos\theta + l_2\cos\varphi \quad (5)$$

The horizontal component of slider's C.G. velocity is,

$$\dot{x}_3 = -l_1\sin\theta\,\dot{\theta} - l_2\sin\varphi\,\dot{\varphi} \quad (6)$$

The geometric constraint for the slider-crank mechanism is,

$$\varphi = \sin^{-1}\left(\frac{l_1}{l_2}\sin\theta\right) \quad (7)$$

4. Small Size Reciprocating Compressor with One Cylinder

The differential equations are derived for each element of the proposed model based on multi-body dynamics of rigid bodies taking consideration of relative motions by using rotating frame of references. This frame of reference translates and rotates simultaneously which is fixed at the C.G. of each body and moves and rotates with the body. This frame may also be called as body fixed frame (BFF). To determine the absolute quantities of acceleration, velocity, and displacement of different bodies, the following equation is considered in the analytical model development [13-15]

$$\dot{\mathbf{V}} = \dot{\mathbf{V}}_{\mathbf{BFF}} + \boldsymbol{\omega} \times \mathbf{V} \tag{8}$$

Here, **V** is any vector and the left hand side of the above equation represents the absolute time derivative of this vector with respect to an inertial frame. And, this absolute time derivative is equal to the time rate of change of **V** with respect to body fixed or translating-rotating frame (simply, absolute velocity expressed in body frame) plus another term which signifies the change of the vector **V** due to the rotation of the body fixed frame. Figure 1, shows OCRC model where two rotating frames (x_1y_1 and x_2y_2) are attached at the C.G. (point o_4 and o_5) of the crank and connecting rod. Vectors $\mathbf{r_1}$ and $\mathbf{r_2}$ are such that are rotating and translating simultaneously. According to relation (8), these vectors may be expressed as,

$$\dot{\mathbf{r}}_1 = \dot{\mathbf{r}}_{1_{\mathbf{BFF}}} + \boldsymbol{\omega}_1 \times \mathbf{r}_1 \tag{9}$$

$$\dot{\mathbf{r}}_2 = \dot{\mathbf{r}}_{2_{\mathbf{BFF}}} + \boldsymbol{\omega}_2 \times \mathbf{r}_2 \tag{10}$$

We also notice that horizontal and vertical forces are considered at joints o_1, o_3 and o_2. Splitting the bodies separately would reveal the forces at these joints which may signify constraint forces acting at these locations. Using bond graph with application of BFF would facilitate to consider these forces. Now, if we consider $\mathbf{r_1}$ and $\mathbf{r_2}$ are the velocity vectors acting at C.G. of crank and connecting rod, the horizontal and vertical components of absolute accelerations of C.G. of both the bodies may be found.

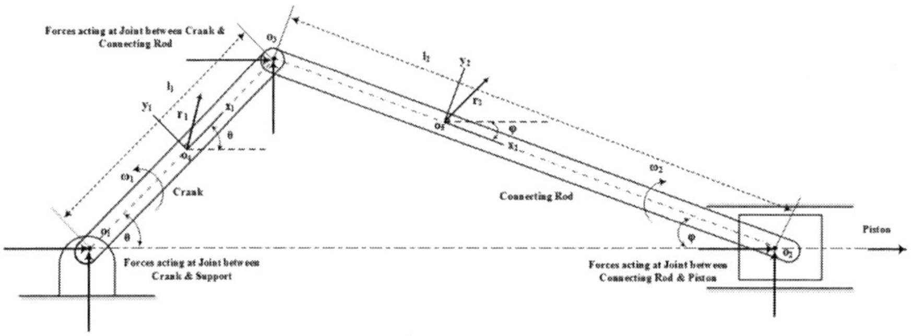

Figure 1: Slider-Crank Mechanism or OCRC highlighting application of rotating frames.

The horizontal component of the C.G. absolute acceleration of the crank is found by [13-15],

$$a_{x_1} = \dot{v}_{x_1} - \omega_1 v_{y_1} \tag{11}$$

Where, the first term on the right hand side of the equation (11) is the acceleration as observed by an observer attached to the rotating frame and the second term accounts for the contribution of the frame rotation to the absolute acceleration which is measured by an observer in the inertial frame.

Newton's law can be applied to the absolute acceleration horizontal component of the C.G., and therefore equation (11) becomes;

$$m_1 a_{x_1} = m_1 \dot{v}_{x_1} - m_1 \omega_1 v_{y_1}$$
(12)

Where, m_1 is the mass of the crank. v_{x_1} and v_{y_1} are the instantaneous velocities along the BFF of the crank. If we rearrange the above equation, this will lead us to the ultimate governing dynamic equation along horizontal direction for the crank.

$$\sum F_{x_1} + m_1 \omega_1 v_{y_1} = m_1 \dot{v}_{x_1}$$
(13)

Where, the sum of the forces along x_1 axis come from any attached systems to the body and the rotational contribution to the acceleration is brought to the force side of the equation. Similarly, we may find the dynamic equations for the rest of the systems. The dynamic equation along vertical direction for the crank is,

$$\sum F_{y_1} - m_1 \omega_1 v_{x_1} = m_1 \dot{v}_{y_1}$$
(14)

The dynamic equations along horizontal and vertical direction for the connecting rod are,

$$\sum F_{x_2} + m_2 \omega_2 v_{y_2} = m_2 \dot{v}_{x_2}$$
(15)

$$\sum F_{y_2} - m_2 \omega_2 v_{x_2} = m_2 \dot{v}_{y_2}$$
(16)

Where, m_2 is the mass of the connecting rod. v_{x_2} and v_{y_2} are the instantaneous velocities along the BFF of the connecting rod. For the piston, it is not needed to consider a BFF as it translates only. Its dynamics equations may simply be expressed in the inertial frame by Newton's second law of motion. The dynamic equation along horizontal direction for piston is,

$$\sum F_{x_{p1}} = m_{p1} \dot{v}_{x_{p1}}$$
(17)

Here, m_{p1} is the mass of piston and $v_{x_{p1}}$ is the linear velocity along horizontal direction. Figure 2 shows the free body diagram of the crank and connecting rod where relative velocities at different points are highlighted. The velocities at point o_1, o_3 and o_2 are needed to be determined to account for the constrained forces acting at different joints within the multi-body system. The location of C.G. of the crank from point o_1 along x_1 axis is half of its length and the location of C.G. of the connecting rod from point o_2 along x_2 axis is two-third of its length.

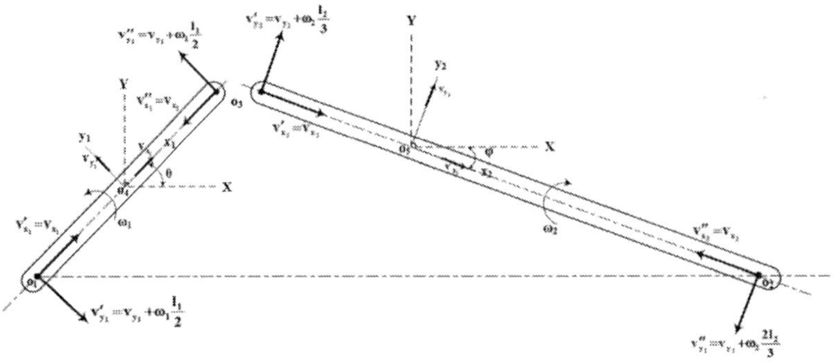

Figure 2: Calculation of relative velocities at different points of the OCRC.

5. Small Size Reciprocating Compressor with Two Cylinders

TCRC is the extension of OCRC with another cylinder. Figure 3 shows the TCRC where two additional rotating frames (x_3y_3 and x_4y_4) are shown that are attached at the C.G. (point o_9 and o_{10}) of the crank II and connecting rods II. Here, l_3 and l_4 refer to the length of the crank II and connecting rod II for the cylinder II. Similar to the OCRC, vectors **r₃** and **r₄** are such that are rotating and translating simultaneously.

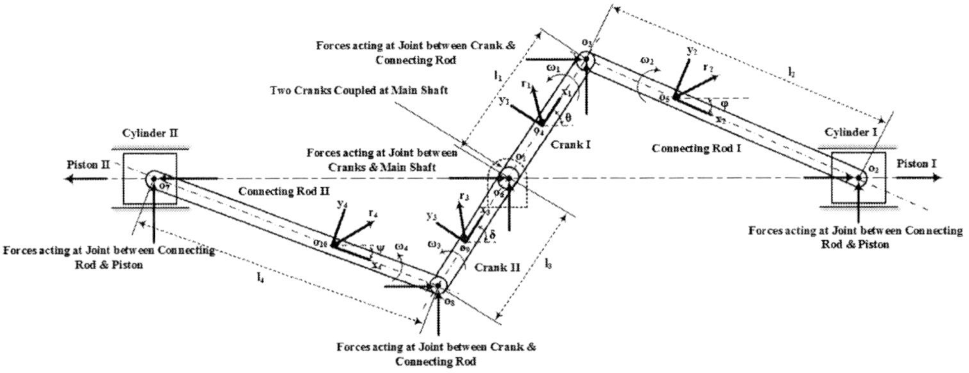

Figure 3: Sketch of two cylinder reciprocating compressors highlighting body fixed frames attached on different parts.

Following previous development of OCRC model, the equations of TCRC for the second cylinder are shown here straightforwardly. The dynamic equations along horizontal and vertical direction for the crank II are,

$$\sum F_{x_3} + m_3\omega_3 v_{y_3} = m_3\dot{v}_{x_3}$$
(18)

$$\sum F_{y_3} - m_3\omega_3 v_{x_3} = m_3\dot{v}_{y_3}$$
(19)

Where, m_3 is the mass of the crank II. v_{x_3} and v_{y_3} are the instantaneous velocities along the BFF of the crank II. The dynamic equations along horizontal and vertical direction for the connecting rod II are,

$$\sum F_{x_4} + m_4\omega_4 v_{y_4} = m_4\dot{v}_{x_4}$$
(20)

$\sum F_{y_4} - m_4 \omega_4 v_{x_4} = m_4 \dot{v}_{y_4}$
(21)

Where, m_4 is the mass of the connecting rod II. v_{x_4} and v_{y_4} are the instantaneous velocities along the BFF of the connecting rod II. The dynamic equation along horizontal direction for piston II is,

$\sum F_{x_{p2}} = m_{p2} \dot{v}_{x_{p2}}$
(22)

Here, m_{p2} is the mass of piston II and $v_{x_{p2}}$ is the linear velocity along horizontal direction. Figure 4 shows the free body diagram of the TCRC where relative velocities at different points are highlighted. The location of C.G. of the crank from point o_6 along x_3 axis is half of its length and the location of C.G. of the connecting rod from point o_7 along x_4 axis is two-third of its length. Velocities at point o_1, o_3 and o_2 have already been identified for OCRC in figure 2. In this part, along with those points, the velocities at point o_6, o_8 and o_7 are shown in figure 4.

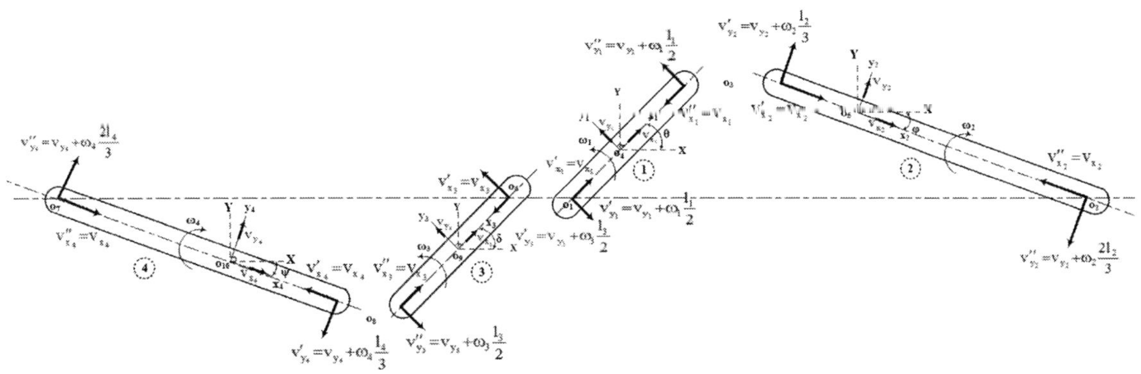

Figure 4: Calculation of relative velocities at different points of the TCRC mechanism.

6. Bond Graph Model of Reciprocating Compressor

We derived the mathematical foundation for the OCRC and TCRC in previous sections. Now it is high time to solve those equations by representing them in bond graph and solve subsequently using powerful and efficient numerical solvers. The bond graph modeling of the OCRC is done in CAMPG [5]. Figure 5 shows the bond graph model of the OCRC in CAMPG. On the model, the dynamic equations for crank, connecting rod, piston and constraint forces are labeled with dashed boxes. Inertias of the rigid bodies are represented by I elements. Constraint forces are modelled by C and R elements. Geometric constraints are represented by TF elements in bond graph. All the equations derived and shown earlier are shown graphically here in figure 5. The beauty of CAMPG software is, it generates the MATLAB 'm file' along with all differential equations ready to be solved by MATLAB. MATLAB solver ODE15s is used to solve nonlinear differential equations.

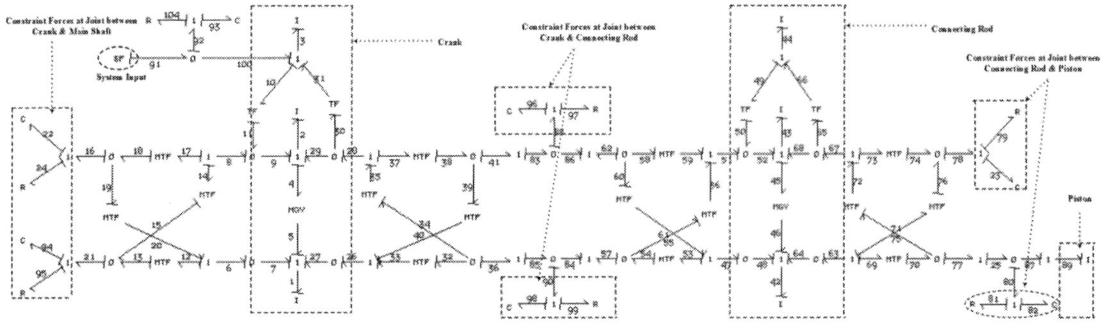

Figure 5: CAMPG model of the OCRC.

7. Model Validation

The validation of our proposed model in done by three ways. First, we do simple kinematic analyses with the model. Then, we compare the results with fundamental kinematic model described in section 3. Next, we build a 3D model in SOLIDWORKS CAD environment and run the model using SOLIDWORKS motion manager. We then use the same properties (mass, moment of inertia etc.) of SOLIDWORKS model in proposed model to simulate and do a comparison of results. Finally, we look into some important effects on the model such as gravity, friction and constraint forces, and critically analyze and compare them with hand calculations.

7. 1. Comparison with Fundamental Kinematics

This section demonstrates validity of the mathematical approach to develop dynamics model of reciprocating compressor by comparing results with kinematics of the mechanism. If simulation results of both methods are same in magnitude, then it would provide a great insight of accuracy of current dynamics model. Comparison with the kinematic model is done for the OCRC model. The lengths of the crank and connecting rod are considered 0.09 and 0.45 m respectively [4]. The input to the mechanism is the constant angular velocity of the crank which is 1 rad/s. The mass of the crank, connecting rod and piston are 36, 125 and 30 kg [4]. The moment of inertia of the crank and connecting rod are 0.1 and 2.3 kg-m^2 [4]. The compliance and damping coefficients for constraint forces are taken very high values which are 1e-9 m/N and 1e9 Ns/m respectively. Results from body fixed frame method for both OCRC and TCRC, and pure kinematics are compared for angular displacement of crank and connecting rod, linear velocity of crank, connecting rod and slider subsequently. Following figure 6 shows comparison of crank angular displacement, figure 7 shows comparison of connecting rod angular displacement, figure 8 shows comparison of crank horizontal velocity, figure 9 shows comparison of crank vertical velocity, figure 10 shows comparison of connecting rod horizontal velocity, figure 11 shows comparison of connecting rod vertical velocity, and figure 12 shows comparison of piston horizontal velocity for both OCRC and TCRC with pure kinematics of corresponding bodies.

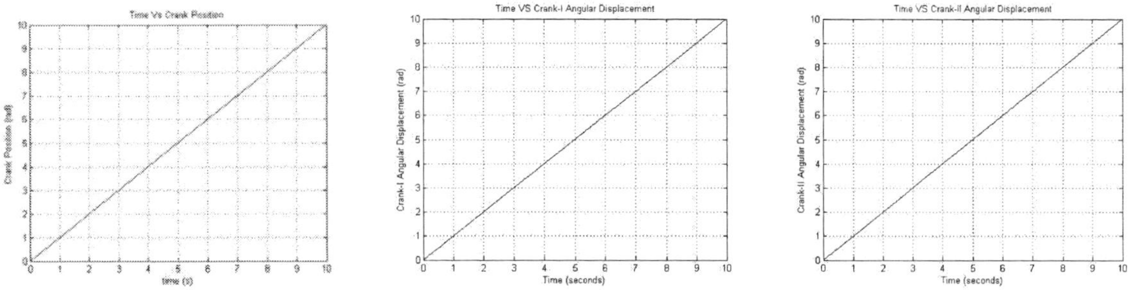

Figure 6: Crank angular displacement from kinematics (left), and from BFF for first cylinder (center), for second cylinder (right).

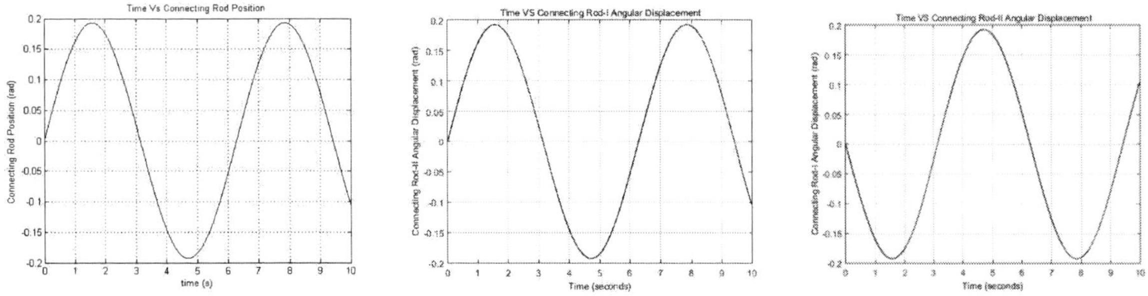

Figure 7: Connecting rod angular displacement from kinematics (left), and from BFF for first cylinder (center), for second cylinder (right).

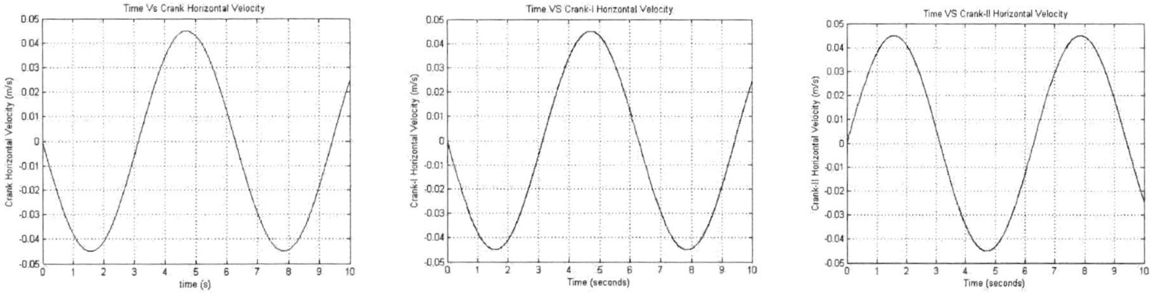

Figure 8: Crank horizontal velocity from kinematics (left), and from BFF for first cylinder (center), for second cylinder (right).

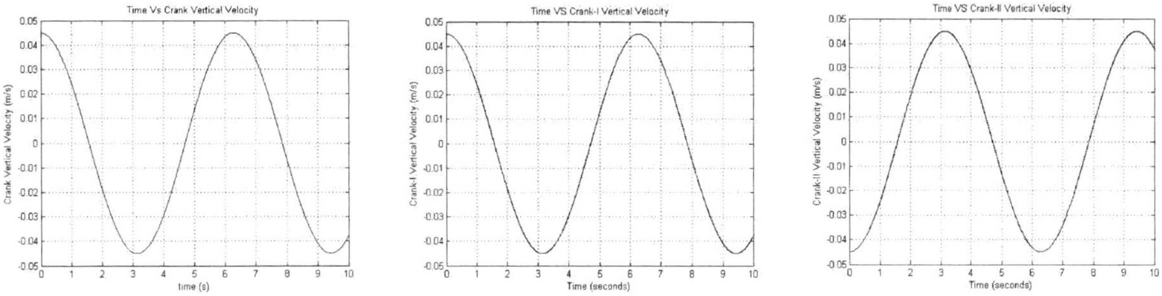

Figure 9: Crank vertical velocity from kinematics (left), and from BFF for first cylinder (center), for second cylinder (right).

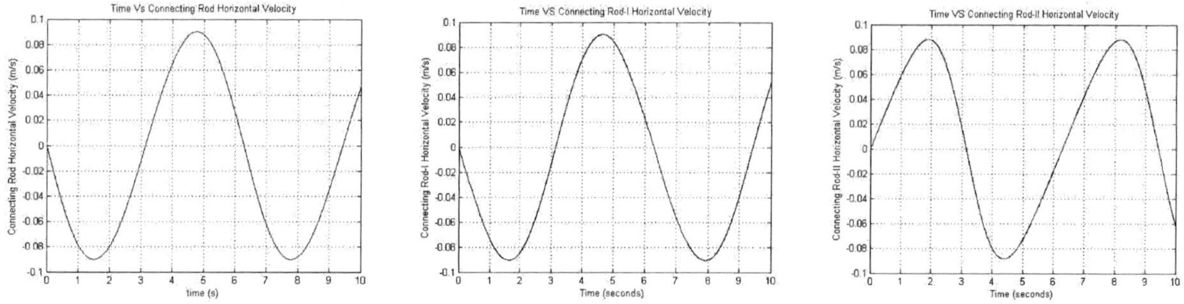

Figure 10: Connecting rod horizontal velocity from kinematics (left), and from BFF for first cylinder (center), for second cylinder (right).

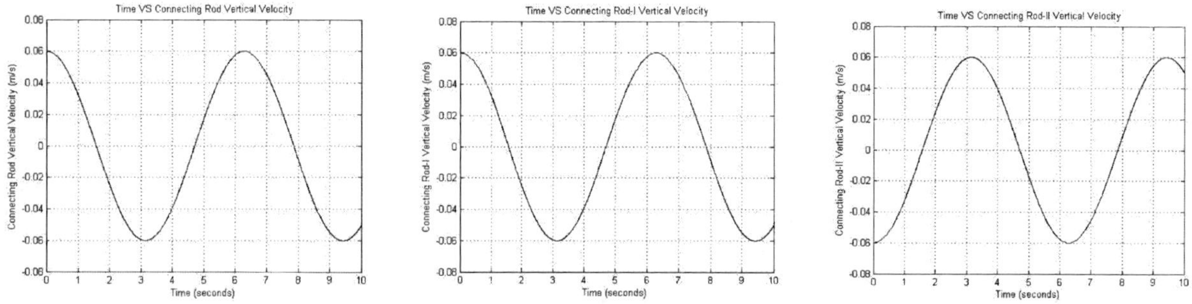

Figure 11: Connecting rod vertical velocity from kinematics (left), and from BFF for first cylinder (center), for second cylinder (right).

Observing figures 6 to 12, it is evident that our proposed model results are perfectly matched with theory. Furthermore, responses for both OCRC and TCRC model are same in magnitude but opposite in phases which make perfect sense as the physical system would respond like these. The dynamics model developed with multi-body dynamics perfectly resembles the kinematics of the mechanism. Thus, these results show accuracy and reliability of the proposed model.

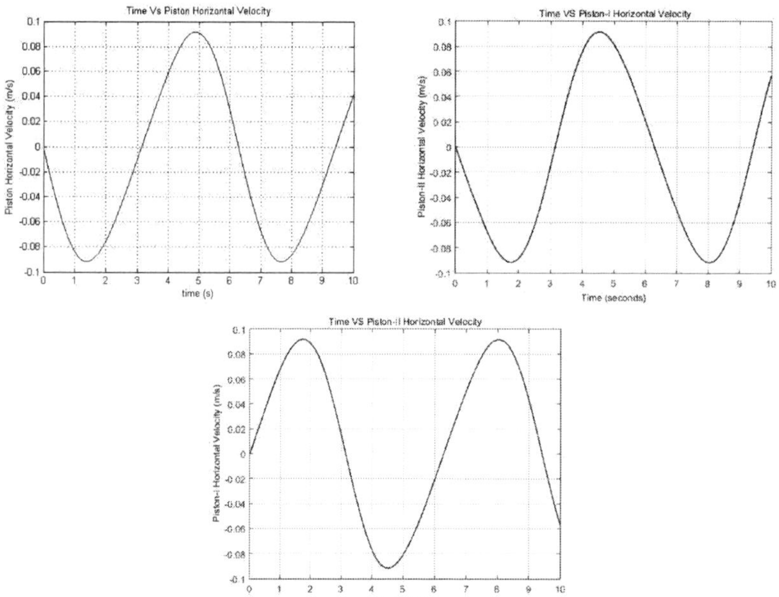

Figure 12: Piston horizontal velocity from kinematics (left), and from BFF for first cylinder (center), for second cylinder (right).

7. 2. Comparison with SOLIDWORKS Model

SOLIDWORKS is a great platform to visualize and analyze the motion of multi-body system. To further check and validate the responses already obtained from bond graph modelling technique in previous section, a TCRC was designed in SOLIDWORKS CAD environment and simulated for analyzing different parts motion later. Figure 13 shows the compressor model in SOLIDWORKS. The properties of different compressor parts are also shown in figure 13.

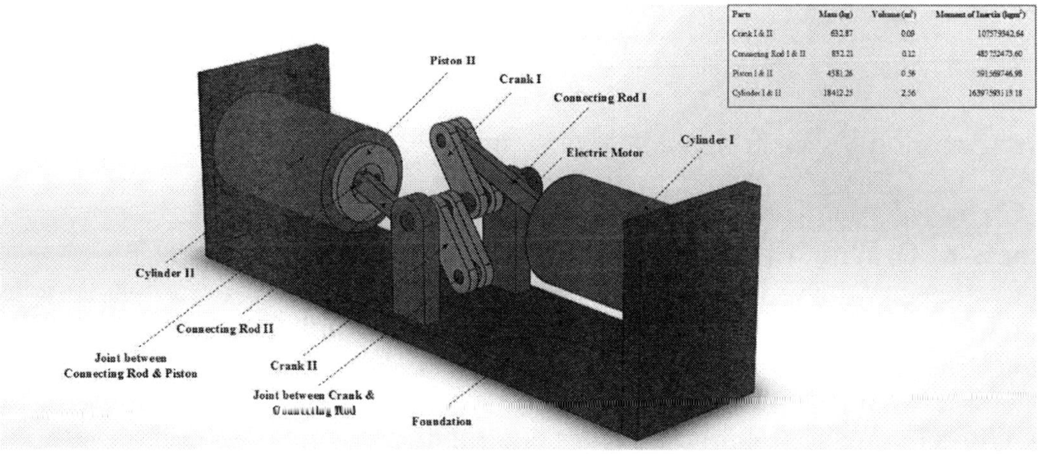

Figure 13: SOLIDWORKS model of the TCRC model.

Figure 14, 15 and 16 show the responses for crank, connecting rod and piston horizontal velocity obtained from both SOLIDWORKS and proposed model. SOLIDWORKS model confirms that the proposed model shows the correct pattern of responses for corresponding bodies.

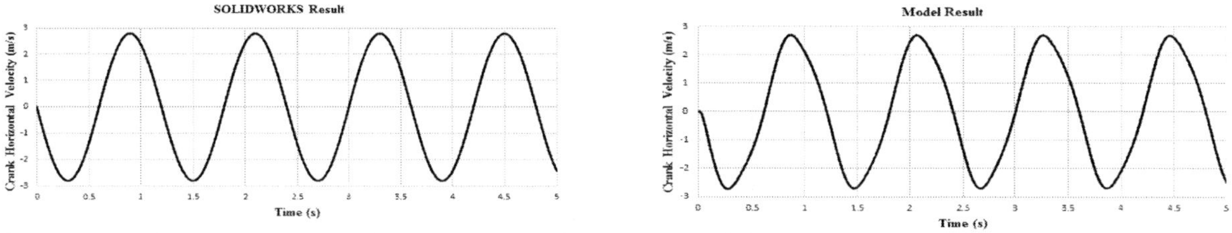

Figure 14: Showing crank horizontal velocity from SOLIDWORKS (left) and bond Graph (right) model.

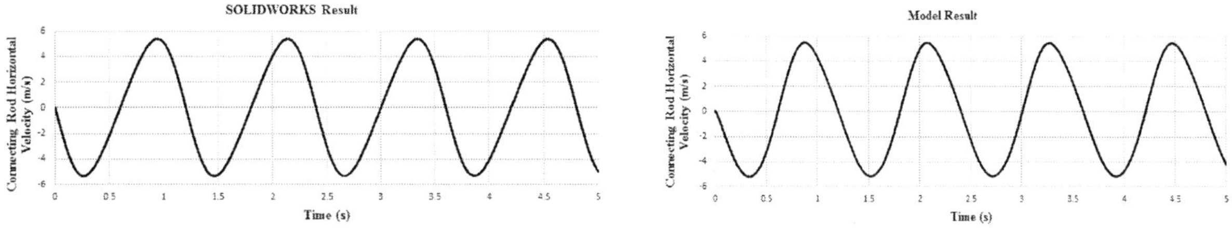

Figure 15: Showing connecting rod horizontal velocity from SOLIDWORKS (left) and bond Graph (right) model.

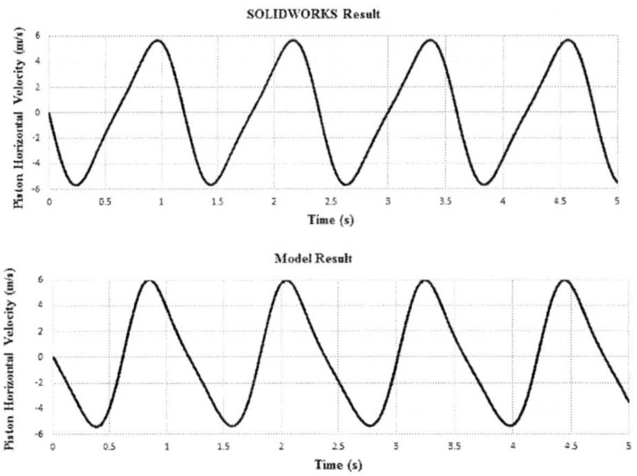

Figure 16: Showing piston horizontal velocity from SOLIDWORKS (left) and bond Graph (right) model.

7. 3. Effects of Gravity, Friction and Constraint Forces

The verification is finally accomplished by looking into effects of gravity, friction and constraint forces. Gravitational forces are added to the crank, connecting rod and piston accordingly. Simple viscous friction model is added between piston and cylinder wall. From the simulation, there is no effect of gravitational and frictional forces on the reciprocating compressor's parts. As the constraint forces between different bodies are extremely high or simply the system is reasonably stiff, effects of gravitational and frictional forces are almost negligible. Simulation results are same as in figures 6 to 12 for which results are not repeated here. Eventually this confirms the model reliability. We are also interested in this part to see the nature and behavior of constraint forces imposed at the connection points between crank and connecting rod, connecting rod and piston and so on. And, we would like to compare model responses with hand calculation results. Hand calculations are the most basic step where from free body diagram of compressor, unknown constraint forces are identified and solved subsequently. This provides the opportunity to compare results quantitatively. Figures 17, 18, 19, 20, 21, and 22 show the comparison of model results with hand calculation for summation of horizontal and vertical components of constraint forces at joints o_1, o_3 and o_2 (fig. 1). The input to the system is crank angular velocity which is 5.236 rad/s. The simulation parameters are same as mentioned in SOLIDWORKS model (fig. 13).

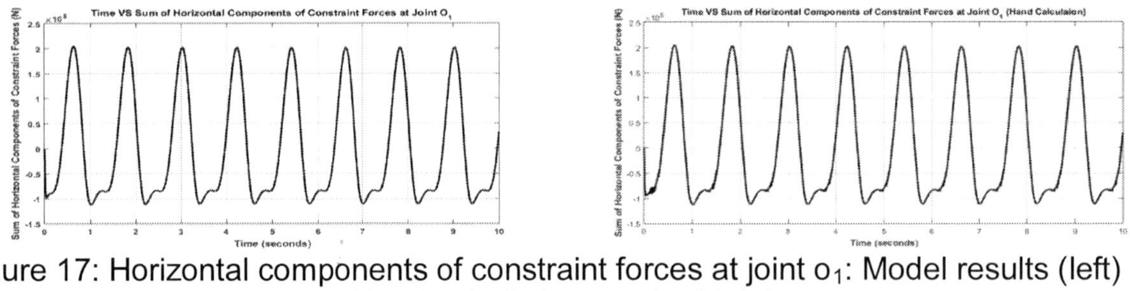

Figure 17: Horizontal components of constraint forces at joint o_1: Model results (left) and Hand calculation results (right).

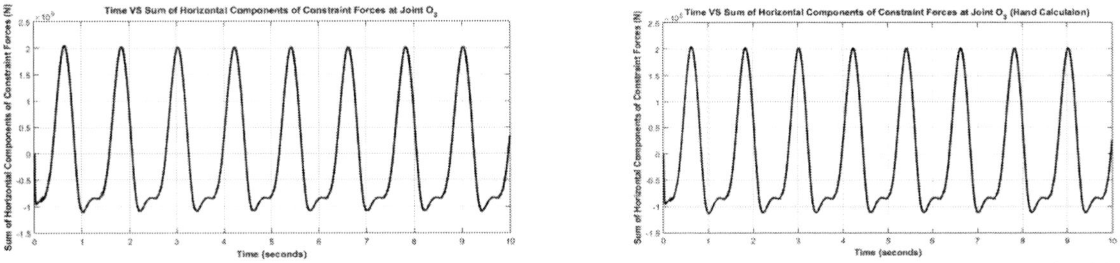

Figure 18: Horizontal components of constraint forces at joint o_3: Model results (left) and Hand calculation results (right).

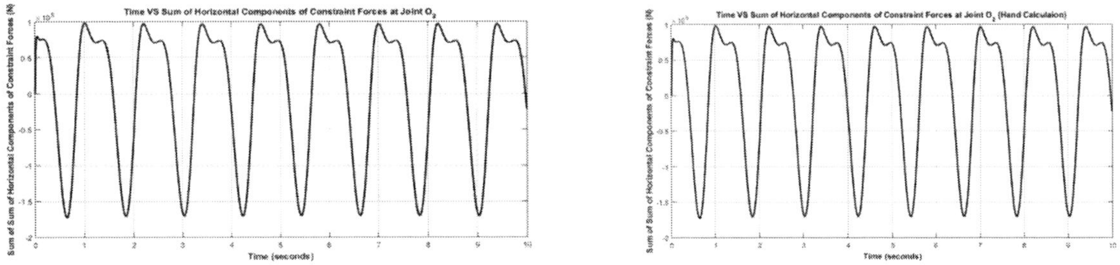

Figure 19: Horizontal components of constraint forces at joint o_2: Model results (left) and Hand calculation results (right).

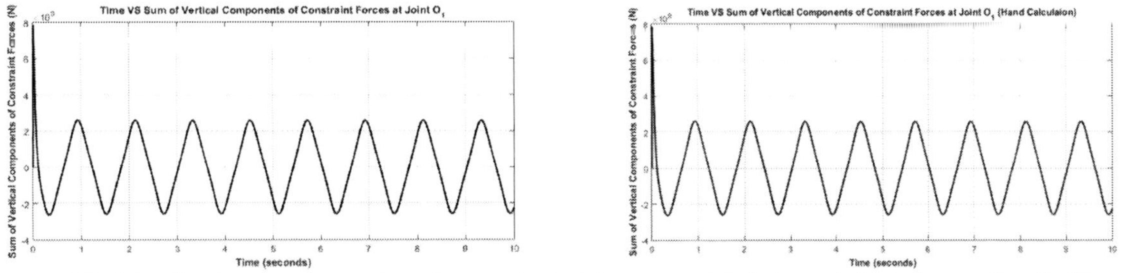

Figure 20: Vertical components of constraint forces at joint o_1: Model results (left) and Hand calculation results (right).

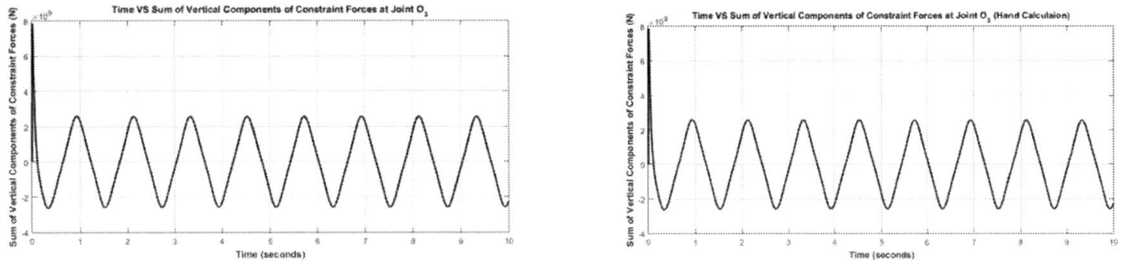

Figure 21: Vertical components of constraint forces at joint o_3: Model results (left) and Hand calculation results (right).

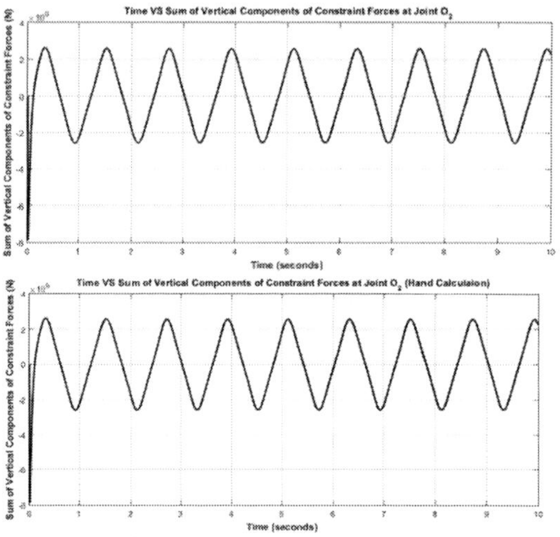

Figure 22: Vertical components of constraint forces at joint o_2: Model results (left) and Hand calculation results (right).

The above model results are in excellent agreement with hand calculations and ultimately validate the proposed model.

8. Conclusion

This paper enlightens the essence of an analytical physics based model of reciprocating compressor that might effectively be implemented in industries for machine condition monitoring. Faults may easily be created in this model and then by comparing results with a faulty compressor in operation, the location and severity of faults may be determined. For mathematical model development, bond graph method is applied. An OCRC model is developed and analyzed and is also extended for a TCRC model. The applicability and validity of this model is demonstrated by couple of ways such as pure kinematics, SOLIDWORKS motion analysis and so on. The effects of external forces such as gravitational and frictional forces are highlighted which shows the efficacy of analytical model over traditional data driven model. The model also gives us privilege to observe the presence of constraint forces which are typically important in multi-body systems.

References

[1] S. M. Schultheis, C. A. Lickteig, R. Parchewsky, Reciprocating Compressor Condition Monitoring, Proceedings of the Thirty-sixth Turbomachinery Symposium, 2007.

[2] E. Mollasalehi, S. A. Z. Jahromi, M. Forcinito, W. Hu, Q. Sun, Structural Fault Diagnosis of a Reciprocating Compressor using Bond Graph Approach, Proceedings of ASME International Design Engineering Technical Conferences & Computers and Information in Engineering Conference, 2012.

[3] B.V. Viscomi, R.S. Arye, Nonlinear dynamic response of elastic slider-crank mechanism, ASME Journal of Engineering for Industry 93 (1971) 251–262.

[4] F.J. Lin, R.F. Fung, Y.S. Lin, Adaptive control of slider-crank mechanism motion: simulation and experiments, International Journal of System Science 28 (12) (1997) 1227–1238.

[5] J.-L. Ha, R.-F. Fung, K.-Y. Chen, S.-C. Hsien, Dynamic modeling and identification of a slider-crank mechanism, Journal of Sound and Vibration 289 (2006) 1019–1044.
[6] A.M. Bos, M.J.L. Tiernego, Formula manipulation in the bond graph modelling and simulation of large mechanical systems, Journal of the Franklin Institute, Vol. 319, pp.51-65.
[7] H. Engja, Bond Graph Model of Reciprocating Compressor, Journal of the Franklin Institute, Vol. 319, No. l/2, pp. 115 124, January/February 1985.
[8] Seyed Abdolali Zareian Jahromi, Ehsan Mollasalehi, Modeling Vibration of a Reciprocating Compressor through Bond-Graph Technique, MITAC Report (2011).
[9] CAMP-G (Computer Aided Modeling Program with Graphical Input) USER'S MANUAL, Cadsim Engineering, Davis, Ca 95617 U.S.A., 2000-2007.
[10] H. M. Paynter, Analysis and Design of Engineering Systems, MIT Press, 1961.
[11] J.F. Broenink, Introduction to Physical Systems Modelling with Bond Graphs, Technical report, University of Twente, Netherlands, 1999.
[12] Wolfgang Borutzky, Bond Graph Methodology: Development and Analysis of Multi-disciplinary Dynamic System Models, Springer-Verlag London Limited 2010.
[13] D. C. Karnopp, D. L. Margolis, and R. C. Rosenberg, System Dynamics: Modeling and Simulation of Mechatronic Systems, John Wiley and Sons Ltd. (Fourth Edition).
[14] J.L. Meriam, L.G. Kraige, Engineering Mechanics, Volume 2: Dynamics, John Wiley and Sons Ltd. (Seventh Edition)
[15] R. C. Hibbeler, Engineering Mechanics: Dynamics, Pearson Prentice Hall (Twelfth Edition).

Biography:

Enaiyat Ghani Ovy was born in Dhaka, Bangladesh, in 1988. He obtained his B.Sc. and M.Sc. degree in Mechanical Engineering at the Islamic University of Technology (IUT) in 2009 and 2012 respectively. Currently, he is pursuing a M.Sc. degree at the University of Calgary, working mainly on dynamic system modelling.

ON THE CONTROL OF AEOLIAN VIBRATION OF TRANSMISSION LINES: CONDUCTOR SELF-DAMPING

Oumar Barry and Ry Long
Central Michigan University
College of Engineering and Technology
Mt Pleasant, MI 48858
barry1o@cmich.edu and long2r@cmich.edu

Abstract

Aeolian vibration is a wind induced vibration with high frequency and low amplitude vibration. It is usually controlled by attaching Stockbridge dampers or by reducing the conductor tension. The conductor tension is dependent on the self-damping of the conductor. In the present paper, experiment is conducted to determine the damping coefficient of the conductor. The performed test is based on the forced response method and is in accordance with IEEE and IEC standards. The conductor, Drake 795 Kcmil, is suspended between two steel reinforced concrete blocks 27.25 m apart. The shaker is placed at mid-span to excite the conductor at various tensions. Data are recorded for each forcing frequency, excitation force, free loop amplitude, and phase angle between the force and the amplitude of vibration. Using the regression analysis method, an explicit expression is presented for the damping coefficient of the conductor. This expression is found to be proportional to the excitation frequency and the amplitude of vibration and inversely proportional to tension of the cable. The conductor self-damping expression can be used by transmission design engineers to predict the amplitude of vibration of the conductor and put a limitation on the conductor design tension for vibration control.

Keywords: Aeolian; Conductor; Vibration; Self-damping; Stockbridge damper.

Introduction:

Overhead transmission lines are continuously exposed to wind forces which make them prone to the development of repetitive cyclic motions that are sources of tremendous damage to power networks. One of the most common types of wind motion is called Aeolian vibration. This type of vibration arises from alternating forces caused by vortex shedding. Vortex shedding is associated with the flow of air across a bluff body. Aeolian vibration is specific to single conductor and it is characterized by low amplitudes and high frequencies vibration. The frequency of Aeolian vibration ranges from 3 to 150 Hz and the wind speed ranges from 1 to 7 m/s [1, 2, 3]. The amplitude of vibration is usually less than the diameter of the conductor. The conductor diameter ranges from 6 to 50 mm. When left uncontrolled it causes fatigue failure at the point of contact between the cable and the suspension clamps. Aeolian vibration is usually controlled

by attaching Stockbridge dampers or reducing the cable tension [4]. The role of the Stockbridge damper is to reduce the vibration amplitude to a safe level. The reduction of tension also minimizes the vibration level.
The efficiency of Stockbridge dampers is characterized by the number of resonant frequencies they possess. There are two types of Stockbridge dampers, symmetric and asymmetric. Symmetrical Stockbridge dampers have identical weights and cable length at each end and they are characterized to possess up to two resonant frequencies. Asymmetric Stockbridge dampers have different weights and cable length at each end and are characterized to possess up to four resonant frequencies.

The most common conductor used in overhead transmission line is called aluminum conductor steel reinforced (ASCR). It consists of aluminum outer layers and steel inner layers. The outer layers are for electric power transmission and the inner layers are for strength required to support the weight of the cable without stretching the aluminum.
For decades, Aeolian vibration has been a subject of study by numerous authors. The most common method to predict the vibration level is the energy balance method [4, 5, 6, 7]. The vibration level is evaluated by determining the balance between the energy imparted to the conductor by the wind and the energy dissipated by conductor self-damping and the added Stockbridge dampers. The energy balance method is simple to implement; however, it tends to overestimate the vibration level of the cable because it does not account for the conductor flexural rigidity.

An attempt to depart from the energy balance method (EBM) was reported in [2, 3, 8, 9]. In [2, 3] the finite element method was used to study the vibration response of the conductor without including the self-damping of the conductor. In [8, 9] an analytical approach was employed to evaluate the response of the conductor. It should be noted that for a complete model and better accuracy, both analytical and finite element method require the conductor damping coefficient. In the present paper, experiment was conducted to determine the damping coefficient of the conductor. The experiment was based on the forced response method and was in accordance with IEEE and IEC standards. An explicit expression is presented for the damping coefficient of the conductor using the regression analysis method.

Experiment Set up and Procedure:

The experiment setup and procedure are based on the forced response method and in accordance with the IEEE and IEC guides [10, 11]. A schematic of the test set-up is depicted in Figure 1. The conductor was attached to two dead-end clamps 27.25 m apart. The clamps were connected to linear bearings mounted on intermediate abutments made of steel reinforced concrete as shown in Figure 2. The conductor was then locked solidly to an insulator clamp at each end. The clamps were suspended on steel reinforced concrete towers as depicted in Figure 3. A hydraulic ram (cylinder) as depicted in Figure 4 was installed at the south-end of the test-span to string the conductor at a desired tension. A cantilever weight (pulley system) was used at the north-end of the test-span to maintain a constant tension throughout the span. A maximum torque of 40 Nm was used to secure the damper on the conductor.

One load cell was placed at the end to monitor the tensile load. The other load cell and an accelerometer were used to measure the input force and velocity from the shaker to the conductor, respectively. The shaker shown in Figure 5 was placed at mid-span. Another accelerometer was placed on an antinode of the corresponding forcing frequency to measure the vibration displacement.

The conductor was then vibrated at various frequencies and power levels based on the Alcoa wind power curve derived from wind tunnel tests on flexible cable [12]. The voltage signal from the load cell and accelerometer were sent through charger amplifiers (low-pass filter) by means of coaxial cable and then to a digital data acquisition system for recording.

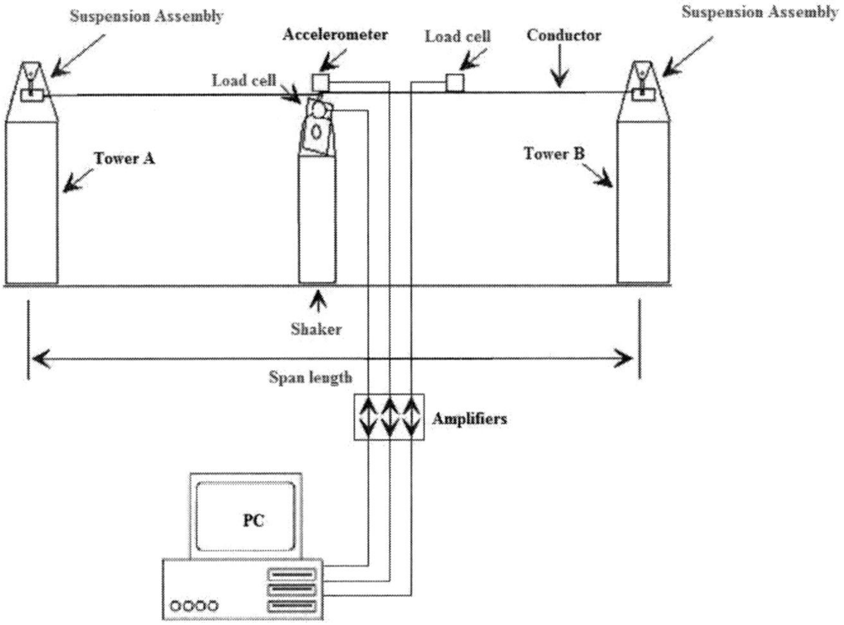

Figure 1: Schematic of experimental set-up.

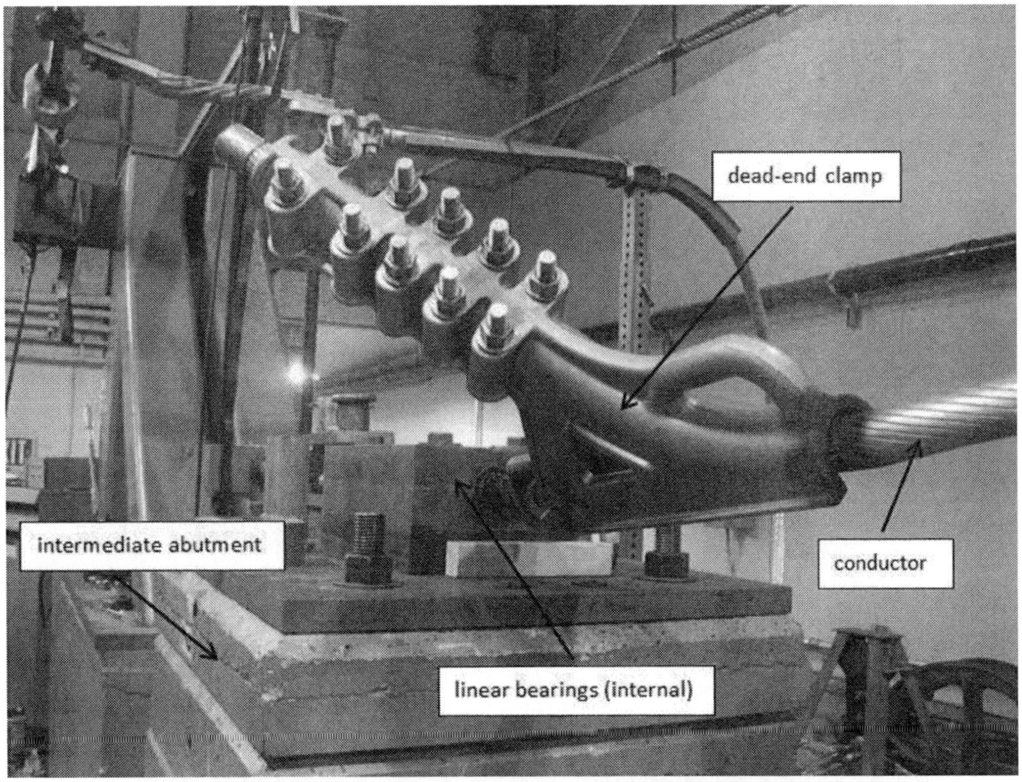

Figure 2: Intermediate abutement dead-end clamp.

Figure 3: Steel reinforced concrete towers and suspension clamp.

Figure 4: Hydraulic cylinder.

Figure 5: Cantilever weight (pulley system).

Figure 6: Shaker (B&K 4802) with conductor, load cell, and accelerometer.

Results and Discussion:

The conductor, Drake 795 Kcmil, was tested at 20 and 25% RTS (rated tensile strength) to determine the conductor self-damping. It was noted that after reaching a stationary condition, all the energy imparted to the span by the shaker is equal to that dissipated by the conductor over one vibration cycle; that is, all the energy introduced by the shaker is dissipated by the conductor self-damping. As such, the energy dissipated by the conductor is given as

$$E_{diss} = \pi F y_f \sin\theta_d \tag{1}$$

where F is the driving force from the shaker, y_f is the displacement of the conductor at the location shaker (0 to peak), θ_d is phase angle between displacement and driving force.

The maximum kinetic energy of the cable is given as

$$E_{k,max} = \frac{1}{4} m_c L_c \omega^2 y_0 \tag{2}$$

where m_c is the conductor mass per unit length, L_c is the span length, y_0 denotes the antinode vibration amplitude and ω is the corresponding resonant circular frequency. The non-dimensional damping ratio is defined as

$$\zeta = \frac{1}{4}\pi \frac{E_{diss}}{E_{k,max}} \tag{3}$$

The expression of the damping ratio of the conductor is now obtained by substituting Eqs. (1) and (2) into Eq. (3)

$$\zeta = \frac{F y_f \sin\theta_d}{m L \omega^2 y_0} \tag{4}$$

The parameters of the tested conductor and Stockbridge damper are tabulated in Table I.

Table I: Conductor Parameters

Parameter	
E_cI_c	1602 Nm^2
m_c	1.628 kg/m
D	28.143 mm
RTS	139.2 kN

The recorded data indicate that the damping ratio is significantly dependent on the excitation frequency, vibration displacement, and the conductor tension. This conclusion is in agreement with [13]. The results show that the conductor self-damping ratio can be expressed as

$$\zeta = \frac{Cf^{\alpha}Y^{\beta}}{T^{\gamma}} \tag{5}$$

where f is the excitation frequency in Hz, Y is non-dimensional peak-to-peak displacement (non-dimensionalized with respect to the conductor diameter), T is the conductor tension, C, α, β, and γ are constants. Using linear regression analysis, the constants in Eq. (5) were determined to be C = 20760, α = 1.406, β = 0.298, and γ = 1.835. It should be noted that measurements were repeated twice to minimize the effect of random error.

Figures 7 and 8 show plots of the damping ratio against the non-dimensional peak-to-peak amplitude. At every resonant frequency, a line of best fit is plotted. Both figures indicate similar trends in that the damping ratio increases with increasing displacement. However, the damping ratio values in Figure 7 (i.e., for T = 20% RTS) are higher than those in Figure 8 (i.e., for T = 25% RTS). This implies that increasing the tension reduces the self-damping of the conductor.

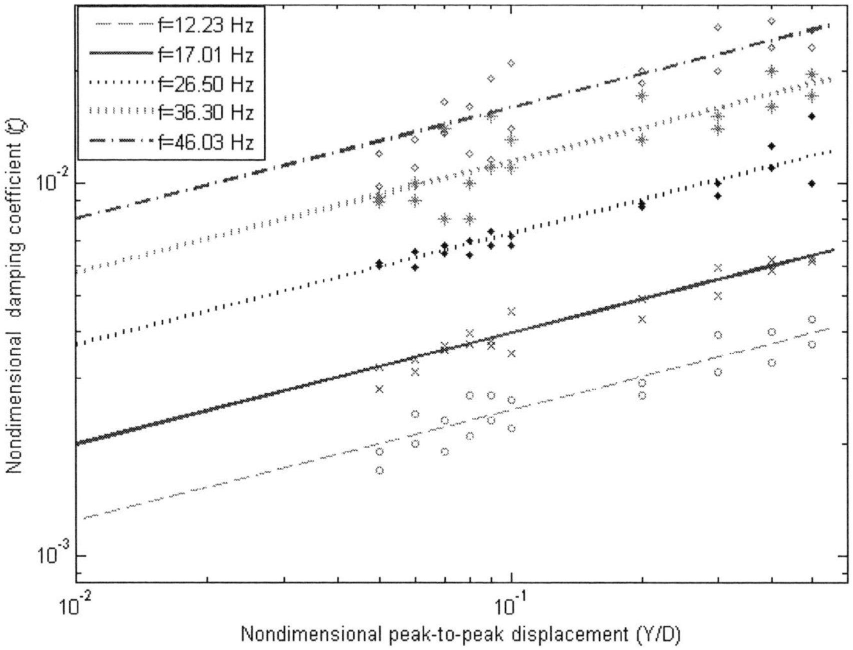

Figure 7: Conductor damping constant for fixed frequency for T = 20% RTS.

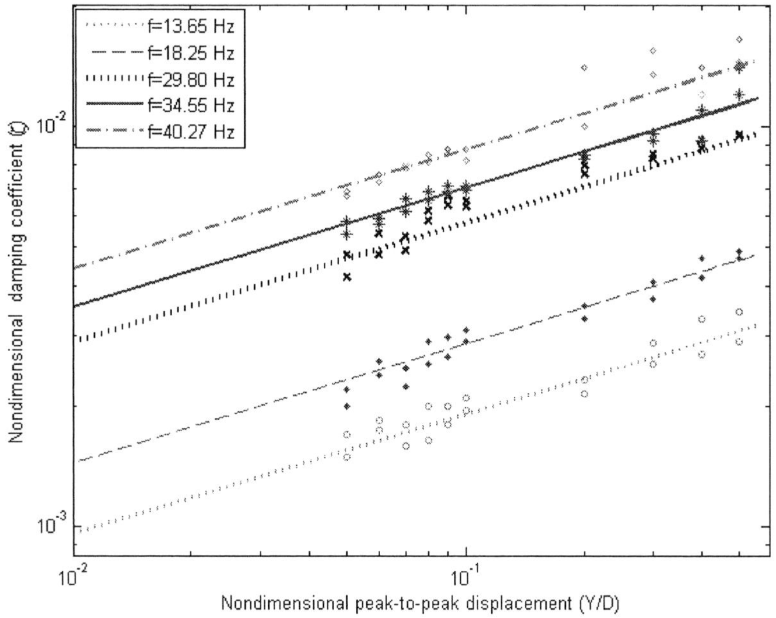

Figure 8: Conductor damping constant for fixed frequency for T = 25% RTS.

For given relative vibration displacement (vibration amplitude over conductor diameter) of Y/D = 0.1 and Y/D = 0.3, plots of damping ratio against resonant frequencies are depicted in Figures 9 and 10 for T = 20% RTS and T = 25% RTS, respectively. These plots indicate that the conductor-self damping increase with increasing frequency and the figure with the lower tension (i.e. Figure 9) exhibits higher conductor self-damping.

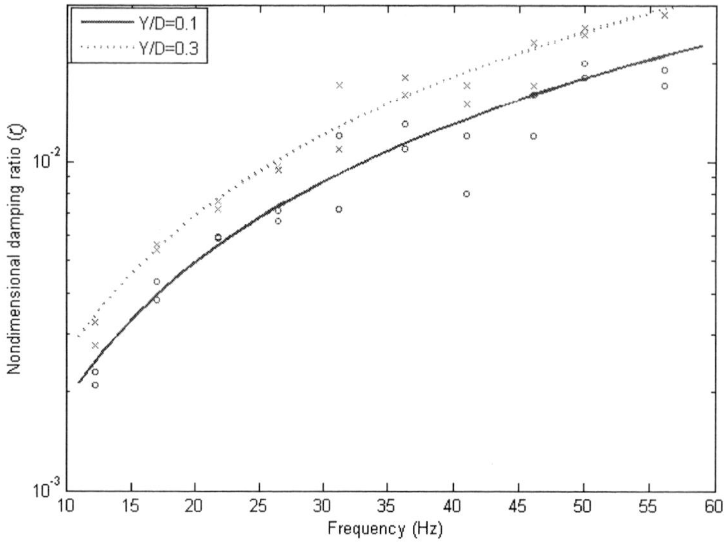

Figure 9: Conductor damping constant for fixed vibration amplitude for T = 20% RTS.

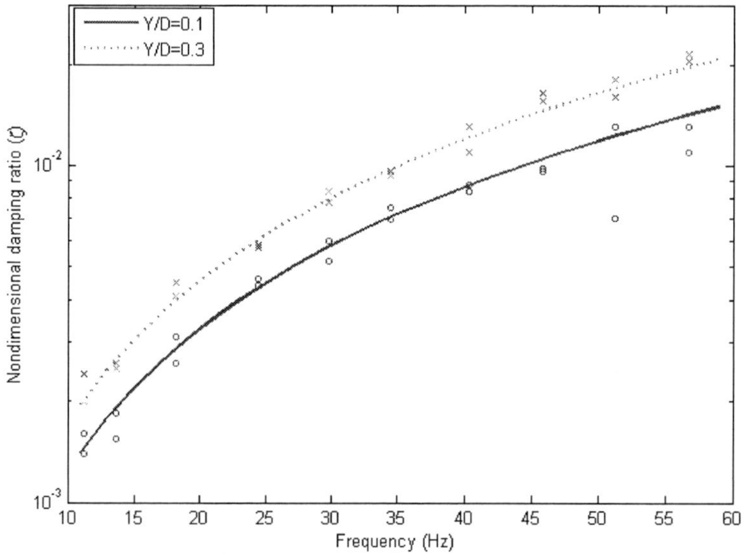

Figure 10: Conductor damping constant for fixed vibration amplitude for T = 25% RTS.

Conclusion:

Conductor damping coefficient is required for a complete analytical or finite element modeling of a single conductor vibration. This paper discusses the method used to determine conductor damping. Experiments were conducted for 795 ASCR conductor using the forced response method. Measurements were recorded for the vibration response, forcing frequency, phase angle between the response and the excitation force. The measured data were employed in Matlab along with the regression analysis to obtain a line of best fit. An explicit expression for the damping ratio of the conductor was then deduced from this line of best fit. Albeit some of the resonant frequencies were found to have a larger variation from this line of best fit, the deduced expression is a good estimator of the damping coefficient of the conductor. Most importantly, the findings deduced from this expression corroborate those in the literature in that the damping coefficient of the conductor is significantly dependent on the tension, excitation frequency and vibration amplitude. The obtained expression for conductor damping can be very useful to design engineers for predicting the vibration response of a conductor and putting a limitation on the conductor design tension to control Aeolian vibration of transmission lines.

Bibliography:

[1] EPRI. Transmission Line Reference Book, Wind Induced Conductor Motion, Palo Alto, California: *Electrical Power Research Institute*, 2006.
[2] O. Barry. *Finite Element Modeling of a Single Conductor with a Stockbridge Damper under Aeolian vibration*, MASc Thesis, Ryerson University, Toronto, ON, 2010.
[3] O. Barry, D.C.D. Oguamanam and D.C. Lin. Aeolian vibration of a single conductor with a Stockbridge damper, *IMechE: Part C, Journal of Mechanical Engineering Science*, 227(5), pp. 935-945, 2013.
[4] A. Oliveira and D.G. Preire. Dynamical Modeling and Analysis of Aeolian Vibration of Single Conductors, *IEEE Transaction Power Delivery*, 9(3), pp. 1685-1693, 1994.
[5] H. Verma and P. Hagerdorn. Wind Induced Vibration of Long Electrical Overhead Transmission Line Spans: A Modified Approach, *Journal of Wind and Structures*, 8(2), pp. 89-106, 2005.
[6] B. Shafer. Dynamic Modeling of Wind Induced Vibration of Over Head Lines, *Int. Non Linear Mechanics*, 19(5), pp. 455-467, 1984.
[7] M. Kraus and P. Hagedorn. Aeolian Vibration: Wind Energy Input Evaluated From Measurements on an Energized Transmission Lines, *IEEE transaction*, 6(3), pp. 1264-1270, 1991.
[8] O. Barry. Vibration Modeling and Analysis of a Single Conductor with a Stockbridge Damper, PhD. Thesis, University of Toronto, Toronto, ON, 2014.
[9] O.Barry, J.W.Zu, and D.C.D.Oguamanam, Forced Vibration of Overhead Transmission Line: Analytical and Experimental Investigation, *Journal of Vibration and Acoustics*, 136 (2014) d.o.i 10.1115/1.4027578.
[10] IEEE Guide on the Measurement of the Performance of Aeolian Vibration Dampers for Single Conductors, *IEEE Std*. 664:1993.
[11] IEC Guide on the Requirements and tests for Stockbridge type Aeolian vibration dampers, *IEC Std*. 61897:1998.
[12] C.B. Rawlins. Wind Tunnel Measurement of The Power Imparted to a Model of a Vibrating Conductor, *Trans. IEEE Power Apparatus and Systems*, 102(4), pp. 963-971, 1983.
[13] G. Diana, M. Falco, A. Cigada and A. Manenti. On The Measurement of Over Head Transmission Lines Conductor Self-Damping, *IEEE Transactions on Power Apparatus and Systems*, 15(1), pp. 285-292, 2000.

Biographies:

Dr. Oumar Barry graduated with a Bachelor degree in Mechanical Engineering with Honors in 2008. In June 2008, he joined Hydro One Inc. where he worked as design Engineer for seven years. During his employment at Hydro One Inc., he obtained a Master of Applied Science (MASc) and a PhD degrees in Mechanical Engineering in 2010 and 2014 from Ryerson University and the University of Toronto, respectively.

Dr. Barry's research has been focused on mechanical vibrations; including wind induced vibrations, energy harvesting using piezoelectric materials, optimization of engine mounting systems for noise and vibration reduction. In July 2015 he joined the University of Central Michigan as an assistant professor in the School of Engineering and technology.

Ry Long, a lecturer in the School of Engineering and Technology at Central Michigan University received her Bachelor degree in Mechanical Engineering with Honors in 2008 and her Master of Applied Science (MASc) degree in Mechanical Engineering in 2010, both from Ryerson University, Toronto Canada. Following that, for five years she worked at Bombardier Inc. as a System's Integrator.

Her research interests include vibration of composite "smart" materials, and the control of vibration on overhead power lines.

DAMPING WITH ADAPTIVE RESONATORS IN 1D ACOUSTIC SYSTEMS

Siddharth Bhagwat and M. Austin Creasy
Purdue University, School of Engineering Technology

Abstract: Acoustic and mechanical systems can be modeled by analogous models that consist of mass, damping, and stiffness of the system. A driving force can cause either of these systems to experience resonance. Passive devices can be used to absorb energy in these systems that are being driven at resonance. 1D acoustical systems are similar to distributed parameter vibrating systems and have multiple resonance frequencies that are dependent on the geometry of the acoustical systems and the boundary conditions. Acoustic pipes can be viewed as a 1D acoustical system where the geometry and the boundary conditions of the pipe will dictate the resonances modeled and measured within the pipe. Helmholtz resonators are passive devices that can absorb acoustical energy from within these acoustical systems because they act as tuned vibration absorbers in acoustical systems. These resonators can be manufactured to be adaptive to varying conditions with the system where damping is being applied and will be explained in the paper. This paper will show how to model Helmholtz resonators, show how an adaptive resonator can be manufactured, and compare the frequency of the Helmholtz resonator model to the measured frequency of the adaptive resonator as the resonator is changed to absorb energy at varying frequencies.

Key Words: Acoustics; Damping; Modeling

Introduction: All acoustical and mechanical systems will resonate when a driving force is applied to the system where the force has a frequency content that is at or near the resonance frequency of the system. The resonance frequency of the system is a function of the mass, stiffness, damping, and boundary conditions of the individual components of the system. Understanding the individual properties (the mass, stiffness, damping, and boundary conditions) of each component will allow analysis of the resonances for the components and group them together in a system to analyze the entire system [1]. Distributed parameter systems allow modeling of the bulk system, but finite element analysis of the distributed parameter systems returns the analysis to the individual properties (the mass, stiffness, damping, and boundary conditions) of each element [1]. Absorbers can be used with these systems to absorb energy from the system around a specific frequency that is based on the design of the absorber. Vibration absorbers are a single degree of freedom system that can be modeled using a second order differential equation to absorb energy of a vibrating system [1]. Helmholtz resonators are the acoustic equivalent of the vibration absorber in acoustical systems

[2]. These resonators can be modeled as well with a second order differential equation where the components of the equation are related to the design of the resonator.

Acoustic systems are more closely aligned with distributed parameter system from vibration analysis. There are multiple resonances in the acoustic system that are dependent on the geometry, boundary conditions, and properties of the gaseous medium. A basic acoustic system is a 1-D pipe where the system resonances are defined from the boundary conditions, the driving frequency, and the gaseous medium [2, 3]. The resonances of these 1-D pipes are normally introduced in physics courses by introducing students to the wavelength of sound waves and the interaction of waves with boundary conditions of the pipe [4, 5]. These resonances are caused by a driven sound wave being in-phase with sound wave reflections from the boundary conditions. Another way to model these resonances of 1-D pipes is by the use of impedance modeling of the system [2, 6]. Each component is seen as an equivalent electrical circuit that can be used to approximate the resonance frequency of the system [3].

The damping in these 1-D acoustic systems is limited to the damping associated with the gaseous medium and the boundary conditions of the structure [2]. Passive Helmholtz resonators are ideal at adding additional damping to these systems. The resonator can be designed to add damping to a specific frequency where the design is based on the speed of sound and density of the gaseous medium and the geometry of the resonator. Therefore the geometry can be used as the design variable to specify the frequency where damping is being applied with a resonator. This work will show the results of modeling a 1-D acoustic system, it will show modeling of Helmholtz resonators, it will show how the geometry of these resonators can be adjusted to make and adaptive resonator, and it will compare the adaptive resonator frequencies to theoretically modeled resonator frequencies.

Background: An impedance model for a 1-D acoustical pipe has been performed in reference [3] where a speaker is driving the system and causing resonances to occur. An impedance model is a relation between the sound pressure and the particle speed of the air. This model allows the boundary conditions to be used to estimate the sound pressure level for a specific driving input and will capture the resonance frequencies of the driven system. This model is not repeated here and was compared to measured data of a 1-D acoustic pipe in the reference. For a 1.5 m 1-D acoustic pipe with a speaker exciting the system on one end and a rigid end cap on the other, the system has 9 resonance frequencies below 1 kHz. Figure 1 shows an example of modeling a 1-D acoustic pipe using the procedure from reference [3] where the sound pressure level (SPL) is modeled as a function of the driving frequency at three different locations.

The modeled locations in the 1.5 m pipe are at the speaker, at 0.75 m (the middle of the pipe), and at the end cap. The results show the 9 resonance frequencies of the system.

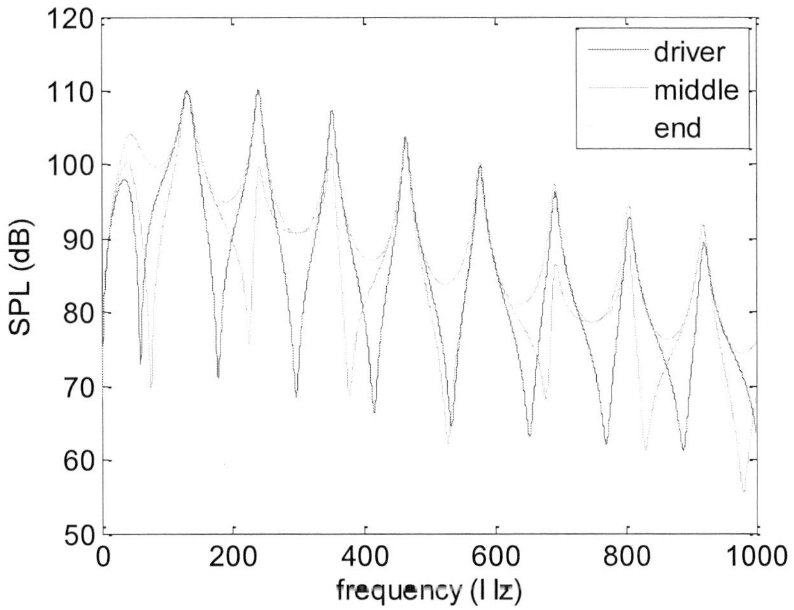

Figure 1. SPL of a 1-D acoustic pipe with a speaker driving the system on one end and a rigid end cap on the other.

Helmholtz resonators are passive devices that can absorb acoustical energy from a system. The main components of a resonator design are the volume of the resonator, the area of the neck of the resonator, and the length of the neck of the resonator. Figure 2 shows an example of a resonator to show the variables at the disposal of the designer that can be used to tune the resonator to a specific frequency. The volume of the resonator is independent of shape. The only constraint on these resonators is the comparison of these three variables to the wavelength of the frequency that is being targeted. The wavelength has to be much greater than the length of the neck, much greater than the volume raised to the 1/3 power, and much greater than the neck area raised to the ½ power [2]. The density of the gaseous medium within the resonator and the geometry can be used to model the resonator as a second order system. The equivalent mass, *m*, of the resonator is

$$m = \rho A L \qquad 1$$

where ρ is the density of air, *A* is the area of the neck, and *L* is the length of the neck with an additional factor that is based on if the neck is flanged or unflanged [2]. This equation shows that the equivalent mass is the mass of the air in the neck of the resonator. The equivalent stiffness, *k*, of the resonator is

$$k = \frac{\rho c^2 A^2}{V} \qquad 2$$

where c is the speed of sound in air and V is the volume of the resonator [2]. The resonance frequency, ω, of this second order system will follow the same equation from the natural frequency equation of a second order system in vibrations,

$$\omega = \sqrt{\frac{k}{m}}\,. \qquad 3$$

Substituting equations 1 and 2 into equation 3 will provide the resonance of the resonator to be

$$\omega = c\sqrt{\frac{A}{LV}}\,. \qquad 4$$

Equation 4 defines the frequency that the resonator can be used to absorb energy in an acoustic system. As stated, a designer has three values that can be used to change the frequency to a resonance of the acoustic system where the resonator will be used. The damping of the resonator will provide a small bandwidth where the resonator absorbs energy.

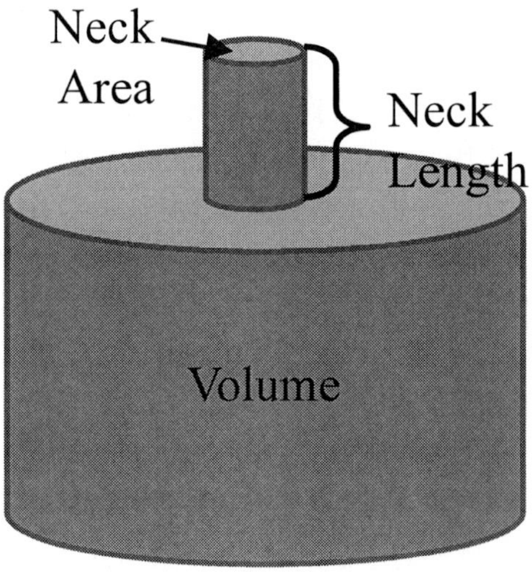

Figure 2. Schematic of a Helmholtz resonator to show the volume and neck properties.

These resonators have been used in numerous applications, but especially in enclosed structures where passive devices can reduce internal noise [7, 8]. In most of these applications, the resonators are designed specifically for a resonant frequency of the acoustic enclosure. In some applications, adaptive resonators have been designed that can be used on systems with varying resonances or to target different resonances at different times [9, 10]. These systems and other systems of similar nature use active control to vary the area of the neck or volume of the resonator in some fashion.

Resonator: The example resonator is made from polyvinyl chloride (PVC) where the neck is a constant. The volume is made with a piston to vary the volume of the resonator. The tolerances of the piston to the interior diameter of the PVC pipe were not emphasized in construction and some losses are assumed to occur around the edge of the piston. This design allows for a very cheap adaptive resonator to be built that can be varied for different resonances. Figure 3 shows a schematic of this design where the piston can be moved by a rod to adjust the interior volume of the resonator. This structure allows for the frequency to be adjusted quickly by using the rod to position the piston within the volume.

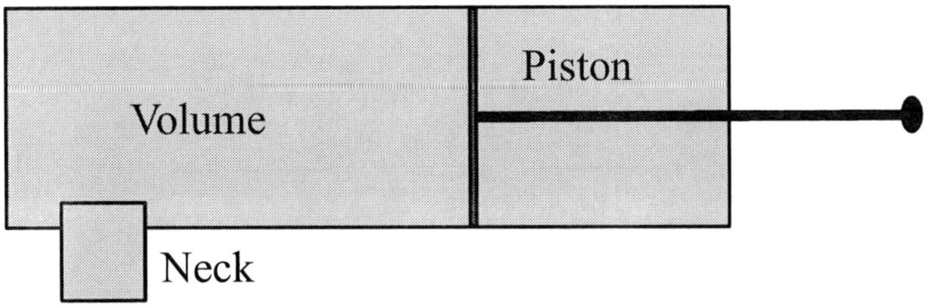

Figure 3. Schematic of the resonator with a varying volume using a piston.

A comparison of the targeted resonances of the example resonator are compared to the theoretical values found using Equation 4. To perform this comparison, two separate PVC resonators were constructed where the neck was 5.08 cm (2 inch) in length and the area of the neck is found from the neck's 2.54 cm (1 inch) diameter. The volumes of the resonators are cylindrical where one resonator had a 5.08 cm (2 inch) diameter and the other had a 7.62 cm (3 inch) diameter cross section. The piston surface was made from renboard and therefore was flat. The other end of the PVC was a circular cap that was sealed onto the resonator end. The volume of the resonators were found by

$$V = \frac{\pi}{4} D^2 L' \qquad (5)$$

where D is the diameter of the resonator and L' is the length from the end cap connection of the resonator to the piston. A small half spherical addition was added to equation 5 to include the volume of the end cap. The rod of the piston was designed to change the length (L') of the resonator from 10 cm (4 inches) to 60 cm (23.5 inches).

The resonators were tested every 1.25 cm (0.5 inches) to measure the resonance frequency of the resonator and compare the value to the theoretical value found using Equation 4. The test was performed by placing the resonator on a 1-D pipe, using a speaker on one end of the pipe to excite all frequencies to 1 kHz using a chirp signal, and measuring the acoustic response at various locations throughout the pipe. The details of the signal and measurement procedures are similar to those used in reference [3]. Figure 4 shows the experimental setup with a resonator placed on the pipe. The rod extruding from the resonator on top is marked to show the equivalent length with the resonator. As the length of the rod is adjusted, at some point the resonator frequency is equal to the resonance frequency of the 1-D pipe. Figure 5 shows an example of this measured data comparison. The resonator is targeting the third natural frequency and reducing the peak by several dB. The experimental resonance frequency of the resonator is assumed to be the value found at the bottom of the valley of the damped third resonance of the measured pipe from Figure 5.

Figure 4. 1-D pipe with an adaptive resonator placed on top to absorb acoustic energy from within the pipe.

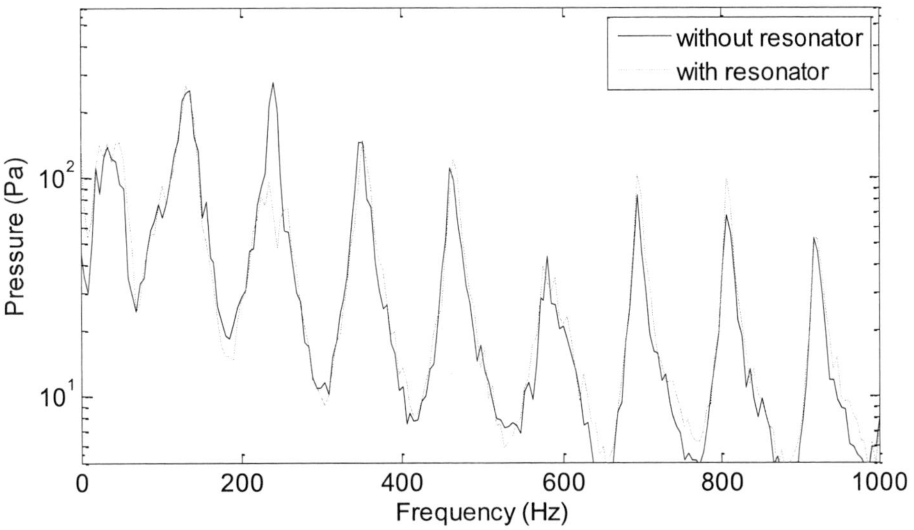

Figure 5. Comparison of the measured sound pressure without the resonator to the system with the resonator.

The relationship between the theoretical and the experimental results are extremely close. The results shows that the simple piston used to change the resonator has very little effect on the resonator frequency compared to the theoretical value. Figure 6 shows the results for both the 5.08 cm (2 inch) diameter and 7.62 cm (3 inch) diameter cross section resonators where the x-axis is the length (*L'*) of the resonator and the y-axis is the frequency of the resonator in Hz. The figures show that the experimental values are on average within 7 percent of the theoretical values.

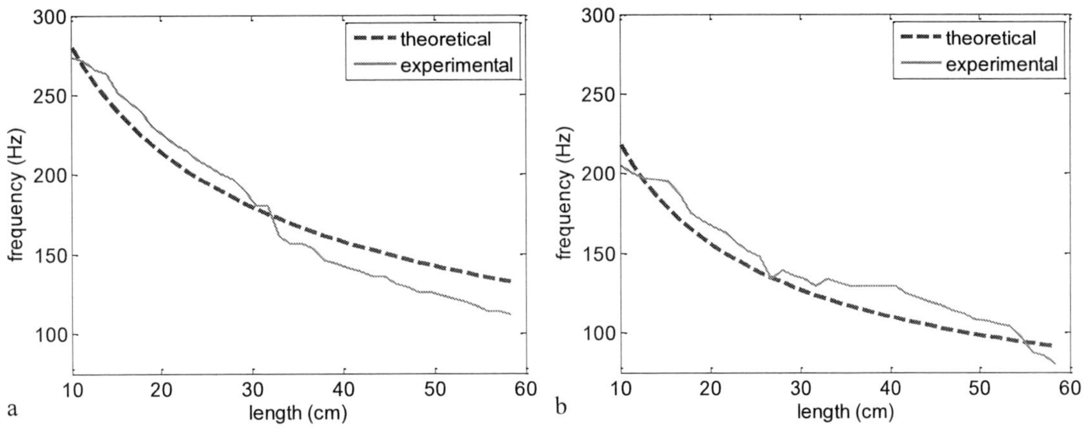

Figure 6. The comparison between the theoretical Helmholtz resonator frequency and the experimental frequency for the a) 5.08 cm (2 inch) diameter and b) 7.62 cm (3 inch) diameter cross section.

Conclusion: This work shows how to model Helmholtz resonators, shows how an adaptive resonator can be manufactured, and compares the resonator model to the measured frequency of the adaptive resonator at varying configurations. The results show that the resonators can be made to easily adapt to changes in the acoustic system to target varying resonance frequencies and add damping to those resonances.

References:

[1] D.J. Inman. Engineering Vibrations, Pearson Education, 2013.

[2] L.E. Kinsler, A.R. Frey, A.B. Coppens, J.V. Sanders. Fundamentals of Acoustics, 4th Edition, John Wiley & Sons, Inc, 2000.

[3] M.A. Creasy. Impedance modelling of pipes, European Journal of Physics 37 (2015) 025001.

[4] A.J. Buffa, J.D. Wilson, B. Lou. College physics, Prentice Hall, 2002.

[5] J. Walker, R. Resnick, D. Halliday. Fundamentals of physics, Wiley, 2008.

[6] L.L. Beranek. Acoustics, Acoustical Society of America, New York, 1996.

[7] S.A. Lane, R.E. Richard, S.J. Kennedy. Fairing Noise Control Using Tube-Shaped Resonators, Journal of Spacecraft and Rockets 42 (2005) 640-646.

[8] D. Li, J.S. Vipperman. Noise Control of Mock-Scale Chambercore Payload Fairing Using Integrated Acoustic Resonators, Journal of Spacecraft and Rockets 43 (2006) 877-882.

[9] S.J. Esteve, M.E. Johnson. Adaptive Helmholtz Resonators and Passive Vibration Absorbers for Cylinder Interior Noise Control, Journal of Sound and Vibration 288 (2005) 1105-1130.

[10] D. Zhao, C. A'Barrow, A.S. Morgans, J. Carrotte. Acoustic Damping of a Helmholtz Resonator With an Oscillating Volume, AIAA Journal 47 (2009) 1672-1679.

VERIFYING MATHEMATICAL MODELS OF THE 2016 FSAE RACECAR DESIGN TO MONITOR AND ESTABLISH SAFE RANGES FOR ROLL AND YAW ANGLES DURING DYNAMIC CORNERING AND ACCELERATION EVENTS

Jennifer Shaffer, Stone Williford
Department of Mechanical Engineering, Georgia Southern University

Abstract:

The 2016 Formula Society of Automotive Engineers (FSAE) racecar was designed and fabricated by Georgia Southern Eagle Motor Sports (EMS) members for the 2016 collegiate competition. Previous researchers have collected data on the 2012 model using Data Acquisition System and Logger Pro software based on certain dynamic conditions of the racecar during certain driving events, such as straight line accelerations, constant radius cornering, and road bumps and pot holes. Previous researchers were able to create a computer model to compare with the experimental data. The major work for this current research involves the development of the computer model of the 2016 car in Adams/Car software to reflect modifications to the current FSAE racecar. Special emphasis has been given to the yawing of the vehicle due to differential slip. One of the major objectives of this research is to create a stability status indicator of the vehicle based on the results of angular displacement, velocity, and acceleration of the vehicle in yaw and roll. Three stages will be assigned based on those parameters observed during modeling and testing. Future work will include designing a self-sustaining control system for the vehicle which will be triggered based on the stability status of the vehicle.

Keywords:

Adams/Car; FSAE; Lateral acceleration; Model verification; RaceCapture/Pro; Yaw

Introduction:

The automotive industry is a crucial area of study for many engineers, as there are numerous aspects that require extensive research and careful design to safeguard the safety of consumers while increasing the vehicles' performance. One system that has come under tougher research scrutiny in recent years is the suspension system, as it contributes significantly to the handling and stability of the vehicle. Increasing a car's stability allows for better handling, as it maintains the maximum contact for the tires on the road, on all four corners, at all times, which reduces tire slip. A suspension system can have many different configurations depending on the needs and performance requirements of vehicle in certain conditions. Independent suspensions are most often dual control arm or multi-link systems. Both systems allows for each corner of the car's suspension to move freely without affecting one another, while minimizing chassis movement. The suspension configuration that is the focus of this research is the dual control arm system, as this configuration is affordable and effective at maintaining vehicle stability. The suspension system helps the tires remain in contact with the ground during extreme driving conditions, as well as minimizing unnecessary vehicle movements. Each corner of an independent suspension has springs and dampeners built in to absorb external forces on the car. This increases rider comfort and allows for a more stable ride. One of the most important components of a high-performing vehicle is a suspension system which will maintain tire contact with the road at all times, minimizing yaw and roll, and thus increase the overall stability and traction of the vehicle. One way to approach this is to attempt to create an inexpensive active or semi-active suspension system that would alert the driver of unsafe driving conditions and adjust suspension parameters accordingly.

The Society of Automotive Engineers (SAE) holds a collegiate event annually that allows student teams to design and build a formula-style open-wheeled racecar and compete them in on-road competitions on a national scale. The FSAE competitions are designed to test the ability of the student racecars in speed, maneuverability, and engine design, and develop a potential production item that caters to a non-professional weekend autocross racer [1], whereas their mini Baja competition is to test agility, strength, and toughness of chassis design. Each school spends the school year assembling its team to design and manufacture the assorted subsystems to comply with the rules listed by the SAE Formula rule book.

Literature Review:

Active and semi-active suspensions have been used for many years to increase stability and handling of all types of vehicles [2]. Many well-known car manufacturers, such as Daimler Chrysler, Land Rover, Bose, and Delphi have all successfully applied these types of systems either as factory or aftermarket. This has led to further research of their applications in the automotive industry, as improving a car's stability through its roll, yaw, and pitch through the suspension is especially beneficial for vehicles with a high center of gravity (e.g. SUVs, ATV, trucks). They can also be helpful in high performance vehicles to improve traction [6]. The main parameter of these systems is dominated by the intention; if it is meant to be used for anti-rolling, the roll angle is controlled. The main cause of rollover in any vehicle is the lateral force which acts on the inner and outer tires. Once the lateral force exceeds certain levels which the car is designed to withstand, rollover conditions initiate [3][4]. The threshold to be aware of is generally set as 5-10 degrees of roll, as it is found that wheel lift-off is commonly found near 23 degrees of roll [3]. Common tests for rollover conditions are the double-lane change and fish-hook maneuvers. Both exert high lateral forces on the tire, although cornering will also have a similar effect [3][5][6]. Variables that are frequently scrutinized to determine rollover conditions are the rolling angle, lateral accelerations, time to rollover, and the rollover coefficient [3][4]. One researcher has shown that the use of a simple hydraulic actuator to generate a warning and increase the stiffness of the anti rolling bar at a certain degree of roll (nearing the vehicle's established threshold for failure) can aid in stabilizing the vehicle [3]. Furthermore, computer simulations have been found to be very useful tools for predicting the rollovers of vehicles, and thus aid in the creation of a dynamic rollover prevention system. Equations for the roll, yaw, and pitch can be solved by complex models in Adams and Matlab Simulink [4] software for many aspects of the suspension. These programs help better understand the effects of configuration modifications to a vehicle's suspension, which can aid a team in designing a racecar suspension, and spare the costs of post-fabrication trial runs and expensive modifications.

Data Acquisition Interface and Instrumentation:
The RaceCapture/Pro unit from Autosport Labs was selected to collect experimental data to be compared against the data generated from the Adams/Car simulation software. The RaceCapture/Pro houses an internal inertial measurement unit (IMU) and magnetic compass. The IMU includes a three axis accelerometer and gyroscope which has six degrees of freedom in measurement, i.e., it can measure the displacement in three orthogonal directions as well as the roll, yaw, and pitch of the vehicle. A diagram describing these six degrees of freedom is provided below in Figure 1.

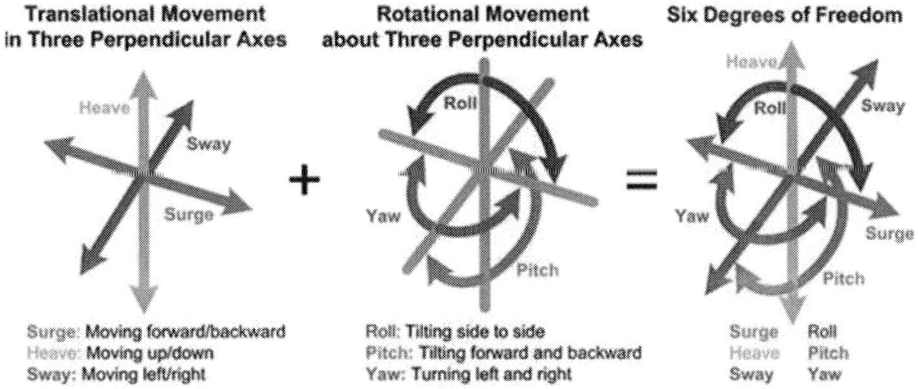

Figure 1- Diagram of the Six Degrees of Freedom (Credit: Honeywell Sensing and Control)

The data is stored in a standalone unit, shown below in Figure 2, which can be accessed wirelessly through a dashboard application via Bluetooth from any Android device. It is also equipped with 2 Control Area Network (CAN) Bus channels with 14 CAN filters per channel. The device also has the option for GPS and satellite telemetry, with a maximum sample rate of 1000Hz, and records lap times and engine data as well. It is powered by a 5V battery and weighs approximately 9 ounces.

Figure 2 – Photograph of the RaceCapture/Pro Data Acquisition Device (Credit: Autosport Labs)

To properly record data from the RaceCapture/Pro device, each piece of hardware and the power supply were affixed to the FSAE vehicle. Several steps were taken to mount the data acquisition device to the chassis of the FSAE vehicle, near the vehicle's approximate center of gravity, along the longitudinal centerline of the vehicle. Mounting the device perfectly along this line was not feasible due to the seating position of the driver, so the device was offset by approximately 10 inches. Supplementary precautions were also added to help eliminate vibrational noise that could potentially distort the signals read by the RaceCapture/Pro device. Neoprene rubber washers were used to add a layer of cushioning between the chassis of the racecar and the RaceCapture/Pro.

Simulation Model Verification:
Full vehicle model simulations in Adams/Car provide an in-depth analysis of the dynamic reactions of vehicle subsystems and components. Adams/Car accounts for the complex geometries, inertias, masses, and forces that affect the Formula SAE racecar, shown below in Figure 3, that a quarter- or half- car model cannot.

Figure 3 – Model of the Front and Rear Dual Control Arm Suspensions of the 2016 FSAE car

For this reason, validating a computer model in Adams/Car with experimental data from the actual racecar is especially valuable to the university's race team, as they can use the validated model to closely predict the behavior and performance of their racecar components under dynamic events, and adjust the vehicle accordingly. This saves both time and funds in the design process of the vehicle, as trial runs of the racecar often require extensive planning and can be costly in terms of fuel, man-hours, and the resulting adjustments to the car. This research specifically sought to examine the effects of the suspension configuration on the vehicle's yaw angle, yaw rate, and roll angle during constant radius cornering events, and the pitch angle of the vehicle during straight line acceleration events. These are events that the FSAE racecar constantly undergoes during a normal collegiate competition and those parameters can greatly affect the performance of the vehicle during competition [6].

To create the 2016 model for the FSAE car, each hard point of a previous 2012 model of the university's FSAE racecar was updated to reflect modifications to the current 2016 car design using comparisons of their models in both SolidWorks and Optimum Kinematics. The Adams/Car model of the 2016 FSAE racecar fabricated by Eagle Motorsports (EMS) is based on this previous solid model, which was updated to accurately reflect the locations and physical characteristics of all components in the chassis and suspension. The past 2015 physical model of the university's FSAE racecar design is shown below for reference in Figure 4.

Figure 4 – The 2015 Georgia Southern University FSAE Racecar Design

Certain suspension parameters were kept constant in the simulations, such as the coil spring rate, spring preload, static toe angle, camber angle, and the tire spring rate, as these were found to be similar to those on the 2012 model. Important parameters that were transferred to the new model were the location of the center of mass, total vehicle weight, and the balance of weight between the front and back tires (weight bias). The completed front and rear dual control arm suspension designs can be seen in Figure 5 and Figure 6 below.

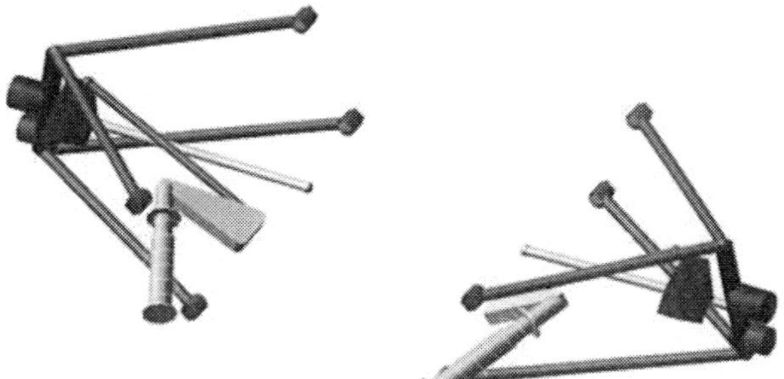

Figure 5 – Front Dual Control Arm Suspension Model

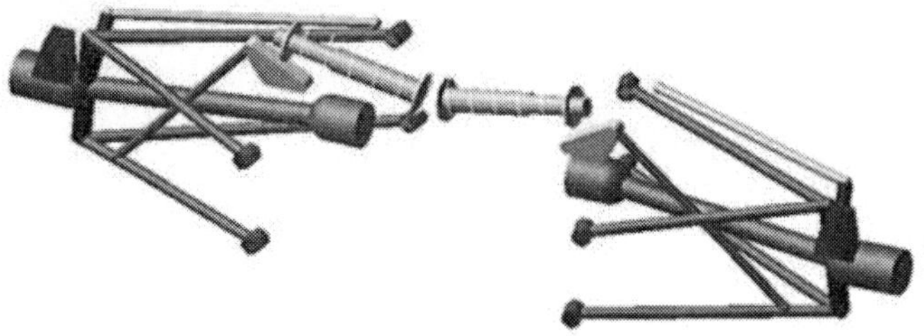

Figure 6 – Rear Dual Control Arm Suspension Model

After each component was positioned correctly in MSC Adams/Car with respect to the origin, based on the Optimum Kinematics model, and the front and rear suspensions placed in the correct configurations, other subassemblies of the car, i.e. the chassis, wheels, steering subassembly, etc., were placed in the full vehicle assembly and configured to match coordinates and values taken from the 2016 car design.

After the computer model of the vehicle was produced using MSC Adams/Car software, a specific set of dynamic conditions were simulated, including straight line acceleration tests and constant radius cornering events at different speeds. The data from these simulations can be used to verify the computer model. It can be compared to the experimental data gathered from the RaceCapture/Pro device after the racecar is fully fabricated. If the experimental data from the vehicle closely matches the simulation results, important assumptions and correlations from the model to the actual racecar can be made, leading to an improved racecar design for competition based on the needs of the race team.

Procedure/Method:

Once the vehicle has been properly equipped with the RaceCapture/Pro data acquisition device, the driver will run a strict series of tests which can be reproduced under similar conditions in Adams/Car simulations, such as constant radius cornering and straight line acceleration. Some events have limitations based on the simulation software, such as the straight line acceleration events will have to be tested with an initial velocity of at least 10km/hr; otherwise, there is a problem with stall speed and the transmission in MSC Adams/Car software gives invalid results and errors.

The 2016 FSAE racecar is currently in the final stages of fabrication after experiencing significant budget setbacks. It should be ready to gather the experimental data no later than April 2016 to verify the computer model simulations. If the experimental data and simulation data are similar, then the computer model can be used to predict the behavior of the racecar's subsystems in other driving events, such as split-mu braking or pothole/road bump events that can occur on the racetrack during the FSAE competitions.

Lastly, preliminary calculations were computed to solve for the expected lateral accelerations of the experiments from data recorded during the actual road tests of the constant radius cornering events. Using the diameter of the circular test course and the time needed to complete a set number of laps, the speed average of the laps can be calculated by dividing the distance traveled around the circumference by the time taken to complete the laps. The average lateral acceleration can be solved for by using the formula $a_c = v^2/r$. The lateral accelerations (in m/s^2) were then divided by $9.81 m/s^2$ to calculate the number of "g"s (gravitational forces) acting on the vehicle during a given test.

Simulation Results:

Several results of constant radius cornering simulations and straight line acceleration events are graphed below. Figures 7-11 are concerned with constant radius cornering events and examine the roll angle, yaw angle, yaw rate, steering wheel angle, and side slip angle of the vehicle during tests run under various radius and velocity parameters. Figures 12-13 are observations of the longitudinal acceleration and pitch angle over time during a straight line acceleration test, as the pitch can play a significant role in the performance of the FSAE formula car.

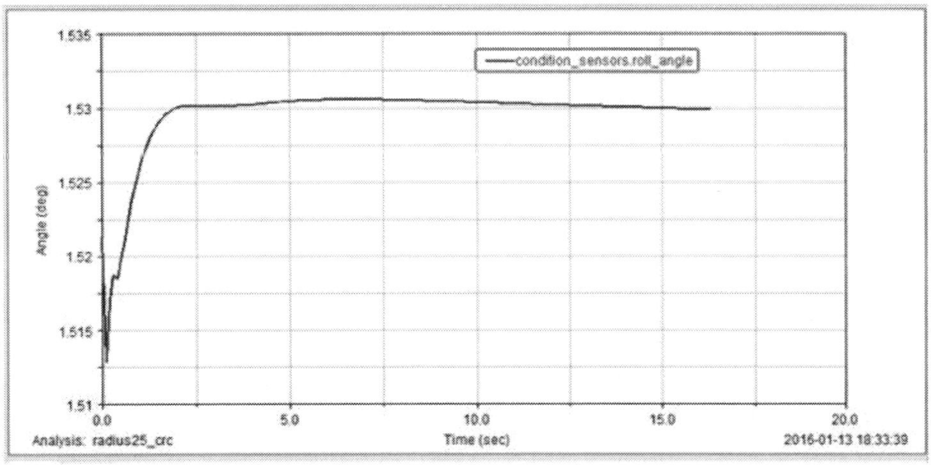

Figure 7 – Roll Angle v. Time (Constant Radius Cornering, 15m radius, 31.55km/hr)

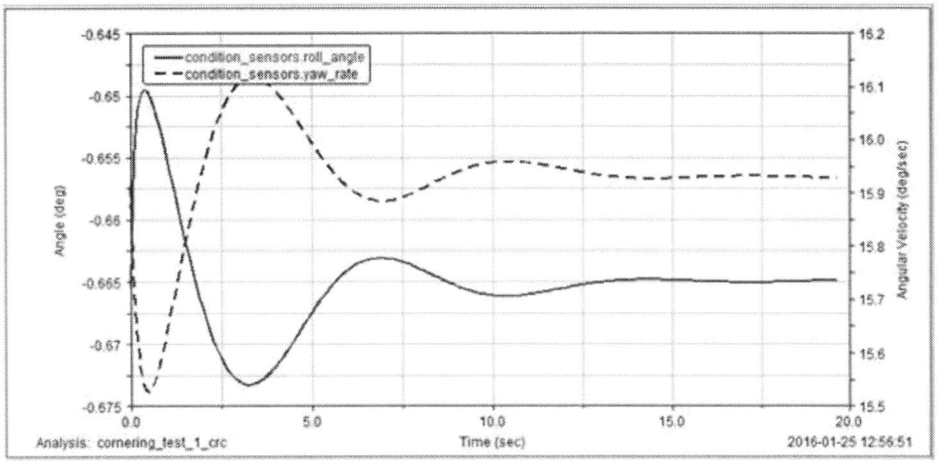

Figure 8 – Yaw Angle and Rate v. Time (Constant Radius Cornering, 12.5m radius, 12.5km/hr)

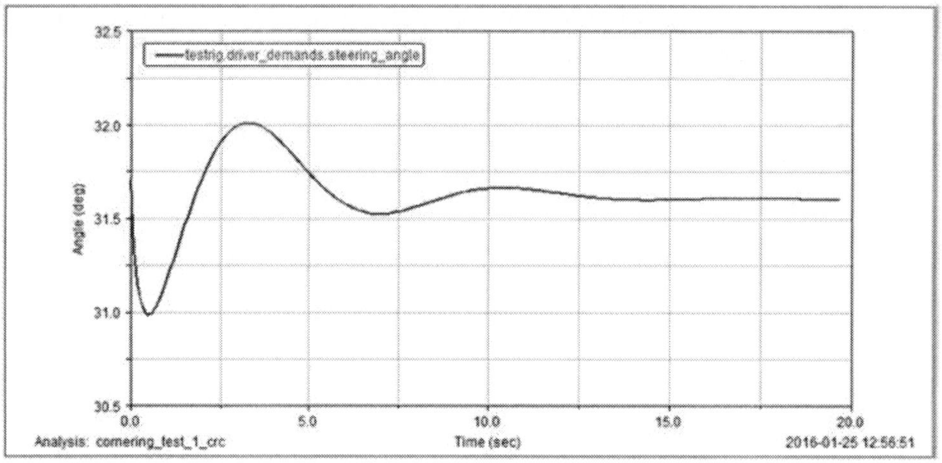

Figure 9 – Steering Wheel Angle v. Time (Constant Radius Cornering, 12.5m radius, 12.5km/hr)

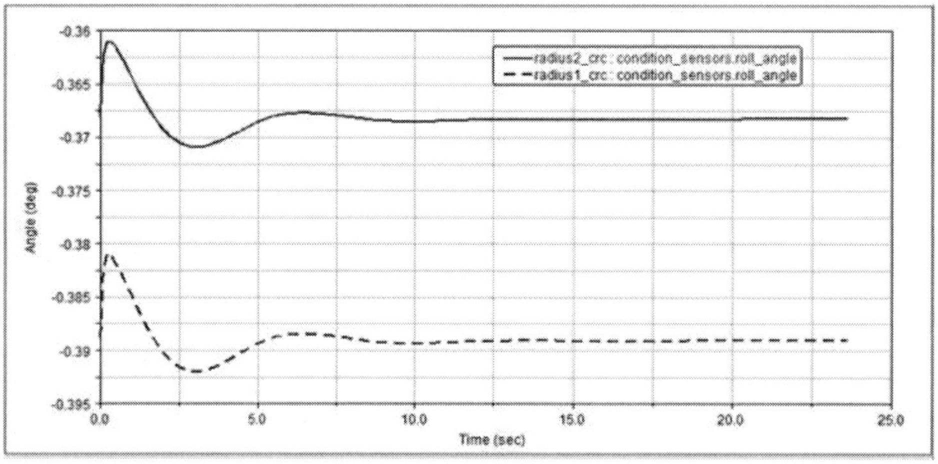

Figure 10 – Roll Angle v. Time (Constant Radius Cornering, 25m radius, 15km/hr)

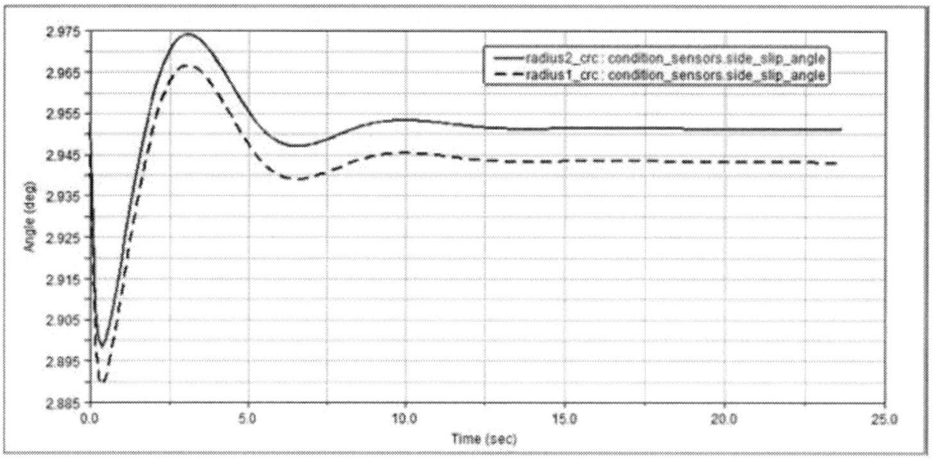

Figure 11 – Side Slip Angle v. Time (Constant Radius Cornering, 25m radius, 15km/hr)

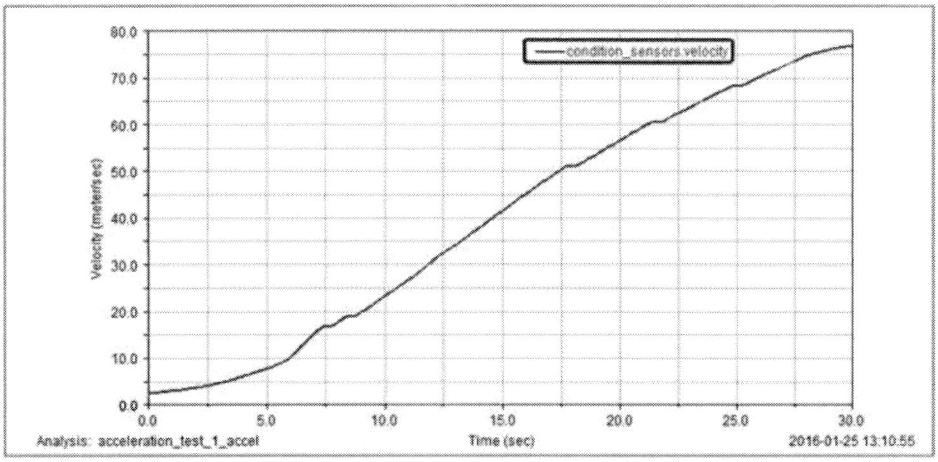

Figure 12 – Acceleration (x-direction) v. Time (Straight Line Acceleration)

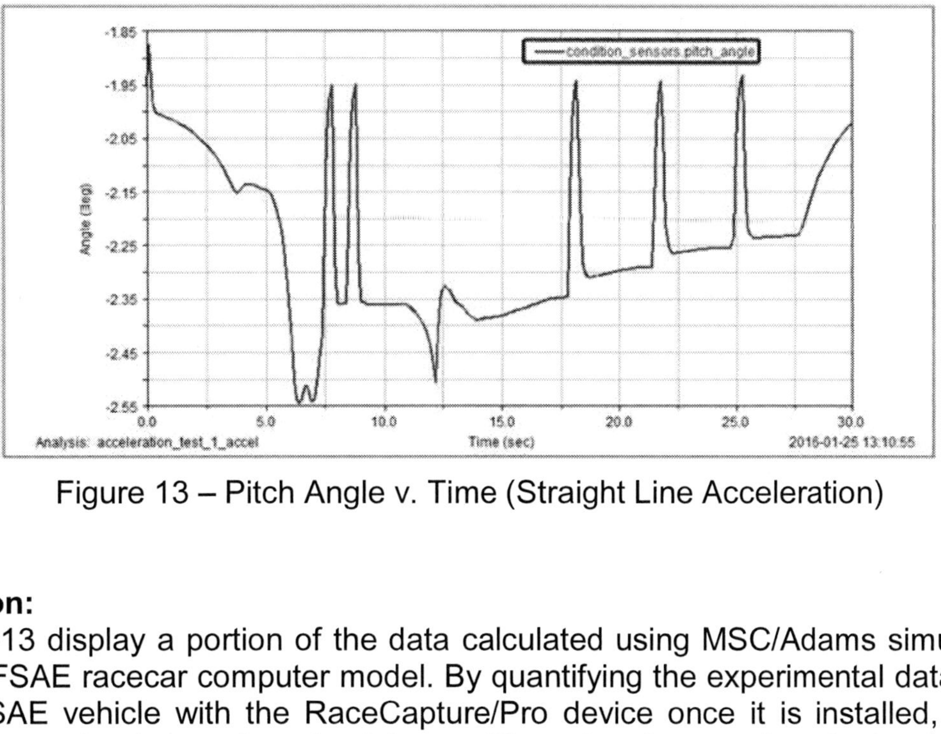

Figure 13 – Pitch Angle v. Time (Straight Line Acceleration)

Discussion:

Figures 7-13 display a portion of the data calculated using MSC/Adams simulations of the 2016 FSAE racecar computer model. By quantifying the experimental data from the Formula SAE vehicle with the RaceCapture/Pro device once it is installed, it can be compared to simulation data in Adams. Thus far the results obtained from the simulations seem to be reasonable estimates of what would happen in the real-world, with relatively low roll angles of around one to two degrees during the cornering simulations. These low roll angles are expected in an FSAE racecar design such as the model represents. Hopefully it will be shown that the real-world data and the simulation data are similar in future experiments, and thus validate the computer model so that it can be used to predict the behavior of the actual racecar during certain dynamic driving events without performing physical tests of the vehicle, which require extensive amounts of planning, time, and cost.

Ideally, a validated model would allow the following year's model to be designed and optimized by altering parameters on the computer model to gather desired results, and then fabricating the physical racecar to match the model as closely as possible. Being able to alter the simulation model in MSC Adams/Car to reflect modifications is exceedingly useful as minute alterations can be made in the model design until the desired results are achieved, and these changes can then be transferred to the physical racecar during fabrication. Designing vehicles in this manner should prove to be less costly and more time efficient in obtaining desirable results and optimizing the racecar design.

Additionally, once the computer model can be reasonably validated in the future with experimental information, and once tire data is updated with each year's model, the computer model can be used to predict and measure certain parameters that are important, but not as easily measured during physical vehicle testing, such as side-slip angle during a turn. The side-slip angle is the angle between the direction the vehicle is moving and the direction the wheel is pointing. This angle is important in predicting vehicle dynamics as a high sideslip angle is a warning of impending traction loss. The problem is, this angle can be quite difficult to ascertain without expensive monitoring systems in road tests. However, with computer simulations this data can be calculated much more easily, as shown in Figure 11. The solid line represents side slip data of the FSAE racer model, while the dotted line is data from the same vehicle, but with the center of mass shifted vertically upwards by 5 cm. The change in roll angles from lowering the mass center are also recorded in Figure 10; the dotted line represents the roll angle of the vehicle with a lower center of mass, while the solid line denotes the roll angle of the FSAE car given a higher center of mass. This data can be useful in determining the configuration of the racecar components such that the end result is a lower center of mass, thus lowering the roll angle and the tendency of the FSAE car to experience rollover and reducing traction during turns. Though the actual change in the roll angle is small, it does not take away from the fact that quick, small changes can be made to the computer model which can reflect the behavior of the actual racecar under similar conditions. If the change in data is favorable to the needs of the driver and race team, the actual vehicle may be modified to match the simulation model.

These simulations can also provide important data for the driver of the vehicle, which can be used to help train a driver how to best control the vehicle during certain dynamic driving conditions. For example, using constant radius cornering simulations, the maximum speed at which the vehicle can safely take a turn can be found. Furthermore, the required angle of the steering wheel to achieve certain turns at specified velocities can be calculated using MSC Adams/Car simulations. Figure 9 provides data on the angle the steering wheel must be at to maintain a 12.5 m right hand turn at 12.5 km/hr. Simulations such as this can essentially teach the driver how to control the vehicle during certain maneuvers to gain the best results.

For future avenues of research, the development of an active control system for the vehicle using "safe", "caution", and "danger" zones for certain values of these parameters (such as roll angle, yaw angle and rate, pitch angle) would be a further application of the results of these simulations. Simulations could be used to determine the three ranges of safety for these parameters, and then sensors could be placed in strategic locations around the vehicle to measure and detect when these parameters exceed the “safe” range of values, alert the driver via a warning light, and engage an active control mechanism to steady the vehicle when the vehicle begins to exceed the established “caution” zone of values.

Conclusion:

The 2016 FSAE racecar currently being fabricated by Eagle Motor Sports is scheduled to undergo a variety of dynamic driving tests under similar conditions to those simulated with the computer model of the racecar. During the road tests, data will be collected using a RaceCapture/Pro device with built-in inertial measurement unit (IMU), GPS, and telemetry capabilities. Pending the results of the experimental data collection from the FSAE racecar, the computer model of the 2016 can be validated if the experimental data closely matches that of the computer simulations run using MSC Adams/Car software. Validation of the computer model will allow the race team to be able to test further modifications to the racecar design and predict the behavior of the racecar, without having to physically alter the racecar until favorable results are met with the simulations. Optimizing the racecar design in this manner will reduce the time needed for fabrication of the final racecar as well as reduce overall costs by eliminating the need for extensive road testing. Additionally, studies of the simulation data compared to the experimental road tests should also give an understanding as to how a vehicle reacts to dynamic situations as well as some insight as to what point an active control system for a vehicle should initiate action to stabilize the vehicle. Though the main purpose of this research was to provide a means for the Eagle Motorsports team to design and test new vehicles, the results gathered can prove themselves valuable on multiple levels by being able to simulate realistic vehicle behavior under dynamic conditions without having to perform extensive road testing.

References:

[1] SAE International. “SAE Collegiate Design Series”. Society of Automotive Engineers International. Web. http://students.sae.org/cds/formulaseries/about/ (accessed November 15, 2015)

[2] Hubert, K., and A. Kumar. “Anti-Roll Stability Suspension Technology.” Society Of Automotive Engineers Papers 3522 (2005).

[3] Aykent, B. “Effects of Sway Acceleration Control on Rollover Propensity and Assessment of Lateral Specific Forces” (No. 2010-01-132.5). Society of Automotive Engineers Technical Papers (2010).

[4] Van der Westhuizen, S. F., and P. S. Els. “Slow active suspension control for rollover prevention.” Journal of Terramechanics 50(1) (2013): 29-36.

[5] Gillespie, T.D. Fundamentals of Vehicle Dynamics. Warrendale, PA: Society of Automotive Engineers, 1992. Print, 147-48.

[6] Milliken, William F. and Douglas L. Milliken. Race Car Vehicle Dynamics. Warrendale, PA: Society of Automotive Engineers, 1995. Print.

Biography:

Stone Williford
Undergraduate Student, Department of Mechanical Engineering, Georgia Southern University

Stone Williford is a senior mechanical engineering honors student at Georgia Southern University. He was born and raised in Albany, Georgia, where he later interned at Peters Performance Speedshop rebuilding engines and transmissions for high performance track cars. This passion evolved into assisting in research for Dr. Mitra where he was introduced into the computer simulation side of the racecar research for his university.

Jennifer Shaffer
Undergraduate Student, Department of Mechanical Engineering, Georgia Southern University

Jennifer Shaffer is a senior mechanical engineering honors student at Georgia Southern University. She was born in Atlanta, Georgia, and recently interned abroad at the Ruhr Universitaet Bochum in Bochum, Germany in the field of materials science and technology. She has been involved with the computer simulation research of Dr. Mitra and his team for a little over a year, and is hoping to eventually pursue a career in the automobile industry.